DRAPSA:

THE VEDIC CYCLE OF ECLIPSES

DRAPSA:

THE VEDIC CYCLE OF ECLIPSES

A KEY TO UNLOCK THYE TREASURES OF TH VEDAs

R. SHAMA SASTRY

GYAN

Published by

Gyan Publishing House
5, Ansari Road
Daryaganj, New Delhi-110002
Phone: 011-47034999, 9811692060
E-mail: books@gyanbooks.com

Distribution Network
gyanbooks.com
India, USA, Canada, UK, Australia

ISBN: 978-93-7169-157-4 (PB)

First Published, 1938

2nd Impression 2025

Printed at: Gyan Press, Delhi.

DRAPSA: THE VEDIC CYCLE OF ECLIPSES
Arthur: R. SHAMA SASTRY

DRAPSA:
The VEDIC CYCLE OF ECLIPSES

a Key to unlock the treasures of the Vedas

By

MAHAMAHOPADHYAYA, ARTHASASTRAVISARADA
VIDYALANKARA, PANDITARAJA,

DR. R. SHAMASASTRY, B.A., PH.D.,

Retired Curator of the Government Oriental Library and Director of Archaeology in Mysore.

Mysore:
Sree Panchacharya Electric Press
1938

DEDICATED

WITH GRACIOUS PERMISSION

TO

HIS HIGHNESS

SRI KRISHNARAJA WODEYAR BAHADUR, IV,

G.C.S.I., G.B.E.,

MAHARAJA OF MYSORE,

AN UNRIVALLED PATRON OF ORIENTAL

LEARNING AND ARTS.

His Highness Sri Krishnaraja Wodeyar Bahadur. IV,
G.C.S.I., G.B.E.,
MAHARAJA OF MYSORE.

PREFACE.

What is ascribed to the Vedic poets in the following pages is not a knowledge of mathematical astronomy which they never had, nor prediction of eclipses with mathematical accuracy which they never did. All that is claimed for them is a knowledge of the famous eclipse-cycle which they seem to have been using in common with the Chaldeans, Israelites, and other contemporary nations. Like those nations they had luni-solar cycles expressed not merely in terms of a thousand and multiples of a thousand days with an intercalary month per thousand days, but also in terms of 5, 17, 19, 20, 30, 33, 57 and 60 years with two intercalary months per cycle of five years. The cycle of 60 years commencing with the year called Prabhava and ending with the year called Akshaya or "Akshayya," as it is spelt in the Taittirīya Araṇyaka (1, 1) is split into three cycles of 20 years each. What is called "Aṭṭāra" and "Ashṭāchkra" in the Tait. Ar. is the same as the 20 years cycle with eight intercalary months, presided over by eight suns called Adityas, or sons of Aditi. According to the Vedic poets a Parva consisted of $14\frac{3}{4}$ days and a lunation $29\frac{1}{2}$ days. The fractions such as $\frac{1}{4}$, $\frac{2}{4}$, $\frac{3}{4}$, and $\frac{4}{4}$ after 14 days of a Parva were called Ekapāth, Dvipāth, Tripāth, and Chatushpāth. As the Vedic poets used to count days in terms of syllables of a verse, assigning one syllable to one day, the first three fractions equivalent to one-fourth, two-fourths and three-fourths of a syllable they rightly regarded as inexpressible and four-fourths which is equivalent to one as expressible. This is clearly stated in the following Vedic hymn (A.V. 7, 10, 27).

" Four are the divisions measurable by a syllable and its parts ; those Brāhmans who are wise know them ; of them, three divisions are kept in the cave ; and are not utterable ; only the fourth division men express."

In other words one-fourth, two-fourths, three-fourths measuring similar divisions of a day are inexpressible, and only four-fourths equivalent to a complete day is what can be uttered. As pointed out in the following pages, Kali, Dvāpara, Tretā, and Krita were originally the names of Parvas ending in one-fourth, two-fourths, three-fourths, and four-fourths of a day. Five hundred such Parvas or seven thousand days in round number constituted a cycle of eight intercalary months and seventy-one or seventy-two eclipses as a rule. In course of time the cycle of 20 years with eight intercalary months was first reduced to a cycle of 20 or 19 years with seven and half intercalary months (Saptārdhagarbhā Aditih), and then to 19 years with seven Manus. In the Amarakośa one thousand Yugas or Parvas are stated to constitute a Day or Night Kalpa of Brahma and are split into 14 Manu-periods or Manavantaras of seventy-one Yugas each. Thus Manu is a name given to 71 or 72 Parvas with eclipses. It is not only found in the Amara and classical literature, but also frequently met in the Vedas undoubtedly in the above sense, when it is used in connection with eclipses.

The Atharvaveda mentions 10000 Yugas. In the Purāṇas and astronomical Siddhānta works the number of Yugas is invariably stated as twelve thousand. These Yugas, as pointed out above and also in the body of the work, are Parvas with eclipses. Hence if we divide 12000 by 72, we get 166 and $\frac{2}{3}$ cycles of eclipses called Manvantaras. This multiplied by 20 gives 3333 as the number of years counted from the beginning of the cycle. The number of year-gods counted in the cycle of 33 years is 3339 which is almost the same as the above number. According to Aryabhaṭa the 60 years' cycle made 60 revolutions by about A.D. 500. This means that by A.D. 500 there had elapsed 3600 years in that era and that it began in B.C. 3101. It follows therefore that the other two eras began at the same time and came to a dead stop at about A.D. 242. The cycle of 60 years was alone continued in the name of Kali Era which is still current. Thus the three Vedic Eras are based on astronomical principles. Of these, the Era of Manvantaras is so misinterpreted as to mean 4,320,000 years and converted as a monstrous background for the Kali era and the era of 33 years' cycle passed for a progressively rising number of gods upto 3339 in the S. Yajurveda and 3349 in the Tait. Brāhmaṇa.

As regards the changes made in the number of gods of the 33 years' cycle, it must be noted that the priests in charge of the Vedas are liable to no charge of tampering or meddling with the reading of the Vedic texts. For necessary changes in case-endings, personal pronouns, and numbers in the Mantras or Vedic formulas, as required by time, place, and agent in connection with any rite, are sanctioned under the name "Ūhas," reasonable alterations. Accordingly the number 3339 year-gods mentioned in the S. Yaj. should not be construed as the date of the Sukla Yajurveda. It is merely the date of the change made in the number of the year-gods, but not the date of the formation of the whole text. Likewise the number of 3349 year-gods found in the Tait.

Brāhmaṇa indicates the date of the change made in the number. It does not at all show the date of the formation of the Brāhmaṇa text which is far earlier than that date.

Since composition of verses in praise of Indra and other gods with thousand and other numbers indicative of eclipses is found only in the Samhitas and no new compositions are seen in the Brāhmaṇas, it may be taken for granted that the practice came to a close with the Samhitas. No definite reasons can be given for this sudden stop. It is probable that there arose serious differences of opinion as to the cause of eclipses and the necessity for the performance of sacrifices to appease the wrath of Yakshas or Gandharvas supposed to be the cause of eclipses. It is also probable that in accordance with the law of change in the track of eclipses, the disappearance of an old series and the rise of a new series of eclipses in the course of about 1150 years, there came a new series necessitating some rearrangement in the matter of beginning and closing the cycle, and thus causing some perplexity to the Vedic poets. It is therefore likely that not knowing what to do, the poets stopped there and added no more verses to the Vedas in praise of Indra and other gods for their success in war with Vritra and other demons of eclipses. It is also probable that after some time there came the Brāhmaṇic period when the old sacrifices with the sacrificial songs were regarded as the most sacred models worthy of performance over again and again and for ever. Thus the system became a mere skeleton with no life. The words and the numbers lost there meaning and were recited with no reference to anything like the eclipses and the gods and demons of eclipses, to which they were rightly applicable. The cyclic numbers were taken to be descriptive of the magnificence of the gods or the mystic virtue of the sacrifices and the light and shade of eclipses were passed for those of day and night. This explains the failure of such celebrated commentators and founders of philosophical systems of sacrificial theology as Yāska, Jaimini, Guru, Bhaṭṭakumārila, and others to rightly interpret the Vedas, notwithstanding their unrivalled learning and monumental works known as the Mīmāmsādarśana which forms a lifelong study to Sanskrit students now.

But for the clue furnished by ancient Jaina Astronomical works and the Hand-book of Astronomy by Chambers, I could have hardly ventured upon such an interpretation of the Vedas. In interpreting the seven Vedic Metres as Gnoman's shadow-measures of seven ordinary or intercalary months and of the solstices, my chief authority is the Jaina astronomical work entitled the "Kālalokaprakāśa" on the Gnoman's shadow-measure of months and solstices, with which the Vedic metrical shadow-measures are identical. In his letter dated January 1, 1938 Dr. F. W. Thomas, Professor of Sanskrit in the University of Oxford, writes to me on this subject as follows—"I think that you must be quite right in considering the matter of the gnoman, as an interesting lecture by Dr. Lerz, which I heard in Germany in 1936, shows the great importance of the gnoman among early peoples in connection with seasonal observations, the time for planting etc., etc, etc. You have done so much in connection with the earliest Vedic astronomy that you are in a very good position for carrying research further in this field."

Though I have not read Dr. Lerz's lectures, I think that whatever Dr. Lerz has said on the use of the gnoman to ascertain the arrival and departure of months and seasons in ancient days will support my interpretation of the Vedic Metres and Vishnu's strides in connection with months.

Likewise in my interpretation of 1000 and other Vedic number puzzles as cyclic numbers of luni-solar days and years with to 7 intercalary months of the eclipse cycle, I owe my inspiration to no less an authority than Chambers who, in his Hand-book of astronomy, interpreted exactly similar number-puzzles of the Bible as those referring to the cycle of eclipses and other self-correcting luni-solar cycles. As the numbers in the two works are so identical, I think that there is no likelihood of my having committed any errors in this matter. In other matters connected with these two main themes it is probable that in some cases my views are farfetched and unsupportable. Even if such be the case, it goes without saying that the main themes remain unaffected. Considerations of age and ill-health and consequently hurry to see the work through the press prevented further revisions of the work which was under my consideration for more than 30 years.

It only remains for me to express my thanks to Mr. H. G. Narahari, M.A., for going through the proofs. My heartiest thanks are due to Pandit Kashināthaśāstri, the Proprietor of the Sree Panchāchārya Electric Press, for his kindness in expediting the work and the neat get-up.

MYSORE,
10—2—1938.

R. SHAMASASTRY.

CONTENTS.

Introduction—1-11.

CHAPTER III

Manvantara Cycle ;—83-130.

lunar years called 33 gods are equal to 32 solar years, 114; Gods of the Apri-verses, 115; The Kālalokaprakāśa on the shadow-measure of months, 115-120; Vishṇu's strides are gnoman's shadows on the days of solstices, 121-122; Epic stories of Vishṇu's incarnations explained 126; Worlds are months, 126; Lengths of days and nights in the Ayanas, 127; Cycles of 17, 33, and 34 years, 128; The Era of the cycle of 33 years, 129; The cycle of 60 years, 129; The Era of Brahmakalpa, the Era of the cycle of 60 years, and the Era of the cycle of 33 years, 130.

CHAPTER IV

The cycle of Eclipses;—131-190.

The cycle of 30 years, 131; The cycle of a thousand and multiple of a thousand days, 132; The cycle of 56 years, the cycle of 18 years and 10 months, or of 7 intercalary months, 133; Swamikannu Pillay on three eclipse-cycles of of 19, 29, and 58 years, 133-134; Hand-book of Astronomy by Chambers on the cycle of 2300 years, the Metonic cycle of 19 years, the cycle of 1000 days, 12000 days, the Messianic cycle of 33 years, the cycle of 1260 years equal to 3½ times 360 years, Daniel's prophecies, the self-corrected cycle of 30 years in 30 times 42 years or 1260 years, 42 and 84 suns, 135-137; The Sūryaprajnapti on 3½, 7, 42 suns and moons, 138; The Tait. Araṇyaka on 7 and 8 suns, 139; Vedic names and description of eclipses, 140; Sheep, goats, bulls, and other names of eclipses, 141; Sun's love to the moon, 143; Thousand troops of Vāyu, a cycle of 1000 days, 144; Manu a name of 71 or 72 eclipses and Tryambaka or Phālāksha, a name of a set of three

CHAPTER V

Srichakra ;—191-224

ERRATA.

Page	*Line*	*Incorrect*	*Correct*
3	22	Idāvtsara	Idāvatsara
3	30	from	form
5	5	familier	familiar
5	26	Ruth	Roth
6	3	comare	compare
6	5	grammer	grammar
6	17	volums	volumes
8	5	new a	a new
8	7	Calledar	Calendar
8	10	fo	to
8	19	penctuation	punctuation
9	9	busines	business
9	18	putāṇas	purāṇas
10	17	geurdon	guerdon
13	17	Calender	Calendar
17	12	celebrate symbolical	celebrate the symbolical
20	10	ग्रयोदसो	त्रयोदशो
21	14	symbolial	symbolical
28	13	Devadasha	Dvādāśāha
39	18	Abhipliva	Abhiplava
40	19	twelth	twelfth
40	22	consists	consist
40	25	tweleve	twelve
40	30	Athervaveda	Atharvaveda
63	8	वा त्रास्मिन्नन्तर्वी-र्यमिति	वास्मिन्नन्तर्वी-र्यमिति
76	5	discribed	described
76	22	thosand	thousand
78	18	choriot	chariot
80	14	षुत्रैरदिति	पुत्रैरदिति
84	21	evedent	evident

Page	*Line*	*Incorrect*	*Correct*
112	27	Punarrūrja	Punarūrjā
112	20	quinoctial	equinoctial
113	13	motnhs	months
113	24	vernral	vernal
117	5	then	the
117	21	derease	decrease
118	18	Karana	Karaṇa
118	24	unevan	uneven
119	20	Hesce	Hence
123	32	demarkation	demarcation
130	26	Brhmp kalpa	Brahmakalpa
131	17	perfoms	performs
131	20	diference	difference
138	26	says	say
140	24	dimmly	dimly
146	27	Pākayajnyas	Pākayajnas
161	20	यत्त	यत्ते
164	1?	spaheres	spheres
164	18	moveable and immoveable	movable and immovable
164	23	colled	called
174	6	diety	deity
190	29	considerstion	consideration
191	26	Jpeshṭha Brama	Jyeṣṭha Brahma
192	8	sixt wins	six twins
194	21	speek	speak
194	21	loarned	learned
200	30	granduer	grandeur
204	14	मु परि	मुपरि
209	18	disassociated	dissociated
220	14	discribed	described

INTRODUCTION.

OF all the world's extant literary records, the Vedas have been admitted to be the most ancient and the earliest expression of the thoughts of civilized society. No nation outside the limits of India has been found possessing such old records; nor is it likely that the archæological explorations made in different parts of the world will bring to light any such religious literature. As the earliest expression of human thought, the Vedas have been of inestimable service for the correct conception of the first principles of philology, mythology, and the origin and growth of religion.

Admitted as is the antiquity of the Veda, scholars are still puzzled about its age. The vedic commentators such as Jaimini, Sāyaṇa, Mahīdhara, and others regard the Vedas as cœval with the creation of the world. They hold that the Vedas form part of the Creator's person; that they are manifested or revealed at the dawn of every Creation, and withdrawn into the Creator's body along with the whole world at each destruction of the world; and that the sacrifices and other ritualistic observances enjoined in the Vedas are so many injunctions imposed upon class of the twice-born. But modern oriental scholars, whether Indian or European, regard this theory of revelation as no better than a nursery tale. These scholars are of opinion that, like any other literary work, the Vedas must be a product of civilization and that as civilization itself is of recent growth, the Vedas, as one of its products, must be still more recent. With this conviction, they have carried on their investigations into the historical records of various nations and founded their conclusions about the chronology of the various kinds of the world's activity on available internal and external evidences. Prof. H. Jacobi,[1] for instance, has found in one of the Ṛigvedic hymns (X, 85, 13) a clear reference to the position of the solstitial colure in Uttara Phalguṇi (*b.* Leonis) and Uttara Bhādrapada (*a.* Andromedæ), the year beginning with the summer solstice in the rainy season (Ṛig. VII, 193, 9), and thereby determined the age of the Ṛigveda to lie somewhere between 4500 and 2500 B.C. He says:—"This period of civilization extended accordingly from about 4500 to 2500 B.C. and we shall, perhaps, not be far wrong if we put the collection of hymns which has come down to us in the second half of this period." Likewise Prof. B. G. Tilak[2] has found independently in the Vedas clear references to the shifting of the position of the equinoctial colure from Ardra (Beteigeuze) to Kṛittikas (*h.* Tauri) and fixed the Vedic age to be the same as that of Prof. H. Jacobi. But other scholars could not bring themselves to believe that the Hindus of such a remote age were so far advanced in their knowledge of astronomy as to make observations of the positions of the colures. Hence, setting aside the above internal evidence, they had recourse to all available external evidence to determine the age of the Vedas. Taking the beginning of the Buddhistic period (500 B.C.) as the land-mark of Indian chronology, Prof. Max Mūller, for instance, divided the previous Vedic period into five parts, the Ṛigvedic period, the Yajurvedic period, the Brāhmaṇic period, the

1. P. 154 Vol., XXIII *Indian Antiquary.*

2. See *Orion* by Prof. B. G. Tilak.

Upanishad period and the Sūtra period; and allowing about two centuries for the growth of each of the literary periods, he fixed the commencement of the Ṛigveda at about 1500 B.C.

It is really surprising that, while making these and other surmises about the Vedic age, both the mediæval commentators and modern critics, notwithstanding their well-established claim to possess a thorough grasp of the contents of the Vedas, should have utterly failed to notice what the vedic poets themselves professed as their aim. The Vedic poets seem to be counting their years in terms of Yugas of the era of Brahmakalpa, or in terms of the cycle of five years or sixty years, or in terms of the cycle of thirty-three years spoken of as thirtythree gods. In A. V. 8, 2, 21 the poets speak of their intention of making their one hundred, two hundred, three hundred, and four hundred (that is, 1000) Yugas ten thousand years. Expressions such as "tenth Yuga", "three Yugas", or "former Yuga" are frequeutly met with in the Vedas. As will be seen, the era of Brahmakalpa is counted in terms of Yugas. Samvatsara, Parivatsara, Anuvatsara, Idvatsara, and Iḍāvtsara are the names of the five years composing a cycle of five lunisolar years and are frequently mentioned in the Vedas (A. V. 6, 55). In A. V. 12, 3, 41 the cycle of sixty years which is a multiple of twelve cycles of five years is clearly mentioned. The thirtythree gods are mentioned in Sukla Yajus. 20, 11; and 20, 36. Of these three eras the first two are still current, not however in their original form, but in a fabulously magnified from given to them by later astronomers of the Siddhānta period. The counting of years in terms of a cycle of thirtythree year-gods fell into disuse, so much so that about six or seven centuries before the Christian era, scholars were hardly able to give the exact meaning of a number of words of the Āpri verses in which the era of the cycle of thirtythree year-gods is enshrouded.

Yāska, one of the Vedic commentators, believed to have lived in the sixth century B.C., "gives the names of no fewer than seventeen predecessors whose explanations of the Veda are often conflicting. The gap between the poets and the early interpreters was indeed so great that one of Yāska's predecessors, named Kautsa, actually had the audacity to assert that the science of Vedic exposition was useless, as the Vedic hymns and formulas were obscure, unmeaning, or mutually contradictory. Such criticisms Yāska meets by replying that it was not the fault of the rafter if the blind man did not see it."[1] It is needless to say that the few specks of clouds which covered the Vedic horizon at the time of Yāska grew so much thicker and darker by the time of Sāyaṇa that it may be said without the fear of contradiction that, notwithstanding the inestimable service which Sāyaṇa has rendered by his exhaustive commentary embodying the mass of tradition current at his time in the world of Vedic scholars, its value as a genuine interpretation of a number of passages is almost *nil.*

Prof. Roth, the founder of Vedic mythology, "propounded the view that the aim of Vedic interpretation was not to ascertain the meaning which Sāyaṇa, or even Yāska, who lived eighteen centuries earlier, attributed to the Vedic hymns, but the meaning which the ancient

1. P. 61, Prof. Macdonell's *History of Sanskrit Literature.*

poets themselves intended. Such an end could not be attained by simply following the lead of the commentators. For the latter, though valuable guides towards the understanding of the later theological and ritual literature, with the notions and practices of which they were familier, showed no continuity of tradition from the time of the poets; for the tradition supplied by them was solely that which was handed down among interpreters, and only began when the meaning of the hymns was no longer fully comprehended. There could, in fact, be no other tradition; interpretation only arising when the hymns had become obscure. The commentators, therefore, simply preserved attempts at the solution of difficulties, while showing a distinct tendency towards misinterpreting the language as well as the religious, mythological, and cosmical ideas of a vanished age by the scholastic notions prevalent in their own."[1]

"Roth,[2] then, rejected the commentators as our chief guides in interpreting the Ṛigveda, which, as the earliest literary monument of the Indian, and indeed of the Aryan race, stands quite by itself, high up on an isolated peak of remote antiquity. As regards its more peculiar and difficult portions, it must therefore be interpreted mainly through itself; or to apply in another sense the words of an Indian commentator, it must shine by its own light and be self-demonstrating. Ruth further expressed the view that a qualified European is better able to arrive at the true meaning of the Ṛigveda than a Brāhman interpreter. The judgement of the former is unfettered by theological bias; he possesses the historical faculty, and

1. P. 60, Macdonell's *History of Sanskrit Literature.*

2. P. 63, Macdonell's *History of Sanskrit Literature.*

he has also a far wider intellectual horizon, equipped as he is with all the resources of scientific scholarship. Roth therefore, set himself to comare carefully all passages parallel in from and matter, with due regard to considerations of context, grammer, and etymology, while consulting, though perhaps, with insufficient attention, the traditional interpretations. He thus subjected the Rigveda to a historical treatment within the range of Sanskrit itself. He further called in the assistance rendered from without by the comparative method, utilising the help afforded not only by the Avesta, which is so closely allied to the Rigveda in language and matter, but also by the results of comparative philology, resources unknown to the traditional scholar. By thus ascertaining the meaning of single words, the foundations of the scientific interpretation of the Vedas were laid in the great Sanskrit Dictionary in seven volums, published by Roth in collaboration with Bohtlingk between 1852 and 1875 "

Valuable as is this dictionary to the scholar of the Vedas, it has so far failed to help the student to proceed a step beyond what is chalked out by Sāyana. The reason is not far to seek. The Vedas abound with a mass of technical terminology of sacrifices which were periodically following the revolutions of the sun and the moon. It is not clearly known at what intervals of time from each other the Vedic poets performed the different kinds of Soma sacrifices, at what recurring periods of time they performed the Horse-sacrifice, of how many years their Yuga consisted, how and when they were periodically adjusting the luni-solar year, and how they kept the account of elapsing years. Unless these and other facts connected with time are clearly understood, no attempt at an interpretation of the terminology of the periodical sacrifices and of the special deities connected with them can be successful. Sāyana, as is well known, lived at a time when the astronomy of the Hindus had long taken a permanent and, as is belived, unalterable shape with its theory of the four Yugas amounting to 4,320,000 years and when the periodical sacrifices were performed at any time convenient to the sacrificer. His was the time when the Horse-sacrifice as well as the so-called *Sattras* or sacrificial sessions extending over a number of days, usually from 12 to 360 days and upwards, had long disappeared from the list of sacrifices in observance. His was the time when the method of keeping up an account of elapsing years and of adjusting the luni-solar year entirely differed, as will be seen, from what was in use during the Vedic period. Hence his failure to find a satisfactory interpretation of many Vedic words and sentences. Hence also the failure of oriental scholars to arrive at a consistent explanation of Vedic thoughts and myths. What the mediæval commentators termed a theological maxim modern critics style a mythological story. Where the former called in the aid of faith to belive, the later saw superstition. The Vedic poets, for example, now and then speak of the failure of Speech[1] to bring the moon, of the loss a feather or nail sustained by the bird-like Gāyatrī,[2] a verse of twenty-four syllables, in bringing the moon, and of cows[3] sitting at a sacrifical session extending over a number of days. The commentators attribute these and other exploits narrated in

1. Book 1, Chapter V, Section 27, *Aitareya Brāhmana.*
2. Book III, Chapter III, Section 25, *Ibid.*
3. Kanda VII, 5, 1, *Krishna Yajurveda.*

the Vedas and Brāhmaṇas to the tutelary gods or goddesses of speech, of metre, and cows, while modern European critics regard these stories as mythological legends, having their origin in the wild imaginations of the poets. Thus new a word, "mythology," is substituted for theology and superstition for faith.

My attempt is to unveil the mystery in which the sacrificial callendar of the Vedic poets is enshrouded, to recover that lost and forgotten era which the poets themselves had invented and continued from 3101 B.C. fo about A.D. 500, and to secure thereby a key to explain a number of theological or mythological stories of the pattern mentioned above.

But before proceeding with the subject I have undertaken, it is necessary that I should assure the scholarly public of the stringent precautions which I have taken against the tendency to "put new wine into old bottles." With the help of grammar and lexicons, it is possible, as has been the practice of many scholars, to twist and split words, to alter the penctuation and shape new sentences in order to extract the desired meaning. A simple thought which an ancient singer wanted to convey by his poetry is thus sometimes converted into a profound, philosophical or mystical idea, and a high-soaring, significant utterance is at other times mistaken for childish prattle. Such are the difficulties which beset the interpretation of the Vedic hymns, when the customs, tradition, and mode of thought of the authors of those hymns have long been forgotten.

But such difficulties and risks are foreign to the subject which I propose to discuss at length. It is a subject which is arithmetical in its character and needs, therefore, no grammatical hair-splitting. The numbers contained in the Vedic passages are plain enough and speak for themselves. What those numbers are ,what their purpose, what their basis, and what their aim, is all that I have to point out. Sāyaṇa's interpretation of the Vedas, Prof. Wilson's translation of the Ṛigveda based upon Sāyaṇa's interpretation, Prof. Whitney's literal translation of the Atharvaveda, none of these affects or disproves my theory. My busines ends with pointing out what the Vedic poets alluded to and what is the chief purport they meant to convey.

The question of Vedic chronology is not however, of primary importance now. It will solve itself, if only what the Vedic poets held before themselves as the main theme of their songs and sacrifices is clearly understood. The one fundamental theme from which the Vedic sacrificial religion and its myths, the stories of the Epics and the Purāṇas, the Jātaka tales of the Buddhists and the Jainas, the Tāntric cult, and the fairy tales of the Panchatantra, in short, the whole of our Indian culture is derived is the cycle of eclipses. It is accepted on all hands that the cult of sun and moon worship is to a greater extent the source of all religions of the world than feticism or spirit-worship. The periodical recurrence of solar and lunar eclipses with mathematical accuracy added to the growth of the cult in various forms. It gave rise to predictive astrology and to the cult of prophecy of the perodical reappearance of the gods to destroy the demons of eclipses and save themselves and the world from their clutches. The Vedas are nothing but a record of the religious spells performed and the prayers sung to help the gods to free themselves from the demons of eclipses.

Of the two planets, soma, the moon, is regarded as the father and the sun, Sāvitrī, as the mother. Their union on the day of new moon is believed to be the cause of the productivity of the world. The moon is the Chaturmukha-Brahma, the four-faced creator of the animate and inanimate world. He is also regarded as the food of departed souls and also of the gods. Accordingly the Vedic poets feared that in the solar eclipses a new beast or demon might outrage the Sāvitrī and become the cause of mixed blood. This they could not avert and the only course left for them was to make an oblation into the fire with prayer to Agni to cut off the second seed. In the annual ancestral rites such an oblation is even now made. The lunar eclipse is sometimes regarded as the capture of king Soma while hunting in distant forests by wild tribes (*Vyādhas*). Hence Soma-krayaṇa or geurdon is symbolically paid to the Soma-seller for the restoration of the moon. As there are only seven lunar eclipses in each of three divisions of the cycle with seven more in what is called the head (Śīrsha) of the cycle, the geurdons are also only seven. As there are, however, eleven lunar eclipses forming part of eleven Rudra eclipses, there may likewise be eleven or ten Soma-krayaṇas corresponding to the eleven Gandharvas or Soma-captors. Sometimes a solar eclipse is also spoken of as a battle between Indra and Vṛitra or Śambara for Soma liquor. At other times it is spoken of as an act of snatching off the Soma-food of the gods and departed souls by a Kravyāda or a flesh-eating demon. On other occasions the moon is regarded as a woman or a she-goat and the lunar eclipse is described as an attack of the she-goat by another male-goat or a demon. The captors are either seven hunters in the country of the Daśārṇas; or they are seven wild beasts in the mountain regions of Kālāñjana; or they are the Chakravāka birds sucking the moon in the Śara island; or they are seven swans eating the lotus-like moon in the Mānasa lake. Thus numberless are the stories devised to account for the phenomena of eclipses.

The eclipses are also enumerated as eight Vasus, eleven Rudras, twelve or seven Ādityas, fourteen Viśvedevas, or sixty-four kalās or delusive arts of the demons, and the like. The sixty-three Śalāka Purushas of the Jainas with their followers called Yakshas, Gandharvas, and Kinnaras are all so many solar and lunar eclipses recurring in each cycle of eclipses.

This is the subject that I propose to expound in the following pages. As a preliminary thing, however, and for the correct understanding of the subject, I shall deal in the next few chapters with the sacrificial calendar of the Vedic poets, their cycle of luni-solar years with seven or eight intercalary months, and their cycle of Manvantaras on which the cycle of eclipses is mainly based.

THE VEDIC CALENDAR.

The Intercalary Month.

THE term " Vedic Calendar " may appear at the outset to be an anachronism, for the reason that there are no clear references to any kind of calendar in the Vedas proper. Even in the Brāhmaṇas, references to a calendar are so vague that it is hardly possible to form a clear conception of the precise nature of the calendar that was in use. But coming to the Sūtras, especially those of the Sāmaveda, we find precise data to determine the various systems of calendar in observance during the Sūtra period. One might, therefore, be led to think that the term " Sūtraic Calender " would be preferable to that of " Vedic Calendar." But it should be borne in mind that the various systems of calendar described in the Sūtras are not the result of an observation of the heavens in a day, but are the outcome of the experience gained and adjustments made by many successive calculators of time. Nor are allusions to a calendar altogether wanting even in the Vedas. The description of the New Year's Day as occurring on the Ekāshṭaka day, *i.e.*, the eighth day of the dark half of the month of Māgha (corresponding to December-January) in the times with which we are dealing, as well as the distinct references to a thirteenth month which must necessarily have been an additional month intercalated for the purpose of keeping the beginning of the year more or less close to its natural time, go a long way to prove that the Vedic poets kept a systematised calendar based upon scientific principles.

The beginning of the year on the Ekāshṭaka day is thus described in the Atharvaveda, III 10:—

"(1) It is she that first shone out; having entered into this (*earth*), she goes about; (*like*) a bride, newly married (*to the* NEW YEAR), she has become the generatrix (*of the days that follow*); three are the great lights that associate with her.

"(2) Extolled in metres, these two shining dawns, coming out of the same womb, and being the wives of the sun, go about all-knowing, making a flag, free from old age, and impregnated with abundant seed.

"(3) Three dawns have reached the path of the sacrifice; three lights [the fire, the sun, and the moon] have also approached it; of them, one protects the offspring, one vigour, and one the rite of those who like to please the gods.

"(4) She who is the fourth has passed into the four sets of Sāma-chants [nine-versed. fifteen-versed, seventeen-versed, and twenty-one-versed chants], maintaining the two wings [halves] of the sacrifice [*i.e.*, the year] as known to the sages, and giving rise to the Great Litany composed of Gāyatrī, Tṛishṭubh, Jagatī, and Anushṭubh metres; and she has preserved this heaven [the solstice].

"(5) With five (*days*) the Creator has made this; he has also created five and five sisters of them; taking various forms and being clothed in sacrificial splendour, five of them run with great speed.

"(6) Thirty sisters [days] partake of the rite, spreading out the same flag; they make the seasons; being wise and all-knowing and residing in the metres, they go about with great splendour.

"(7) Clothed in splendour, this shining night takes to herself the rites addressed to the sun above; even the various kinds of beasts, on awakening, see her on the lap of this mother [the earth].

"(8) This eighth day, bearing the troubles of pregnancy, has brought forth this great Indra; with his help the gods repelled the enemies; in virtue of his own might, he has become the destroyer of the Asuras.

"(9) O sole Ashṭakās, ye gave a sister to me hitherto without a sister; ye speak the truth; listen to this prayer: just as ye are pleased with the behaviour of this (*Indra*), so may ye be pleased with mine; do not send me away to any one else!

"(10) This all-knowing dawn stepped into my mind and has taken a firm hold of it; just as ye are pleased with this (*Indra*), so may ye be pleased with me; do not send me away to any one else;

"(11) The five mornings, the five milkings, and the five seasons follow the cow with five names; the five quarters regulated by the fifteen-versed chant and possessed of the same characteristics as the five mornings follow this single light [the dawn].

"(12) (*Of the five mornings*) the first is the womb of the dawn: one bears the magnificence of the waters; one presides at the rites addressed to the sun; one presides over the heat; and one the sun controls.

"(13) She that first shone out has become a cow at Yama's: let her, rich in milk, yield to us each further summer!

"(14) Foremost among the lights, clothed in brilliant splendour, has arrived this illuminating dawn with various colours, like a flag of the sacrificial fire ; O ever youthful Dawn, conducive to the performance of unchanging rites, and grey with old age, thou hast arrived!

"(15) The wife of the seasons, the first (*Dawn*) has arrived, leading the days and being the mother of creatures; though one, thou hast become many; free from old age, thou causest the rest to grow old."

Likewise the Tāṇḍyamahābrāhmaṇa describes the Ekāshṭaka as the wife of the year:—V. 92.

एषा वै संवत्सरस्य पत्नी यदेकाष्टका. एतस्यां वा गतां रात्रिं वसति. साक्षादेव तत्संवत्सरमारभ्य दीक्षन्ते.

"What is called the Ekāshṭakā (*day*) is the wife of the year; when the night of this day arrives, (*prajāpati*) lies with her. Hence, commencing with the (*true*) beginning of the year, (*sacrificers*) observe the rite of initition."

The important points to be particularly noticed in the above passages are (1) the beginning of the sacrificial year, probably solar, on the eighth day of the dark half of the month Māgha; (2) the designation of this day by such names as 'a cow,' 'dawn,' 'Prajāpati's daughter,' and 'Sūryā'; (3) the association or a kind of secret marriage of the dawn with three lights, the moon, and the sun, as pointed out by Sāyaṇa in his commentary on verse 1; (4) the birth of the days of the following year or cycle of years, as well as of Indra and Soma from the marriage of the dawn with the sun; (5) the celebration of the dawn by the four well-known Sāma-chants; namely, the nine-versed chant, the fifteen-versed chant, and the twenty-one-versed chant, each of which is, as we shall see, intended to signify as many intercalary days as the number of verses contained in it; (6) the destruction of enemies and Asuras brought about by Indra, the son of the dawn.

As regards the first point, it is true that we are told nowhere in the Vedas themselves that the word Ekāshṭakā means the eighth day of the dark month of Māgha; still, on the authority of Āpastamba and other Sūtra-writers, who have defined it as such, we may take it to mean that particular day. From the next three points we have to understand that, at the commencement of every, year, or cycle of years, it was the usual custom with the Vedic poets to celebrate symbolical marriage of the New Year's Day with the sun in order to enable the year to beget its 720 children, [1] *i. e.*, its days and nights, or, in other worlds, to perpetuate an auspicious flow of time for themselves. This seems to be the sum and substance of the celebrated marriage hymns,[2] in which the marriage procession of Sūryā or the dawn to be wedded to the sun is the subject of a long and mystic description, and which are even now recited on the occasions of marriages performed as a rule after the winter solstice and before summer solstice.

Thus, while the Yajurveda connects the mornings and the mystic cows with the Chatustomas and makes the number of intercalary days to be twenty-one, the same is distictly stated as three times seven in the Sāmaveda—

तेऽमन्वत प्रथमं नाम गोनाम्
त्रिस्सप्त परमं नाम जानन ।
ता जानतीरभ्यनूषत क्षाः
आविर्भुवन्नरुणीर्यशसा गावः ॥

---

1 R. V. I, 164, 10, 11.   2 R. V. X, 85; and A. V. XIV, 1, 2

"First they (*the sages*) came to know the sacred
name of the cows; they came to know the
sacred names to be three times seven;
knowing them, they extolled the morning (*kshāḥ*)
then the red cows became famous.'

Thre are two more verses which express the same idea—

अयं पुनान उषसो अरोचयत्
अयं सिन्धुभ्यो अभवदु लोककृत् ।
अयं त्रिस्सप्त दुदुहान आशिरम्
सोमो हृदे पवते चारुमत्सरः ॥ Sāma ii, 1 17, 3.

त्रिरस्मै सप्त धेनवो दुदुहिरे
सत्यामाशिरं परमे व्योमन् ।
चत्वार्यन्या भुवनानि निर्णिजे
चारूणि चक्रे यद्दतैरवर्धत ॥ Sāma, vi, 2, 2, 7.

'He, being purified, hath made the mornings shine;
and it is he who gave the rivers room to flow:
making the three times seven pour out the milky
stream, Soma, the cheerer, yields whatever the heart finds
sweet."

"The three times seven milch-kine in the loftiest
heaven have for this Soma poured the genuine
milky draught; four other beauteous worlds
hath he made for his adornment when he
waxed in strength through holy rites."

We find clear referenes to a thirteenth intercalary month not only in the Yajurveda and the Atharvaveda, but also in the Ṛigveda. The Ṛigveda I, 25, 8, thus alludes to the intercalary month:—

वेद मासो धृतव्रातो द्वादश प्रजावतः । वेद य उपजायते ॥

"He, who, accepting the rites (*dedicated to him*), knows the twelve months and their productions, and that which is supplementarily engendered."

In his translation of the Ṛegveda, Professor H. H. Wilson remarked as follows—

"**वेद य उपजायते**, who knows what is *upa*, additionally or subordinately produced. The expression is obscure, but in connection with the preceding, **वेद मासो द्वादश,** who knows the twelve months, we cannot doubt the the correctness of the scholiast's conclusion, that the thirteenth, the supplementary or intercalary month of the Hindu luni-solar year, is alluded to; that 'the thirteenth or additional month which is produced of itself, in connection with the year,' '**यस्त्रयोदशोऽधिमास उपजायते संवत्सरसमीपे स्वयमेवोत्पद्यते**'." The passage is important, as indicating the concurrent use of lunar and solar years at this period and the method of adjusting the one to the other."

Notwithstanding Sāyaṇa's interpretation of the word *upajāyate* in the sense of 'a supplementary month,' it is doubtful whether the word indicates a complete intercalary month or an intercalated period less than a month; for we shall presently see that along with the custom of adjusting the lunar and solar reckoning by the addition of a complete month, there was also the practice of adjusting them, by adding as many days as formed the difference between any two kinds of years or sets of years. Still, it is certain that some sort of intercalation, either in the form of a month or in the form of a period less than a month, is what is implied in the above verse of the Ṛigveda. But coming to the Atharvaveda, we see therein a clear description of a thirteenth intercalary month—

अहोरात्रैर्विमितं त्रिंशदंगं त्रयोदशं मासं यो निर्मिमीते ।
तस्य देवस्य क्रुद्धस्यैतदागो य एवं विद्वांसं ब्राह्मणं जिनाति, ॥

" He who mesuars the thirteenth month, fabricated of days and nights, having thirty members—against that god, angered, is this offence." A. V., XIII, 3. 8.

सनिस्रसो नामासि
त्रयोदसो मास इन्द्रस्य गृहः ।

" Weakling by name art thou, the thirteenth month, Indra's house." A. V., v. 6. 4.

In the Kṛishṇa Yajurveda, I. 4. 14, the twelve months together with a thirteenth intercalated month are thus enumerated :—

मधुश्च माधवश्च शुक्रश्च शुचिश्च नभश्च नभस्यश्च इषश्चोर्जश्च सहश्च सहस्यश्च तपश्च तपस्यश्चोपयामगृहीतोऽसि संसर्पोऽसि अंहस्पत्याय त्वा.

" Thou art the month of Madhu,[5] the month of Mādhava, the month of Śukra, the month of Śuchi, the month of Nabhas, the month of Nabhasya, the month of Isha, the month of Ūrja, the month of Sahas, the month of Sahasya, the month of Tapas, and the month of Tapasya; and thou art caught hold of in a wooden vessel; thou art the month Samsarpa [a creeping month]; and thou art the receptacle of sins."

5 Madhu corresponds to Chaitra; Mādhava to Vaiśākha; and so on.

The Brāhmaṇa portion contained in the Kṛishṇa-Yajurveda, vi. 5, 3, 12, comments on this passage as follows :—

प्रसिद्धमेवाध्वर्युर्दक्षिणेन प्रपद्यते प्रसिद्धं प्रतिप्रस्थातोत्तरेण. तस्मादादित्यष्षण्मासो दक्षिणेनैति षडुत्तरेण उपयामगृहीतोऽसि संसर्पोऽस्यंहस्पत्याय त्वेत्याहास्ति त्रयोदशो मास इत्याहुस्तमेव तत्प्रीणाति.

"Clearly does the Adhvaryu first go to the south; clearly the Pratiprasthāṭri priest to the north. Hence does the sun go to the south for six months. He says: 'Thou art caught in a wooden vessel'; thou art Saṁsarpa [a creeping month] and a receptacle for sins.' They say that there is also a thirteenth month; it is that thirteenth month which he pleases thereby."

The symbolial practice connected with this passage is this—The adhvaryu priest fills thirteen vessels with Soma-juice; and with the help of another priest, called Pratiprasthāṭri, he makes offerings therefrom to the seasons. While performing the rite, the Adhvaryu goes to the south and the Pratiprasthāṭri to the north, imitating the southern and northern movements of the sun respectively. As will be seen, it was in the middle of the year, during the summer or the winter solstice, according as the year began with the winter or the summer solstice, that the intercalary period was inserted, delaying the sun's turning movement so long and occupying that period in performing the initiatory rites. Hence the reference in this passage to the sun's northern and southern movements, and to the thirteenth month during which the commencement of those turning movements is delayed. The fact of representing the months by Soma-vessels is clearly stated in the Maitrāyaṇiya-Saṁhitā, iii, 10. 4. 5.

द्वादश पात्राण्युपांशुसवनस्त्रयोदशं यत्त-
न्मीमांसंते पात्राउन्नपात्रामिति मीमांसंते
हि त्रयोदशं मासं मासाउन्नमासाउ इति.

"Twelve are the vessels; the pressing stone, called *Upāṁśusavana*, is the thirteenth; the discussion they hold, whether there ought to be a vessel or no vessel (*to represent a thirteenth month*) is a discussion as to whether there is a thirteenth month."

Regarding the sacrificial function observed during a thirteenth month, the Tāṇḍyamahā Brāhmaṇa, X, 3. 2, says—

पिता नोऽरात्सीदिति माता उपासीदन् । ते दीक्षयैवाराध्नुवन् । उपसत्सु त्रयोदशमदीक्षयन् । सोऽनुव्यमभवत्. तस्मादुपसत्सु दिदीक्षाणोऽनुव्यं भवत्येव च हि त्रयोदशं मासं चक्षते नैव च.

"The months observed the vow of *Upasads* [sessions] with the intention that their father [the year] might prosper. They, however prospered merely by observing the initiatory rites, and initiated the thirteenth month during the period of the vow of *Upasads* [session]. Therefore the thirteenth month became their follower. Hence whoever undergoes the rite of initiation during the period of the vow of *Upasads* [sessions] becomes the follower (*of the rest of the priests*). Accordingly they declare a thirteenth month as existent and also as non-existent."

Again, the Maitrāyanīya-Samhitā i. 5. 5. 6, says—

अग्नीषोमीयया त्रयोदशी उपस्थेयोऽस्ति.
मासस्त्रयोदशः तमेवैतयाप्त्वावरुंधे.

"The thirteenth lunar day is to be propitiated by the immolation of a beast sacred to Agni and Soma. There is thirteenth mouth ; it is that thirteenth month which he catches hold of by this offering."

These and other references to the New Year's Day and the thirteenth month intercalated solely for the purpose of keeping the seasons or the month in their proper places in the year, are enough to show that the Vedic poets kept a calendar with far more scientific precision than we are pleased to credit them with. Whether we will or no, the ·fact cannot be denied that the idea of a thirteenth month, *i.e.*, an interclated month, could not have dawned upon the mind of the Vedic poets unless they had been quite familiar with the true lengths of several kinds of years. There is also reason to believe that, before the system of adjusting the difference between any two kinds of years by the insertion of an intercalary month was begun, the practice was to adjust them by adding sets of intercalary days, such as 9, 11, 12, 21, and so on. That such was the custom, is clear from the following passage of the Kaṭhaśākhā-Brāhmaṇa, quoted in the *Smṛititattva.*[6]

अर्धमासा वै अधस्तात्संतः अकामयत मासास्स्यामेति. ते द्वादशाहं क्रतुमुपायन् त्रयोदशं ब्राह्मणं कृत्वा तस्मिन् मृष्ट्वा उदतिष्ठन् तस्मात्सोऽनायतनः इतरानुपजीवतीति तस्माद्द्वादशाहस्य त्रयोदशेन ब्राह्मणेन भवितव्यमिति कठशाखाब्राह्मणम्.

अस्यार्थो जयस्वामिना व्याख्यात :—

ते चार्धमासास्त्रयोदशं मलमासं ब्राह्मणं कृत्वा द्वादशाहं क्रतुमुपायन् उपाहृतवंतः तस्मिन्मलमासे मृष्ट्वा संमार्ज्य किमित्याकांक्षायां अरातीरित्यध्याह्रियते अरातीः पापानि संमार्ज्य उदतिष्ठन्. पापभारशून्या उत्थिता अभवान्नित्यर्थः तत्र पापनिर्मार्जनार्थवादात्

6 *Smitilatva.* Calcutta ed., 1895, p. 782.

सम्भवत्कालानंतरं कर्म तत्र न कर्तव्यम् न तु निरवकाशमिति. अर्थवादात् विधीकल्पनायाः प्रतीतिवाधेनैवौचित्यात्. अतो नित्यनैमित्तिकशांतिकादेः मलमासेन पर्युदासः सोऽनायतन इति नाव्यस्य चैत्रादिवत प्रतिनियतस्थानमित्यर्थः इतरानुपजीवतीति मासांतरेषु चन्द्रक्षयवृद्धिभ्यः. तस्योपजननात्.

"Being at a lower level [*i.e.*, being less than a month]. the half-months desired that they might grow into months, They approached the twelve days' sacrifice. Having appointed a Brāhmaṇa as a thirteenth priest [in addition to the twelve priests] and having washed off (*the sins*) on him, they got up. Hence it is that he [thirteenth month or the priest who represents it] is homeless and dependent for his existence upon others. Hence there ought to be a thirteenth Brāhmaṇa priest in every twelve days' sacrifice. This is a passage from the *Brāhmaṇa* of the Kaṭha School.

"This passage is thus commented upon by Jayasvāmin—Having represented the thirteenth 'dirty,' month by a Brāhman priest, those half-months collected the twelve-days' sacrifice, [*i.e.*, converted it into a month's sacrifice]. Having wished off in that 'dirty' month,—if it is asked what was that which they washed off on the 'dirty' month, we have to understand the word *arāli*, 'enemies';—having washed off the enemies, *i.e.*, the sins, they got up, *i.e.*, they rose up free from the burden of sin. From the descriptive statement of washing off the sins, it follows that the sacrificial performance which can possibly be observed in the subsequent month should neither be undertaken during the 'dirty' month nor be given up. The inference of a rule from a descriptive statement is reasonable in as much as the sense of the descriptive statement can otherwise have no application whatever.[7] Hence the performance of obligatory, casual, expiatory, and other religious rites is prohibited in a 'dirty' month. It is homeless, *i.e.*, like Chaitra and other months, it has not a fixed place of its own in the year. 'Dependent for its existence upon others, *i.e.*, the thirteenth month comes into existence owing to the waxing and waning of the moon in the intervals of months."

The meaning of the above passage is this—Giving up a practice of adding twelve days to the synodic lunar year of 354 days in order to adjust it to the sidereal solar year of 366 days, the Vedic poets allowed the twelve days to accumulate to the extent of a month in the course of two and a half years, and then performed their sacrifice at the close of the thirteenth month with thirteen priests, of whom the thirteenth priest represented the thirteenth month, the 'dirty' month and took up the sins of the sacrificer for the gold that was presented to him.

In the *Aitarēya-Brāhmaṇa*, i, 12, the thirteenth priest is called Soma-vikrayin, 'seller of Soma.' This passage, with a brief commentary upon it by Raghunandana Bhaṭṭāchārya, the author of the *Smṛitilattva*, runs as follows[8]—

प्राच्यां दिशि वै देवाः सोमं राजानमक्रीणंस्तस्मात्प्राच्यां दिशि क्रीणते त्रयोदशान्मासादक्रीणंस्तस्मात्त्रयोदशो मासो नानुविद्यते पापो हि सोमविक्रयीति. अस्यायमर्थः

---

7 Here the rule is that the sacrifice should neither be performed in the 'dirty' month nor be given up, but should be performed in the subsequent month. The statement is that of washing off the sins in the dirty month.

यतोऽधिमासस्सोमविक्रयी अतोऽसावितरमासवन्नानुविद्यते. विद्यमानोऽपि कर्मानर्हत्वादसन्निवेत्यर्थः सोमविक्रय्यपि ऋत्विगन्तरवत्.

"The gods bought the king Soma in the eastern direction. Thence he is (*generally*) bought in the eastern direction. They bought him from the thirteenth month. Thence the thirteenth month is found unfit (*for any religious work to be done in it*); a seller of Soma is (*likewise*) found unfit (*for intercourse*), for such a man is a defaulter." The meaning of the passage is this—" Because the intercalated month is the seller of Soma, therefore it has no proper existence like other months. Although it has its own existence, it is yet regarded as having no proper existence in asmuch as no rites are performed in it. The seller of Soma is like other priests employed for the performance of sacrifice."

As regards the sinful nature of intercalated months, the author of the *Smṛititattva*, quotes the following passage°—

वत्सरांतर्गतः पापः यज्ञानां फलनाशकृत् ।
नैर्ऋतैर्यातुधानाद्यैस्समाक्रांतो विनामकः ॥

इत्यादि ज्योतिश्शास्त्रे.

विरुद्धनामको विनामकः कुतः मलिम्लुचादिनामकत्वात्.

"(*The intercalated month*) contained in the body of the year, is sinful, is destructive of the good results of sacrifices, is infested by Nirṛita, Yātudhāna, and other evil spirits, and is of a disagreeable name. This and other passages are found in astrological works. The word *vināmaka* means 'that which has a disagreeable name'; for it has Malimlucha and other (*disgusting*) names."

The three passages quoted above throw a flood of light on the nature of the conception which the Vedic poets entertained regarding the intercalary days and months. We have to understand the three important points specified in these passages—

(1) At first the Vedic poets used to adjust their lunar year with the sidereal solar year by adding twelve days to the former, but in course of time they gave up that custom and began to intercalate one month to every third lunar year.

(2) Instead of performing any sacrifice during the intercalated period, they spent that period in performing such accessory rites as are called *Upasads* or sessions and *dīksha* or rites of initiation.

(3) They regarded the intercalated days as being infested by evil spirits and enemies. It is therefore probable that the apparent acts of sorcery undertaken in connection with every kind of sacrifice in order to drive out or to destroy 'those who hated the poets and whom the poets hated,' are acts intended to symbolise the fact of getting rid of an intercalated period. Since an intercalated period is regarded, not only as being burdened with dirt or sin, but also as being infested by Varuṇa, Nirṛti and other good or evil spirits with nooses in their hands to bind their victims, it is probable that, during an intercalated period, the Vedic poets regarded themselves, not only as being burdened with sin but also as being bound with the noose (*pāśa*) of Varuṇa or Nirṛiti.

---

8 *Smṛititattva*, p. 782. 9 *Idid*, p. 778.

It also follows that the removal of sin or of Varuṇa's fetters at the close of a period of twelve or twenty-one days, is a technical expression of the Vedic poets implying the intercalary nature of those days. The removal of guilt at the close of the twelve intercalated days is thus referred to in the *Aitareya-Brāhmaṇa*, IV. 4, 24—

त्रयश्च वा एते ऽत्र्यहा आदशममहरद्वावतिरात्रौ यद्वादशाहो द्वादशाहानि दीक्षितो भवति. यज्ञिय एव तैर्भवति द्वादश रात्रीरुपसद उपैति. शरीरमेव ताभिर्धुनुते. द्वादशाहं प्रसूते धूत्वा शरीरं धूत्वा शुद्धः पूतो देवता अप्येति. य एवं वेद. षट्त्रिंशदह्नो वा एष यद्द्वादशाहः ।

"The Devādasāha consists thrice three days together with the tenth day and two Atirātra days. After having undergone the ceremony of initiation during twelve days one becomes fit for performing the sacrifice. During the twelve days he undergoes the *Upasad* or the vow of fasting. By means of them he shakes off all guilt from his body. He who has such a knowledge becomes purified and clean, and enters the deities after having, during (*these*) twelve days, been born anew and shaken off (*all guilt*) from his body. The Dvādaśāha consists (*on the whole*) of thirty-two days."

The thirty-six days referred to in this passage are three sets of twelve days each, constituting the difference between three lunar and sidereal solar years. The *Atharvavēda*=(Rig. V. VII. 103. 1.) IV. 15. 13, also speaks of the twelve day's vow as follows—

संवत्सरं शशयाना ब्राह्मणा व्रतचारिणः ।
वाचं पर्जन्यजिन्वितां प्रमंडूका अवादिषुः ।

"Having lain for a year, (*like*) Brāhmans preforming a vow, the frogs have spoken forth a voice quickened by *parjanya* [the raining clouds]."

So also the same says IV. 11. 11—

द्वादश वा एता रात्रीर्व्रत्या आहुः प्रजापतेः ॥
तत्रोप ब्रह्म यो वेद तद्वा अनडुहो व्रतम् ॥

"Twelve, indeed, they declare those nights of the vow of Parajāpti; whoso knows the Brahma whithin them—that verily is the vow of the draught ox."

The release from Varuṇa's fetters at the close of twenty-one days is thus referred to in the *Atharvaveda*, IV. 16. 6—

ये ते पाशा वरुण सप्त सप्त त्रेधा तिष्ठंति विषिता रुशंतः ।
सिनंतु सर्वे अनृतं वदंतं यः सत्यवद्यति तं सृजंतु ॥

"What fetters of thine, O Varuṇa, seven by seven, stand triply relaxed, shining—let them all bind him that speaks untruth; whoso is truth-speaking, let them let him go."

I presume that the expression of three times seven milch kine pouring their milky draught, as referred to in the two verses of the *Sāmaveda* (vi. 2, 2, 7) implies the same idea as that of an intercalation of twenty-one periods. From the consideration of these and other similar passages too numerous to be quoted here, we may conclude that expressions such as 'the milking of the kine,' 'the destruction of the evil spirits or of enemies,' and 'the release from the fetters of Varuṇa or of Nirṛiti,' are Vedic expressions implying the passing off of an intercalated period.

It appears that the number of days intercalated differed with different schools of Vedic astronomers, and depended upon the difference between any two kinds of years selected for adjustment with each other. The school which had adopted the synodic lunar year of 354 days seems to have added to every lunar year a Dvādaśāha or period of twelve days, during which they performed a sacrifice with recitation of a Sāma-chant of twelve verses on the last day. With the school which had adopted the sidereal lunar year of 351 days, *i.e.*, the year of thirteen months of 27 days each, and adjusted it with the Sāvana year of 360 days, the number of days added was nine. In this way there seems to have been during the Vedic period a variety of different astronomical schools, whose chief religious function was the performance of a grand sacrifice during each period of their respective intercalary days. A regular account of the 'cows' or intercalary days which each school counted and observed is found preserved under the general title of Gavām-Ayana, "the walk of cows or intercalary periods." The term Gavām-Ayana seems to have been originally intended to be a name of only the intercalary days; but in course of time it appears to have also been used to signify that year which contained intercalary days added to it, no matter whether the number of days so added, or counted as having been added, amounted to a year or more than a year. These and other important points connected with the Vedic calendar are clearly explained both in the *Nidāna-Sutra* and in the *Śrauta-Sūtra* of Lāṭyāyana.

The passage of the *Nidāna-Sutra* in which a few forms of Gavām-Ayana inclusive of the 3 years cyle are defined, runs as follows, v, 11, 12 :—

[10]अथातस्संवत्सरा वर्गाणां. पंचसंवत्सरा वर्गाः [11]तेषु धीरो मनीषया कर्मण उपसदो विद्यात् संस्था वा एषु व्रतानि च. षट्त्रिंशोनो नवोनश्च षडह्नोनोऽथ सावनोऽष्टादशभिर्ज्यायानह्नोभिः सावनात्परो नाक्षत्रमिति मासश्च तस्य चैव त्रयोदश. चांद्रमसस्सावनश्चोभावथाष्टादश्युत्तनोऽष्टा सप्तत्रिंशते पौर्णमास्यां प्रसाधयेत् ।

गवामयनस्योपायांश्चतुरः प्रतिपादयेत्. तेषां नाक्षत्रः प्रथमस्तस्य सप्त विंशिनो मासाः सप्तविंशतिर्नक्षत्राणीति. तस्य कल्पः प्रथमस्य प्रथमस्याभिप्लवस्य स्थाने त्रिकद्रुकत्र्यहं कुर्यात्प्राग्विषुवत उत्तमस्योत्तमस्योर्ध्वं. विषुवतः ते खल्वभिप्लवतन्त्र एव कृताः स्युरित्येके. एते चाधिकृता न चापि निवर्तयत्यथापि दृश्यते त्र्यहस्त्र्यहतन्त्रे कृप्तो यथा स्वरसामानस्त्रिकद्रुकपंचाहश्चाभिप्लवतन्त्रे. सप्तदशरात्रे । स्वरतन्त्रा इत्यपरम्. एवं च तन्त्राविलोपः अपि च सत्रेषु त्रिकद्रुकत्र्यहः स्वतंत्रो भवति.

अथ नवोनस्तस्यैवं त्रयोदश मासाः संभार्यंयोर्मासयोर्नवाहं लुंपेच्चतुरहमेव प्राग्विषुवतः पंचाहमूर्ध्वं तस्य कल्पः प्रथमस्याभिप्लवस्य स्थाने ज्योतिषं च गां च कुर्यात्प्राग्विषुवत ऊर्ध्वं विषुवत उत्तमस्याभिप्लवस्य स्थाने ज्योतिषं नात्र विषुवानभिभवत्युत्तरेऽत्र पक्षसि विषुवानुपसंख्यायत इति.

अथ षड्नाश्चांद्रमसाः षट्पूर्णोपक्रमाः ऊनावसानाः पूर्वे पक्षसि मासास्स्युः ऊनोपक्रमाः पूर्णावसाना उत्तरे. तस्य कल्पः प्रथमस्य प्रथमास्याभिप्लवस्य स्थानेऽभिप्लवपंचाहं कुर्यात्प्राग्विषुवत ऊनेषु मासेषु उत्तमस्योत्तमस्योर्ध्वं विषुवतः ।

---

10. From पंच° to सावनात्परो (for °र:)—and perhaps farther—seems to be a metrical quotation from some other work; with one or two words separated, and one omitted.—J. F. Fleet.

11. *Varga* is not fairly to be rendered by 'cycle.' Cycle is *yuga* or *chakra*: *varga* is a 'group, class.'—J. F. Fleet.

व्याख्यातस्सावनः। स एष आदित्यसंवत्सरो नाक्षत्र आदित्यः खलु शश्वदेतावद्भिरहोभिर्नक्षत्राणि समवैतित्रयोदशाहं त्रयोदशाहमेकैकं नक्षत्रमुपतिष्ठत्यहस्तृतीयं च नवधा कृतयोरहोरात्रयोर्द्वे द्वे कले चेति. सांवत्सरास्ताश्चतुष्पंचाशतं कलाः ते षण्नववर्गाः स षट् षष्टित्रिशतः षष्टित्रिशते । श्लोकौ भवतः—

सप्तविंशती राष्ट्रस्य राज्ञो वसतयो मिताः ।
त्रयोदशाहं त्रयोदशाहमेकैकं नक्षत्रमुपतिष्ठति ॥
त्रयोदशाहानि तृतीय मह्नश्चतस्त्रेधा दशतयो विकुर्वन् ।
त्रिणवं पंथानं विततं पुराणं चत्वारिंशता नवरात्रैसमश्नुते॥ इति.

अथाष्टादशभिर्ज्यायानादित्यसंवत्सर एव तैर्यगयनिको भवति. आदित्यः खलु शश्वदेकदा षण्मासानुदङ्ङेति नव चाहानि तथा दक्षिणा. तदप्येते श्लोका भवंति :—

यस्मिन्वै परिवत्सरे
सौम्यो मासोऽथ चांद्रमासो ।
नाक्षत्रो न विलुप्यते
कस्स्वित्तं वेद क स्स्वित् ॥

अष्टासप्तत्रिंशते तस्मिन्
संवत्सरे मिते ।
सौर्यो मासोऽथ चांद्रमसो
नाक्षत्रो न विलुप्यते ॥

सप्तविंशतिमेवैष सप्ताहानेति दक्षिणा ।
तथोदङ सप्तविंशतिमिति ॥

तस्य कल्पः संभार्ययोर्मासयोरष्टादशाहान्युपाहरेन्नवाहमेव प्राग्विषुवतः नवाहमूर्ध्वं त्रिकद्रुकांश्चाभिप्लवं च प्राग्विषुवतोऽभिप्लवं च त्रिकद्रुकांश्चावृत्तानूर्ध्वं विषुवतः।

Then the years of the class: the classes (*are*) of the five years. In them the sage by his wisdom will know the sessions of the ritual,[12] and the basic form (of *the sacrificial rites*),[13] and the vows *or* ceremonies (*to be observed*) in them.

"(*The year*) which is less (*than the Sāvana year*) by 36 (*days*); that which is less by 9 (*days*); that which is less by 6 (*days*); then the Sāvana year (*of 360 days*); then the year which is greater than the Sāvana year by 18 days. The sidereal year (*of 351 days*) has a thirteenth month (*of 27 days*). Then the two kinds of years: the lunar and the Sāvana. Then the year which is greater than the Sāvana year by 18 days: one has to observe (*it*) on every 38th or 37th full-moon.

He has to know the four forms of Gavām-Ayana. Of them, the sidereal year (*of 324 days*) is the first; its months are of 27 days each, because there are 27 *Nakshatras*. The mode of observing it (*is this*): in the place of each first [period of six days called] *Abhiplava* (*of every month of 30 days*) before the central day (*of the year*), one should observe [a period of only three days known as] *Trikadruka*:[14] (*likewise in the place*) of each last (*Abhiplava*) after the central day. Some say that

---

12. *Upasad*: lit. 'the sitting down, waiting for the arrival of the final sacrificial day.'

13. *Saṁsthā*; lit. a staying or abiding together.'

14. *Trikadruka* is the name given to a unit of three days' of which the first day is called *jyotis*, 'light,' the second *go*, 'cow' and the third *āyus*, 'life.' *Abhiplava* is the name given to a unit of six days, of which the first three days are named like the *Trikadruka* days and the last three days are called *go*, *āyus*, and *jyotis*.

they (*the Trikadruka days*) are so devised as to be of the same form as the (*Abhiplava* days; they have their place here; nor is their observance opposed to that of the *Abhiplava* days. It is also known that, like the *Svarasāman* days,[15] the unit of three days [the *Trikadruka* days] is devised as a special period of three days. The three *Trikadruka* days, as well as the five days of the six *Abhiplava* days are observed together in the sacrificial session of seventeen nights. Others say that the *Trikadruka* days are the same as the *Svarasāman* days. And thus the usual form of the calendar days and their rites is not lost; for the *Trikadruka* days have their own independent place in all sacrificial sessions.

"Then (*the year of 351 days*) which is less (*than the* Sāvana *year*) by nine days:[16] thus it has months (*each* of *27 days*). He has to omit nine days in the two intercalary months [*sambhārya*; *i.e.*, the sixth and the seventh month, each of 30 days]; four days (*are to be omitted*) before the central day of the year, and five days after it. This is how it is done: in the place of the first *Abhiplava* (of *the sixth month* of *30 days*) before the central day, only two days known as *jyotis* and *go*, are to be observed; and in the place of the last *Abhiplaua* (of *the seventh month*) after the central day, only one day, known as

---

15. *Svarasāman* is a name given to the three days before and after the central day of a sacrificial session. Special Sāma-chants are sung on these six days. If the *Trikadruka* days were considered as identical with the *Svarasāman* days, which are strictly observed immediately before and after the central day of a sacrificial session, the other days of the session would be counted in periods of six days each. This appears to be the meaning of 'an independent place for the *Trikadruka* days.'

16. See Calendar, Form II. below.

jyotis, is to be observed. No central day occurs in the year (of *351 days*); for it is counted in its later half.

"Then the lunar years (of *354 days*) which are less (*than the Sāvana year*) by six days:[18] in the first half (*of this year*) there are six months, beginning with one which is full [*i.e.*, consists of 30 days] and ending with one which is deficient [*i.e.*, contains only 29 days]; in the latter (*half there are six mouths*), beginning with one which is deficient and ending with one which is full. This is how it is observed: in the deficient months before the central day, in the place of each first *Abhiplava* one should observe (*only*) five days of *Abhiplava*; (*likewise in the place*) of each last (*Abhiplava*) in the deficient months after the central day.

"The Sāvana year (*of 360 days*)[19] has been explained. It is this same sidereal year of the sun.[20] The sun is known to pass through (*each of*) the *nakshatras* in a fixed number of days: he remains in each *nakshatra* for thirteen days, together with a third part of a day and two out of nine *kalās* or parts of a day-and-night [*i.e.*, of a whole day]: these *kalās* or parts amount in a year to 54, and

---

18. See Calendar, Form II, below.

19 See Calendar, Form I., and for the year of 366 days, see Form IV. below.

20 The text treats here in a somewhat obscure manner of *both* the Sāvana year of 360 days *and* the sidereal solar year of 366 days. The latter is not mentioned as one of the five classes of years in the beginning of the passage. The text seems to suggest that the Sāvana year, before being regarded as=30 days × 12, was a sidereal year of 27×13⅓ days, and that the year of 366 days,=$27 \times 13\frac{5}{9}$, was a refinement of it, as a result of experience showing that the sun required $\frac{2}{9}$ of a day more time to pass through each *nakshatra*.— J. F. Fleet.

are equal to six times nine *kalās* [*i.e.*, 6 days]: thus it consists of 366 (*days*) as contrasted with the (*Sāvana year*) consisting of 360 (*days*). There are two verses about this:—

" 'Twenty-seven are the mansions in the king's [*i.e.*, the Sun's] dominion; thirteen and thirteen days he resides in each *nakshatra*: thirteen days and one-third of a day; thus dividing four times ten days into three (*equal*) parts, he traverses the broad and ancient path of thrice nine stations in the course of forty periods, each of nine nights.'

"Then the year of the sun (of 378 *days*)[21] which is greater (*than the Sāvana year*) by eighteen days; this indeed is made by his transverse motion;[22] it is well known that the sun always goes to the North for six months and nine days, and likewise to the south. Accordingly there are following verses—

Who knows that year in which the solar, the lunar, and the sidereal months are not lost, who knows that? In the year measured by 37 or 38 (*full-moons*), the solar, lunar, and the sidereal months are not lost. The sun goes to the South for twenty-seven times seven days.

"This is how this year is observed—In the two intercalary months,[23] one should intercalate eighteen days; nine days before the central day of the year and nine days after it; three *Trikadruka* days and six

21 See Calendar, Form III. below.

22 The motion from South to North and back again, which the sun seems to have in passing from solstice to solstice, is transverse, at right angles, to his actual motion from West to East through the *nakshatras*.—J. F. Fleet.

23 See Note 18, above.

Abhiplava days before the central day, and six *Abhiplava* and three *Trikadruka* days after the central day."

Similar forms of calendar, together with some more varieties, are also described in the *Śrauta-Sūtra* of *Lāṭyāyana* iv. 8, 1—7.—This is what he says—

ज्योतिषामयन विकल्पाः। तत्र यदादितोऽन्ततस्तदूर्ध्वं विषुवतः। मासि मास्याद्यस्याभिप्लवस्य स्थाने त्रिकद्रुकाः। स षट्त्रिंशदूनो नाक्षत्रस्सप्तविंशिनो हि मासाः। षष्ठाद्यस्याभिप्लवस्य स्थाने ज्योतिर्गौश्च ज्योतिरेवावृत्ते स नवोनो नाक्षत्र एव त्रयोदशी युग्ममासेष्वाद्यस्याभिप्लवस्य स्थाने तत्पंचाहः स षड्नश्चांद्रमसः। षष्ठादौ त्रिकद्रुकानाभिप्लवं चोपदध्यात् सोऽष्टादशाधिकः पौर्णमासीप्रस्रवस्तैर्यगयनिक आदित्यस्य।

"Varieties of the movements of the heavenly luminaries. In the calender pertaining to these movements, that which is observed at the beginning (of *each month*) before the central day, is observed at the close (of *each month*) after the central day. In the place of the first six *Abhiplava* days in each month, only three *Trikadruka* days are observed. Thus this sidereal lunar year is less (*than the Sāvana year*) by 36 days, since its months consist of 27 days each.

"In the place of the first six *Abhiplava* days of the sixth month (of *the Sāvana year*), there are observed only two days, known as *jyotis* and *go* and in the second part of the year, which is merely a repetition of the first part, only one day, known as *jyotis*, is observed in the place of the last six *Abhiplava* days (of *the seventh month*); thus it is less (*than the Sāvana year*) by nine days, and is a sidereal year having a thirteenth month.[24]

24 See Calender, Forms I and II.

"In the place of the first six *Abhiplava* days in all the even months, only five *Abhiplava* days are observed; this is less (*than the Sāvana year*) by six days, and is lunar.[25]

"In the beginnig of the sixth month, one should intercalate three *Trikadruka* days and six *Abhiplava* days;[26] thus it is greater (*than the Sāvana year*) by 18 days, and is productive of a full-moon; and it is caused by the transverse motion of the sun."[27]

Besides the three forms of calendar mentioned above, which are similar to those described in the *Nidāna-sutra*, a few more varieties also are noticed in the Lāṭyāyana *Śrauta sūtra*. As some of these varieties are referred to, though only briefly, in the *Kṛishṇa-Yajurvêda*, it can not be said that they existed during the Vedic period, and that they are not the later contrivances of Sūtra writers. It is therefore necessary that we should understand them as clearly as possible. The Lāṭyāyana-Sūtra continues in iv. 8, 8-20:—

उत्सर्जनानि मासि मासि । यथाऽन्त पवमावृत्तानामादिः । पूर्वेष्वभिप्लवेषु षष्ठमहरुक्थ्यं कृत्वाऽग्निष्टोममुत्तमे । तद्वैकत्रिकस्तोमम् । सवनविधं पशुं कुर्वन्नुत्तममभिप्लवपंचाहं कृत्वा षष्ठस्थाने सवनविधः पशुः । प्रथमं चाभिप्लवं पंचाहं कृत्वा मासांते सवनविधः पशुः । सर्वानूनानेके प्रथममभिप्लवपंचाहं कुर्युः । अहनी वा समस्येयुरभिप्लवपृष्ठ्ययोः सन्निपातके । अभिप्लवयोरुत्तमे । तथा सत्येकादश्यां पूर्वपक्षस्य दीक्षित्वा त्रयोदशादीक्षाः कुर्वीरन् । सप्तदश वा । व्यत्यासं वा पूर्णोनानूनपूर्णानावृत्तान् शालंकायनिनः ।

25 *Ibid*, form III.

26 That is, nine days before the central day and nine days after it.

27 See Calender, Form IV.

"Omissions (of *days*) month after month.[28] Just as the last day (*in each month in the first half of the year*) is omitted, so the first day (*in each month*) in the repeated part of the year [*i.e.*, the second part] is omitted. Having treated as *Ukthya* days the sixth day in each of the three *Abhiplava* periods of six days, they observe the sixth day of the last *Abhiplava* [*i.e.*, the fourth *Abhiplava*] as an Agnishṭoma day.[29]

"On the sixth day of the fourth *Abhiplava* period of each month, they have to recite a set of Sāma-verses called *Ekatrika.*[30] In view of immolating a sacrificial animal, they make the last (*i.e.*, *the fourth*) *Abhiplava* consist of only five days, and immolate a sacrificial animal on the sixth day. Having made the first *Abhiplava* consist of only five days, they immolate a sacrificial animal at the close of the month. Some teachers make all the months deficient by one day: they make the first *Abhipliva* of each month consist of only five days.[31] At

---

28 This is what is called *utsarginām ayanaṁ*, which is described in the *Krishṇa-Yajurveda, VII.* 5, 6.

29 It should be noted here that according to this school a month is made to consist of four *Abhiplavas* of six days each and a *Prishṭhya* of six days closiug the month. According to the commeutary of Agnisvāmin on this sūtra, it is the *Ukthya* days that are omitted. Accordingly, three days are omitted in each month, thus making it consist of 27 days. See Calendar, Form V. below.

30 Agnisvāmi quotes a passage on the authority of which the day with the Ekatrikastoma is omitted, Hence, according to this school, the month seems to consist of only 20 days. See Calandar, Form VI. below.

31 See Calendar, Form VII. below.

the junction of *Abhiplava* and *Pṛishṭhya* days,[32] they reckon the last day of the (*fourth*) *Abhiplava* period and the first day of the *Pṛishṭhya* as one day [*i.e.*, they treat the two as a single day]. In the last month [*i.e.*, the twelfth month], they make the last day of the last but one *Abhiplava* the first day of the last *Abhiplava.*[33] If so [*i.e.*, if they omit one day in each month of the year], they should undergo the vow of initiation for their sacrifice on the eleventh day of the bright half of the month, and spend thirteen days in vow (before they perform their sacrifice on the fourteenth, *i.e.*, the *Ekāshṭakā* day of the dark half of the month). Or they have to spend seventeen days in vow.[34]

---

32 Like *Abhiplava*, *Pṛisṭhya* is also a name given to a period of six days which are called: (1) Rathantra, (2) Bṛihat, (3) Vairūpa, (4) Vairāja, (5) Śākvara, and (6) Raivata, after the names of the Sāma-verses recited on those days. In some schools, the last six days of each month are observed as *Pṛishṭhya.*

33 The twelth month, when recast in the form of Gavām-Ayana, consists of three *Abhiplavas of* six days each and a period of *Dvādasāha* or twelve days. In order to make this month also consists of 29 days, they make the last day of the second *Abhiplava* the first day of the third Abhiplava. See Callendar, Form VIII.

34 As each month of the year is made to consist of 29 days (total 348), the deficiency in the year amounts to tweleve or seventeen days according as we take the Sāvana year of 360 days or a solar year of 365 days for comparison. It is clear, therefore, that the twelve or seventeen days regarded as *Dikshā-days* are no other than intercalary days required to make up the year in observance Compare *Aitareya-Brāhmaṇa*, iv. 4, 24; and *Atherva-veda*, iv, 11, 11; iv, 16, 6, quoted above.

Calendar—Form I.

[Abbreviations: J=*jyotis*; G=*go*; A=*āyus*.]

Savana Year of 360 days.

1st Month.

	J.	G.	Ā.	G.	Ā.	J.
Abhiplava I	1	2	3	4	5	6
,, II	7	8	9	10	11	12
,, III	13	14	15	16	17	18
,, IV	19	20	21	22	23	24
,, V	25	26	27	28	29	30[1]

1 Similarly the other eleven months.

2 In order to convert this Sāvana year into the sidereal lunar year of 351 days, the last four days of the first Abhiplava of the 6th month and the first five days of the last Abhiplava of the 7th month are omitted.

N.B.—Instead of being called *Abhiplava*, the last week in each month seems to have been called by others, as *Pṛishṭhya*, the days being named Rathantara, Bṛihati, Vairūpa, Vairāja, Śākvara, and Raivata respectively.

Calendar—Form II.

Sidereal Lunar year of 351 days.[1]

Calendar—Form III.

Synodic Lunar year of 354 days.

1st Month.

	J.	G.	Ā.	G.	Ā.	J.
Abhiplava I	1	2	3	4	5	6
" II	7	8	9	10	11	12
" III	13	14	15	16	17	18
" IV	19	20	21	22	23	24
" V	25	26	27	28	29	30

2nd Month.

	J.	G.	Ā.	G.	Ā.	J.
Abhiplava I	......	1	2	3	4	5
" II	6	7	8	9	10	11
" III	12	13	14	15	16	17
" IV	18	19	20	21	22	23
" V	24	25	26	27	28	29[2]

1 See Note 2, Page 41.

2 Similarly the other months, omitting the first day in the 4th and 6th months, and the last day in the 7th, 9th, and 11th months.

Calendar—Form IV.

Sidereal year of 366 days with an intercalation of 18 days.

1st Month.

	J.	G.	A.	G.	A.	U.
Abhiplava I	1	2	3	4	5	6
,, II	7	8	9	10	11	12
,, III	13	14	15	16	17	18
,, IV	19	20	21	22	23	24
Pṛiṣṭhya I	25	26	27	28	29	30[1]

6th month.

	J.	G.	A.	G.	A.	J.	J.	G.	A.
Abhiplava I	1	2	3	4	5	6	7	8	9
,, II	10	11	12	13	14	15	......	......	......
,, III	16	17	18	19	20	21	......	......	......
,, IV	22	23	24	25	26	27	......	......	......
,, V	28	29	30	31	82	33	......	......	......
,, VI	34	35	36	37	38	39*	40	41	42
,, VII	43	44	45	46	47	48	......	......	......

1 Likewise the 2nd, 3rd, 4th and 5th months, and also the months from 7 to 12.

Calandar—Form V.

	1	2	3	4	5	6
Abhiplava I	J.	G.	A.	G.	A.	U.
,, II	J.	G.	A.	G.	A.	U.
,, III	J.	G.	A.	G.	A.	U.
,, IV	J.	G.	A.	G.	A.	Agni.
Prishṭhya I	R.	B.	V.	Vr.	S.	Rai.

J=gyotishṭoma; G=Goshṭoma A=Āyushṭoma; U=Ukthya.

Note—Similarly the other months, only *Prishṭhya*, taking the place of the first Abhiplava in the second half of the year.

Calendar—Form VI.

	1	2	3	4	5	6
Abhiplava I	J.	G.	A.	G.	A.	J.
,, II	J.	G.	A.	G.	A.	J.
,, III	J.	G.	A.	G.	A.	J.
,, IV	J.	G.	A.	G.	A.	Ekatrika[1]
Prishṭhya I	R.	B.	V.	Vr.	S.	Rai,

1 This day is not counted; similarly the other months.

Calendar—Form VII.

	1	2	3	4	5	6
Abhiplava I	J.	G.	A.	G.	A.	J.
„ II	J.	G.	A.	G.	A.	J.
„ III	J.	G.	A.	G.	A.	J.
„ IV	J.	G.	A.	G.	A.	J.
Prishṭhya I	R.	B.	V.	Vr.	S.	Rai.

Calendar—Form VIII.

Month 11	1	2	3	4	5	6
Abhiplava I	J.	G.	A.	G.	A.	J.
„ II	J.	G.	A.	G.	A.	J.
„ III	J.	G.	A.	G.	A.	J.
„ IV	J.	G.	A.	G.	A.	J.
Prishṭhya I	R.	B.	V.	Vr.	S.	Rai.

Calendar - Form VIII.

Month 12.	1	2	3	4	5	6
Abhiplava I	J.	G.	A.	G.	A.	J.
,, II	J.	G.	A.	G.	A.	J.
,, III	J	G.	A.	G.	A.	.....
The Dvādaśāha	R.	B.	V.	Vr.	S.	Rai.
	C^1	C^2	C^3	C^4	M.	U.[1]

1 C^1 to C^4=Four Chandoma days; M=Mahāvrta; U=Udayanīya day.

Calendar—Form IX.

The Savana year with twenty-one intercalary days inserted between the sixth and seventh months.

6th Month	J.	G.	Ā.	G.	Ā.	J.
Abhiplava I	1	2	3	4	5	6
,, II	7	8	9	10	11	12
,, III	13	14	15	16	17	18
,, IV	19	20	21	22	23	24
Priṣṭhya days	25	26	27	28	29	30

Abhijit	Six Prishthya days						Three svarasama days			Central day	Visvajit	Three svarasama days			Six Prishthya days					
1	2	3	4	5	6	7	8	9	10	11	12	13	14	15	16	17	18	19	20	21

7th Month of Calendar Form IX	J.	G.	Ā.	G.	Ā.	J.
Abhiplava I	1	2	3	4	5	6
,, II	7	8	9	10	11	12
,, III	13	14	15	16	17	18
,, IV	19	20	21	22	23	24
Prishthya days	25	26	27	28	29	30

Names for the different kinds of years as suggested by Dr. J. F. Fleet :

$324 = 27 \times 12$ ...	" Sidereal lunar year of 324 days."
$351 = 27 \times 13$ ...	" Sidereal lunar year of 351 days."
$354 = 30 \times 6 + 29 \times 6$ ...	" Synodic lunar year."
$360 = 30 \times 12$ or $27 \times 13\frac{1}{3}$	The best possible term for this is the original one, " Sāvana year."
$366 = 27 \times 13\frac{5}{9}$ ...	" Sidereal solar year "
$378 = 189 + 189$ ...	" Pseudo-solstitial year of 378 days."
365	" Vague solar year."
$365\frac{1}{4}$	" Julian solar year."—This term involves an anachronism but it is customary and explains at once what is meant.

This is a portion of the article that appeared with Dr. J. F. Fleet's notes in I. A. 1913.

CHAPTER II

THE CYCLE OF FIVE YEARS.

THE nature of the five-years cycle is explained in the *Maitrāyaṇīya Saṁhitā* itself. The passage (1. 10, 8) in which it is described runs as follows:—

प्राणेभ्यो वै ताः प्रजाः प्राजायंत. प्राणा वा एतानि नव हवींषि नव हि प्राणाः आत्मा दैवता ततः प्रजायते. नव प्रयाजा नवानुयाजा द्वा आज्यभागा अष्टौ हवींष्यग्नये समवद्यति. वाजिनो यजति. तत्त्रिंशत्. त्रिंशदक्षरा विराट्. विराजेव प्रतितिष्ठति. विराजो वै योनेः प्रजापतिः प्रजा असृजत. विराजो वा एतद्योनेर्यजमानः प्रजायते. त्रिंशत्त्रिंशद्वै रात्रयो मासः. यो मासः स संवत्सरः. संवत्सरः प्रजापतिः. तत्प्रजापतेश्च वा एतद्विराजश्च योनेर्मिथुनाद्यजमानः प्राजायत. एकैकया वा आहुत्या द्वादश द्वादश रात्रीरयुवत. ता यावतीः संख्याने तावतीः संवत्सरस्य रात्रयः. संवत्सरमेव भ्रातृव्याद्युवते. वैश्वदेवेन चतुरो मासो-ऽयुवत. वरुणप्रघासैः परांश्चतुरः. साकमेधैः परांश्चतुरः. तानेव भ्रातृव्याद्युवत. ऋतुयाजी वा अन्यश्चातुर्मास्ययाज्यन्यः. यो वसन्तोऽभूत्प्रावृडभूच्छरदभूदिति यजते स ऋतुयाजी. अथ यस्त्रयोदशं मासं सम्पादयति त्रयोदशं मासमभियजते स चातुर्मास्ययाजी ऋजूंस्त्रीनिष्ट्वा चतुर्थमुत्सृजेत. ऋजू द्वौ परा इष्ट्वा तृतीयमुत्सृजेत. ये वै त्रयस्संवत्सरास्तेषां षट्त्रिंशत्पूर्णमासाः यौ द्वौ तयोश्चतुर्विंशतिः. तद्योऽमी षट्त्रिंशत्याधि तानस्यां चतुर्विंशत्यामुपसम्पादयति एष वाव स त्रयोमासः तमेवैनत्सम्पादयति. तमभियजते. वैश्वदेवेन यजेत पशुकामः न वरुणप्रघासैर्न साकमेधैः. सर्वो वै पुरुषः साहस्रो जायते यावत्तरसं त्वेवैति. प्रजननं वा एतद्धविर्यद्वैश्वदेवं. यद्वैश्वदेवेन यजते प्रजननाय वा एतद्यजते स्वां मात्रां गच्छानीति. स यदा सहस्रं पशून्गच्छेदथ वरुणप्रघासैर्यजते. यदेवादः सहस्रमगंस्तस्यैतदंहोऽवयजति.

M. S. I, 10, 8.

From vital breaths are those creatures born. Vital breaths are these nine oblations, for nine are the Vital breaths. Atmā [the inner man] is the deity. From him (*the deity*) is (*the sacrificer*) born. Nine fore-offerings, two butter portions, and eight oblations, he puts together for Agni. He makes the oblation of curdled milk (*vājina*). That amounts to thirty. The Virāṭ metre consists of thirty syllables. By means of the Virāṭ, he has a firm footing; for Prajāpati created the creatures from the womb of Virāṭ. From this womb of Virāṭ is also the sacrificer born. Thirty and thirty nights are a month. That which is the month is the year. Prajāpati is the year. From the womb of the couple, Prajāpati and Virāṭ, is the sacrificer born. With each oblation he inserts twelve and twelve nights. There are, when counted, as many oblations as there are nights in a year. He separates the year from the enemy. With the Vaiśvadeva sacrifice he inserts four months; with the Varuṇapraghāsa sacrifice, the next four months; with the Sākamedha sacrifice, the next four. These are the months which he has separated from the enemy. He who sacrifices for the seasons is one, while he who sacrifices for the four months is another. He who sacrifices for the reason that that which was the spring has become the rains, and that that which was the rainy season has become the autum, is a sacrificer for the seasons. But he who gains a thirteenth month, and sacrifices for that thirteenth month, is the one who sacrifices for the four-months. Having sacrificed for three regular (*months*), he should omit the fourth; and then having sacrificed for the next two regular (*months*), he should omit the third. What are counted as three years, there are in them thirty-six full moons; what are counted as the next two, there are in them twenty-four. Those (*days*) which exceed (*an intercalary month*) in thirty-six full moons, he puts in (*the next*) twenty-four full moons. This is, verily, that thirteenth month. This is what he gains and sacrifices for. He who is desirous of cattle should observe the Vaiśvadeva sacrifice, but neither the Varuṇapraghāsa nor the Sākamedha. All the Purusha amounts to a thousand when counted together as far as the flesh-oblation (*Tarasa*). The oblation made in the Vaiśvadeva sacrifice is, verily, the birth (*of creatures*). The reason for which he sacrifices with the Vaiśvadeva is the birth of creatures, for which he sacrifices with the thought that he may attain his own measure. When he comes by a thousand cattle, then he should sacrifice with the Varuṇapraghāsa. When he comes by a thousand of this, then he gets rid of his sin by means of sacrifice."

Omitting the sacrificial technicalities with which the above passage abounds, we may confine our attention to that portion of the passage where a distinction is drawn between the Season-sacrificer and the Four-monthly sacrificer, and where the nature of the three Four-monthly sacrifices, the Vaiśvadeva, the Varuṇapraghāsa, and the Sākamedha is clearly defined. It is clear from this passage that during the Vedic period there were two important schools of priestly astronomers, the Season-sacrificers and the Four-monthly sacrificers. Of these two schools, one seems to have been observing the lunar year of 354 days without adjusting it to the solar or sidereal year, and to have allowed it to fall back by 12 days in every year and to regain its original initial point at the close of 32 or 30 years, making a full rotation through the seasons. This is what is meant by the expression that what was the spring became the summer, and that what was the summer became the autumn. The priests who were sacrificing for such rotating seasons are called Ritu-yājins'. The other school of sacrificers called the Chāturmāsyāyājins, 'Four-monthly sacrificers' did not, like the Season-sacrificers, allow the year to fall back for want of intercalation, but adjusted their lunar year of 354 days to the sidereal year of 366 days by adding two months in five years or four moths in ten years. From the reference made to twelve days in the beginning of the passage, it is clear that it is the sidereal year of 366 days that is taken for adjustment with the lunar year of 354 days. Accordingly the extra days in three lunar years amount to thirty-six days, *i.e.*, one month and six days, says the author, are to be added to the twenty-four days of the subsequent twenty-four full-moons or two years. From the statement that whoever gains a thirteenth month is a Four-monthly sacrificer, it is clear that the three Chāturmāsyas or Four-months are undoubtedly three intercalary periods of four months each. I have pointed out in my *Vedic Calendar* how the vedic poets regarded the intercalary days or months as enemies and as sinful periods infested with demons. This is what the writer means when he says that the sacrificer has to separate the Chāturmāsyas, the Four-months, from the enemy. What is meant by the expression that Purusha amounts to a thousand will be explained later on in connection with the Purusha hymn.

It appears that when the three Four-monthly periods were got rid of by intercalation, the Vedic poets used to renew their sacred fire by churning anew. This idea is conveyed in the following passage of the *Maitrāyaṇīya Saṁhitā* (I. 10, 7):—

त्रेधा विहितानि चातुर्मास्यानि. संवत्सरं वै चातुर्मास्यानि-संवत्सरेणाग्निं मन्थति.

"Three are the Four-monthly sacrifices to be performed. To a year (*amount the three*) Four-monthly periods. In such a year [*i.e.*, once in thirty years] the sacrificer churns the fire [*i.e.*, sets up the sacrificial fire again]."

It is not to be understood that the Vedic poets were adjusting the lunar year to the sidereal year by intercalating four months once in ten years alone. Since a thirteenth month is frequently mentioned in the *Véda*, we may believe that they were adjusting the years once in two and a half years, when one intercalary month occurs. It is, therefore, certain that whenever a thirteenth month is mentioned, half a cycle of five luni-solar years is meant. The following passage of the *Maitrāyaṇīya Saṁhitā* (I. 5, 6) refers to a thirteenth month and the form of the sacrifice performed in it:—

जीर्यति वा एष आहितः पशुर्ह्यग्निः. तदेतान्येवाग्न्याधेयस्य हवींषि संवत्सरे संवत्सरे निर्वपेत्. तेन वा एष न जीर्यति तेनैनं पुनर्णवं करोति. तन्न सूक्ष्यम्. एताभिरेवाग्नेयपावमानीभिः अग्न्याधेयस्य याज्यानुवाक्याभिरुपस्थेयः. तेन वा एष न जीर्यति. तेनैनं पुनर्णवं करोति. द्वादशभिरुपतिष्ठते द्वादश मासाः संवत्सरः संवत्सरमेवाप्त्यावरुन्धे. अग्नीषोमीयया त्रयोदश्युपस्थेयोऽस्ति मासस्त्रयोदशः तमेवैतयाप्त्वावरुन्धे.

"When once set up, he becomes old; for Agni is (*like*) a beast. Hence he should offer, year after year, these oblations of the Agnyādheya rite. He does not thereby grow old. The sacrificer renews him thereby. This (*way of renewing the fire*) is not well-considered. The sacrificer should simply praise the fire with the Yājya and Anuvākya hymns called Āgneya-pāvamāni, used in the Agnyādheya rite. Thereby he does not become old. Thereby the sacrificer renews him. The sacrificer praises him with twelve verses, for there are twelve months in the year. Thus he catches hold of the year and keeps it. He is to be praised with a thirteenth verse dedicated to Agni and Soma, for there is the thirteenth month also. With this verse he catches hold of that month and keeps it."

Since the Atharvaveda (V. 6, 4) assigns the thirteenth month to Indra (**त्रयोदशो मास इन्द्रस्य गृहः**), ("the thirteenth month is the home of Indra,") we shall not be wrong in considering Indra also as one of the chief deities worshipped in a thirteenth month. The following passage of the *Maitrāyaṇiya Saṁhitā* (II. 1, 3) furnishes additional evidence about the same fact:—

अग्नीषोमाभ्यां वै वीर्येणेन्द्रो वृत्रमहन्. स ओजसा वीर्येण व्यार्धत्. स एतमैंद्राग्नमपश्यत्. तेन ओजो वीर्यमात्मन्नधत्त.

"Indra killed Vṛitra with the power of Agni and Soma. Hence he grew with brightness and strength. He saw the power of Agni also. Thereby he kept brightness and strength in himself."

The connection of Vṛitra with Agni and Soma, the gods of an intercalary month, will be explained later on. That intercalary months were being observed, either singly or in sets of two, three, or four months, is clear from the following passage of the *Maitrāyaṇīya Saṁhitā* (I. 11, 10):—

वसवस्त्रयोदशाक्षरया त्रयोदशं मासमुदजयन्. रुद्राश्चतुर्दशाक्षरया चतुर्दशं मासमुदजयन्. आदित्याः पञ्चदशाक्षरया पञ्चदशं मासमुदजयन्. अदितिष्षोडशाक्षरया षोडशं मासमुदजयत्.

"The Vasus conquered the thirteenth month with a verse of thirteen syllables. The Rudras conquered the fourteenth month with a verse of fourteen syllables. The Adityas conquered the fifteenth month with a verse of fifteen syllables. Aditi conquered the sixteenth month with a verse of sixteen syllables."

Since in this passage a year of 12 months is referred to before speaking of the thirteenth, fourteenth, and other months, I take them to be of an intercalary nature. There is no reason to believe that the Vedic poets were counting thirteen or sixteen ordinary months in a year, though they were acquainted with the luni-solar cycle of 5 years, as pointed out above. The following passage of the *Maitrāyaṇiya Saṁhitā* (I. 10, 5) leaves no doubt that Indra is a god of an intercalary month :—

देवाश्च वा असुराश्चास्मिंल्लोके आसन्. स प्रजापतिरकामयत प्रासुरान्नुदेय प्रजाः सृजेयेति. स चातुर्मास्यान्यपश्यत्. चातुर्मास्यैर्वै सोऽसुरान्प्राणुदत चातुर्मास्यैः प्रजा असृजत. तद्य एवं विद्वांश्चातुर्मास्यैर्यजते प्र भ्रातृव्यं नुदते प्र प्रजया च पशुभिश्च जायते अग्निष्टोमाद्वैश्वदेवं यज्ञक्रतुं निर्माय प्रजापतिः प्रजा असृजत. उक्थ्याद्वरुणप्रघासान्यज्ञक्रतुं निर्मायेमाः प्रजा वरुणेनाग्राहयत्. अतिरात्रात्साकमेधान्यज्ञक्रतुं निर्मायेंद्रो वृत्रमहन्. सृष्टाः वा अन्याः प्रजाः आसन्नसृष्टा अन्याः अथ प्रजापतिरकामयत प्रजास्सृजेयेति. संवत्सरो वै यज्ञो यज्ञः प्रजापतिः स एते पयसी आत्मन्नधत्तोधन्यं च बाह्यं अथैताभ्यो देवताभ्यो एतानि हवींषि भागं निरवपत्. तैः प्रजा असृजत. ऋतुभ्यो वै ताः प्रजाः प्राजायंत. ऋतवो वा एतानि पञ्च हवींषि. *M. S.* I. 10, 5.

"The Devas and the Asuras were in this world together. Prajāpati desired that he might drive out the Asuras and create children. He looked to the 'Four-months:' for it is by the 'Four-months' that he drove out the Asuras and created children. Whoever, knowing thus, performs the sacrifice of 'Four-months,' will drive out his enemy and get both children and cattle.

"Creating the Vaiśvadeva sacrifice on the model of the Agnishṭoma, Prajāpati created children; and creating the Varuṇa-praghāsa sacrifice on the model of the Ukthya, he put these children under the clutches of Varuṇa. Creating the Sākameda sacrifice on the model of the Atirātra sacrifice, Indra killed Vṛitra. The children that were created became of one kind, while those that were not created became of another. Then Prajāpati desired that he might create children. The year is the sacrifice, and sacrifice is Prajāpati. He kept in himself this pair, the year and the sacrifice, (*like two kinds of*) milk, that which is produced from the udder and that which is external. Then to these gods he offered the following oblation as a share, and created children from them: from the seasons were those children born. The seasons are the five oblations."

From this passage we can understand the technical sense in which the words *Dēva*, *Asura*, and *Prajās*, are commonly used in the Vedas. In the terminology of the Vedic poets the name of the ordinary days of a year is *prajāḥ*, 'children.'[1] I have pointed out in my notes in the *Vedic Calendar*, how the Vedic poets regarded the intercalary months as Asuras, demons. It follows therefore that the word Deva as opposed to Asura must

1 R. V. I., 164; A. V. IX, 8.

mean an ordinary month or days. Accordingly, we may interpret the conflict between the Devas and the Asuras as denoting some inconsistency between the ordinary and the intercalary months. That the words, *Deva*, *Asura* and *Prajā* have such meanings as the above, is confirmed by the above passage: We are told in the passage that Prajāpati or Father Time repelled the Asuras by means of the Chāturmāsyas, a period of four intercalary months, as pointed out above—and that having done so, he created children. This evidently means that Prajāpati got rid of the extra months by intercalating four months in ten years, and adjusting thereby the lunar to the sidereal year, brought the seasons and days to their usual position which was four months behind before intercalation. We also learn that Indra is a god of an intercalary month, and that the oft-repeated destruction of Vṛitra by Indra is an act of getting rid of the sinful and demon-like eclipse through the worship of Indra and other gods; for we are told in the passage that Indra killed Vṛitra by the Sākamedha, or the sacrifice performed during the third period of the four intercalary months, *i.e.*, at the end of 30 or 60 years.

From a consideration of the passage explained above, we learn that Prajāpati is Father Time, that his children are the ordinary days of the year, that the Asuras are the sinful intercalary months, and that Indra is a god of an intercalary month. We know from the story of Aditi that Indra is one of her sons. Accordingly, we may take Aditi to mean the cycle of five luni-solar years, bringing forth Indra periodically along with her other sons. The other sons also must necessarily be the gods of intercalary months. This idea is, as clearly as the sacrificial terminology of the poets could permit, conveyed in the following passage of the *Maitrāyaṇīya Saṁhitā* (I. 6, 12):—

यस्या राज्याः प्रातरग्निमाधास्यमानः स्यात्तां रात्रीं चतुश्शरावरावमोदनं पक्त्वा ब्राह्मणेभ्यो जीवतण्डुलमिवोपहरेत्. अदितिर्वै प्रजाकामौदनमपचत्. सोच्छिष्टमाश्नात्. तस्या धाता चार्यमा चाजायेतां. सापरमपचत्. सोच्छिष्टमाश्नात्. तस्या मित्रश्च वरुणश्चाजायेताम्. सापरमपचत्. सोच्छिष्टमश्नात्. तस्या अंशश्च भगश्चाजायेताम्. सापरमपचत्. सैक्षतोच्छिष्टं मेऽश्नत्या द्वौ द्वौ जायेते. इतो नूनं मे श्रेयः स्याद्यत्पुरस्तादश्नीयामिति सा पुरस्तादशित्वोपाहरत्. ता अन्तरेव गर्भः संता अवदताम् इदं भविष्यावो यदादित्या इति. तस्याः आदित्या निर्हंतारमैच्छन्. ता अंशश्च भगश्च निरहताम्. तस्मादेतौ यज्ञेन यजंते. अंशप्रासोंऽशस्य भागधेयं जनं भगोऽगच्छत्. तस्मादाहुर्जनो गंतव्यः तत्र भगेन संगछत इति. स वा इन्द्र ऊर्ध्व एव प्राणमनूदश्रयत. मृतमितरमाण्डमवापद्यत. स वाव मार्ताण्डो यस्येमे मनुष्याः प्रजाः सा वा अदितिरादित्यानुपाधावत्. अस्त्वेव म इदं मोघे परापत्तदिति. तेऽब्रुवन् अथैषोऽस्माकमेव ब्रवातै न नोऽतिमन्यता इति स वाव विवस्वानादित्यो यस्य मनुश्च वैवस्वतो यमश्च. मनुरेवास्मिल्लोके यमोऽमुष्मिन् एते वै देवयानान्पथो गोपायंति यदादित्याः त इयक्षमाणं प्रतितुदंते. यो वा एतेभ्योऽप्रोच्याग्निमाधत्ते तमेते स्वर्गाल्लोकात्प्रतिनुदंता उच्छिष्टभागा वा आदित्याः यदुच्छिष्टे विवर्तयित्वा समिध आदधाति तदादित्येभ्योऽग्न्याधेयं प्राह नैनं स्वर्गाल्लोकात्प्रतिनुदंते. संवत्सरमुत्सृजेताग्निमाधास्यमानो नास्याग्निं गृहाद्धरेयुर्नान्यता आहरेयुः. संवत्सरे वृद्धा गर्भाःप्रजायंते प्रजातमेनं वृद्धमाधत्ते. द्वादश रात्रीरुत्सृजेत द्वादश वै रात्रयः संवत्सरस्य प्रतिमा. संवत्सरे वृद्धा गर्भाः प्रजायंते प्रजातमेनं वृद्धमाधत्ते तिस्र उत्सृजेत. त्रयो वा इमे लोकाः. इमानेव लोकानाप्नोति. एकामृत्सृजेत एको वै प्रजापतिः.

M. S. 1, 6, 12.

"During that night on the morrow of which he is going to set up the sacred fire, he should cook four dishes of rice and present them to Brāhmans as fresh rice. Desirous of getting children, Aditi cooked the rice. She ate the remnant (*of what remained after the gods partook of the dish*). Two sons, Dhātā and Aryamā, were in consequence born of her. She cooked another (*dish*) and ate the remnant. Two sons, Mitra and Varuṇa, were in consequence born of her. She cooked another (*dish*), and ate the remnant. Two sons, Aṁśa and Bhaga, were in consequence born of her. She cooked another (*dish*). She thought that in consequence of her eating the remnant, two and two sons are being born of her; and that it would indeed be to her advantage if she would eat it before (*presenting it to the gods*). Accordingly, having previously eaten it, she offered the remnant (*to the gods*). The seeds, still remaining in the embryo form, said: 'We shall become what the Ādityas are.' The Ādityas on the other hand looked for a murderer of those two. Aṁśa and Bhaga struck them. Hence sacrificers worship these two in their sacrifices. Aṁśaprāsa became the portion due to Aṁśa in sacrifices. Bhaga went to the people. Hence they say that if one is desirous of getting wealth, one should go to somebody among men. That Indra, however, got up and recovered his breath. The other egg appeared as dead. He is, verily, the Mārtāṇḍa (*broken egg*) whose children are men. Aditi then went to the Adityas and said: 'Let this one be to me, but not the other which has fallen lifeless. They said: 'Then let it be to ourselves, as we say; do not despise us.' He is, verily, the Aditya, the Vivasvat, whose offspring are Manu, the Vaivasvata, and Yama, the Vaivasvata. Manu is in this world, and Yama in the other. These are the Adityas who guard the paths through which gods move. They drive away that sacrificer who sets up his sacred fire without calling upon them: they drive him away from the heavens. The Adityas are, verily the portions of the remnant. When a sacrificer puts the sacred sticks into the fire after rotating them in the remnant, then he may be taken to have spoken to the Adityas of his setting fire. Him they do not throw away from the heavens. He who is going to set up the sacred fire should *omit* a year (*i.e.*, intercalate a year). He should not bring his fire from a household or from any other place. Embryos [due to the remnant, *i.e.*, the twelve days at the end of the sidereal year of 366 days], developed in the course of the year are born. When the embryo is born and fully developed, the sacrificer sets it up (*while setting up the sacred fire*). Twelve nights he has to omit (*in a year*); for twelve nights are the index (*pratimā*) of the year. Embryos [*i. e.*, the twelve days] developed (*in the form of months*) in the course of (*the cyclic*) year are born. When it is born and fully developed, he sets it up [*i.e.*, intercalates while setting up the sacred fire]. He should omit three, for three are the worlds: these worlds he will thereby attain. He should omit one, for one is the Prajāpati."

From what has been said above, it is clear that the three good twin sons brought forth by Aditi in consequence of her eating the remnant must necessarily be the three pairs of intercalary months occurring in the course of three luni-solar cycles of five years each in consequence of the difference, or remnant as it is called, of twelve days between a lunar and a sidereal year. There is a sufficient clue in the passage itself to interpret the story of Aditi in this way We are told in the passage that the sacrificer should omit or intercalate a year and that then he should set up the sacred fire anew. From the *Maitrāyaṇīya Saṁhitā* I. 10, 7, we also know that the rite of setting up or churning the fire anew was performed at the end of the third intercalary period of four months at the close of thirty years. We are told in the above passage that the sacrificer had to omit twelve days every year and that the embryos developed in the course of the (*intercalated*)year were born. In the parlance of the Vedic poets, embryos or children are, as already pointed out, days of the year, either ordinary or intercalary. If, then, the twelve days at the end of the sidereal year are, as implied in the above passage, the embryo, which, when developed and born, the sacrificer is called to set up, it follows that the remnant which gave to Aditi a pair of sons is the same period of twelve days, giving rise to two intercalary months in the course of five luni-solar years. If this meaning is true, it follows that the three other pairs of Aditi's sons must also be three other pairs of intercalary months, occurring in the course of fifteen luni-solar years. If this is true, it is clear that what are called Dhātā, Aryamā, Mitra, Varuṇa, Aṁśa, and Bhaga, are the gods of the six intercalary months occurring in the course of fifteen luni-solar years. The only riddle that remains to be solved in the above passage is that connected with the birth of the fourth pair of sons, of whom one, called Indra, is said to have been fully born while the other, called Mārtāṇḍa, is said to have been half-born. If we paraphrase the Vedic language in our modern language, and say that three pairs of intercalary months and a seventh one were full and the eighth intercalary month was a broken month, we know where to seek for an explanation of this break. We know that the only year which can keep the seasons, especially the commencement of the much-desired rainy season, in their usual position, is the solar year of 365¼ days, but not the sidereal year of 366 days, which is evidently too long by three-fourths of a day. This excess will amount to $\frac{3}{4} \times 20 = 15$ days in the course of four cycles of five years each or in twenty sidereal years. Accordingly if this greater cycle of twenty years, with eight or rather seven and a half intercalary months to be intercalated at the end of the twenty years, had begun to be observed, as months, as the Vedic poets seem to have done, then the beginning of the year would fall back, not by eight months, as the Vedic poets first supposed, but by seven and a half months; or in other words the Hindu lunar year which begins with Chaitra[10] would then fall back and begin at the middle of Srāvaṇa of the rainy season, instead of at the end of Ashāḍha, as the poets seem to have expected it. How the poets found out the error, is a question with which we are not concerned here. Whatever may be the way in which they detected the break or error, the only explanation that can possibly be given for the half-birth of the eighth intercalary month or son, seems to be the one I have given above. This theory of intercalary months explains the simultaneous arrival of the 'seven streams' of the rainy season, of the

10 But it is only in the later calendar that we have a Chaitrādi year. In the Vedic period the year and the cycle began with Māgha.—J. F. F.

demon, Vṛitra, meaning eclipse, and of Indra, the god of the seventh intercalary month, for the destruction of the demon.

The Vedic poets seem to have entertained two kinds of conceptions about the intercalary months; one evil and another good. Indra, Mārtaṇḍa, and other sons of Aditi seem to have represented the good side of the months, while Vṛitra, Sambara, and other demons of eclipses are regarded as the personification of evil nature of the intercalary months. If there still remains any doubt about this point, the following passage of the *Maitrāyaiṇya Saṁhitā* (II. 3, 4) will probably help to remove it:—

ततो यस्सोमोऽत्यरिच्यत तमग्ना उपप्रावर्तयत् । स्वाहेंद्रशत्रुर्वर्धस्व इतींद्रस्याहैनं शत्रुमचिकीर्षदिंद्रमस्य शत्रुमकरोत्. तथा वाक्स्वयमेव व्यैत्स यं सोमं प्रावर्तयद्यास्मिन्नग्ना उपप्रावर्तयत्ता अग्नीषोमौ देवते प्राणापाना अभिसमभवताम्. स यावदूर्ध्वबाहुः पराविध्यत्तावति व्यरमत. यदि वा प्रवणं तावदासीद्यदि वाग्नेरधि ताव दासीत्स वा इषुमात्रमेवाह्ना तिर्यङ्ङवर्धतेषुमात्रमेवान्वङ्ङाथो आहुरहोरात्रे एवेषुमात्रं तिर्यङ्ङवर्धतेषुमात्रमन्वङ्ङित्यथो आहुरर्धमासमथोमास मथो संवत्सरमिति. स वा इमाः सर्वाः स्रोत्याः पर्यशयत्तस्माद्वा इन्द्रोऽबिभेत्तस्मादु त्वष्टाबिभेत्तस्यैंद्रः प्रत्तिमैच्छत्तमस्य प्रायच्छत्. तस्मै त्वष्टा वज्रमासिंचत्तपो वै स वज्र आसीत्तमुद्यमं नाशक्नोदथ वै तर्हि विष्णुरन्या देवतासीत्सोऽब्रवीद्विष्णा एहि इदमाहरिष्यावो येनायमिदमिति स त्रेधात्मानं विन्यधत्ताभिपर्यावर्तादबिभेदस्यां तृतीयमन्तरिक्षे तृतीयं दिवि तृतीयं. स यदस्यां तृतीयमासीत्तेन वज्रमुदयच्छद्विष्णवनुष्ठितः. स वज्रमुद्यतं दृष्ट्वाबिभेत्सोऽब्रवीदस्ति वा इदं त्यस्मिन्नंतर्वीर्यं तत्ते प्रदास्यामि मा मा वधीरिति. तद्वा अस्मै प्रायच्छत्. तत्प्रत्यगृह्णात् । अथा मा इति तद्विष्णवे-

ऽतिप्रायच्छत्. तद्विष्णुः प्रत्यगृह्णात् । अस्मास्विन्द्र इन्द्रियं दधात्वस्मान्रायो मघवानः सचन्ताम् । अस्माकं संत्वाशिषः । इति सोऽवेदस्ति वा त्यस्मिन्नन्तर्वीर्यमिति. स यदन्तरिक्षे तृतीयमासीत्तेन वज्रमुदयच्छद्विष्णवनुष्ठितः. स वज्रमुद्यतं दृष्ट्वाबिभेत्सोऽब्रवीदस्ति वा इदं त्यस्मिन्नन्तर्वीर्यं तत्ते प्रदास्यामि मा मा वधीरिति. तद्वा अस्मै प्रायच्छत् प्रत्यगृह्णात्. द्विर्माधाः इति तद्विष्णवेऽतिप्रायच्छत् तद्विष्णुः प्रत्यगृह्णात्. अस्मास्विन्द्र इंद्रियं दधात्वस्मान्रायो मघवानः सचंताम् । अस्माकं संत्वाशिषः. इति. सोऽवेदस्ति वा यास्मिन्नन्तर्वीर्यमिति. स यद्दिवि तृतीयामासीत्तेन वज्रमुदयच्छद्विष्णवनुष्ठितः. स वज्रमुद्यतं दृष्ट्वाबिभेत्सोऽब्रवीदस्ति वा इदं त्यस्मिन्नंतर्वीर्यं तत्तें प्रदास्यामि मा मा वधीरिति. संधां नू संदधावहै यथा त्वामेव प्रविशानीति. सोऽब्रवीद्यन्मां प्रविशेः. किं मे ततः स्यादिति. सोब्रवीत्वामेवेंधीय तव भोगाय त्वां प्रविशेयमिति. तद्वा अस्मै प्रायच्छत्. तत्प्रत्यगृह्णात्. त्रिर्माधाः इति तद्राव त्रैधातव्या सहस्रं वा अस्मै तत्प्रायच्छत्. ऋचः सामानि यजूंषि यद्वा इदं किंच तत्त्रैधातव्या तदाप्नोति पशूनेव.

उदरं वै वृत्रः पाप्मा क्षुद्भ्रातृव्यः पुरुषस्य. यत्तप उपैति पाप्मानं वा एतत्स्तृणुते भ्रातृव्यं क्षुधमेव तस्मिन्वा अवदेतां सयमस्या अध्यूर्ध्वा वागवदत्. उभा जिग्यथुर्न पराजयेथे न पराजिग्ये कतरश्च नैनोः ।

M. S. II, 4, 4.

"Then what Söma there remained, he poured it into the fire, and said rather in favour of Indra than Agni: 'Grow with Indra as thy enemy.' He wanted Agni to be Indra's enemy: but he made Indra the enemy of Agni: for his expression itself came out (*with that meaning*). Both the Soma he pressed and the Soma he put into the fire became the two deities Agni and Soma, and also the two vital airs, Prāṇa and Apāna (*air inhaled and air exhaled*). No sooner did this dual god with his arm raised up attempt to strike Indra, than he himself fell down. Whether when the dual deity fell down, or when he was inside the fire (*it cannot be said*),—he, however, began to grow breadthwise by the measure of an arrow in the course of a day, and also lengthwise by the measure of an arrow in the course of day. They say that day and night themselves grew breadthwise by the measure of an arrow and also lengthwise by the measure of an arrow. They say that then the half months (*grew*); then the month; and then the year. Then this dual deity lay covering all these streams. Indra became afraid of him; Tvashṭri also feared him, Indra requested the help of Tvashṭri. The latter promised help: he sprinkled the thunderbolt (*with water*) for him. Tapas [the month so called] is, verily, the thunderbolt. Indra could not raise it. Then there was another god, Vishṇu, near. Indra said: 'Come, Vishṇu, let us catch hold of this by which this (*is done*).' Vishṇu stretched his body in three directions, one-third portion on the earth, one-third in the air, and one-third in the heaven, so that Indra might get rid of his fear from the universal growth of the dual deity. Followed by Vishṇu, Indra raised the thunderbolt agaist the one-third part of the dual deity lying on this earth. Seeing the thunderbolt raised, he became afraid of it, and said: 'There is in me some power and I shall give it to you. Do not kill me!' He gave it to Indra, and the latter took it and thought: 'May Indra put vital force into us; may Indra bring prosperity to us; may there be blessings upon us; for there is internal power in him.' Followed by Vishṇu, Indra raised the thunderbolt against the one-third part that lay in the air. Seeing the raised thunderbolt, he became afraid of it, and said: 'There is some power in me and I shall give it to you. Do not kill me!' He gave it to Indra, and Indra took it and gave it to Vishṇu, saying 'keep this for me a second time.' Vishṇu took it, thinking: 'May Indra put vital force into us; may Indra bring prosperity to us; may there be blessings upon us: for there is internal power in him.' Followed by Vishṇu, Indra raised the thunderbolt against the one-third part that lay in the sky. Seeing the raised thunderbolt, he became afraid of it, and said: 'There is some power in me and I shall give it to you. Do you kill me? let us make peace: I shall enter into you.' Indra said: 'If you enter into me, of what use will it be to me?' He said: 'I shall brighten yourself; I shall enter into you for your own enjoyment.' (*So sayiug*) he gave it to Indra, and Indra took it and gave it to Vishṇu, saying: 'Keep this for me for a third time.' It (the power) is, verily, a thousand of what are called Tridhātus (three colours). He gave it to Vishṇu. The Ṛiks, the Sāmas, the Yajus, and whatever else there is, all that belongs to the three elements. Hence he obtains cattle alone."

"Vṛitra is the belly; and sin is hunger, the enemy of man. When man obtains Tapas, he rends the sin, the inimical hunger. This is what the heavenly utterance said: 'Both of them conquered, but never sustained defeat; and no one defeated either of them (*Indra and Vishṇu.*)"

We are told in the above passage that Vṛitra grew out of the remnant of Soma and that he grew first in the form of a day, then of half a month, then of a month, and at last of a year. Thus Vṛitra is clearly identified with time. Special attention should be paid to those sentences of the passage which clearly declare: 'Vṛitra began to grow breadthwise by the measure of an arrow in the course of a day and also lengthwise by the measure of an arrow in the course of a day. They say that *day and night themselves grew* by the measure of an arrow, and became half-months, months, and a year.' It is clear therefore that Vṛitra is a demon infesting the intercalary months, or rather of the eighth intercalary month, since Indra who destroys him periodically is, as we have seen above, the god of the seventh intercalary month of the luni-solar cycle of five years. Since Vṛitra is made to 'enter into Indra himself,' it is clear that he is the broken eighth month coming after the seventh month.

I have pointed out in the *Vedic Calendar* how the Vedic poets regarded the intercalary days as being sinful and inimical to man. In the above passage Vṛitra is spoken of as a kind of sin and enemy to man. We have already seen how Agni and Soma are considered as the gods of the light half of an intercalary month. In the following passage of the Taittirīya Saṁhitā (II. 5,. 2) Agni and Soma are clearly described as the life-principles of Vṛitra. It follows therefore that Vṛitra must be the light half of an intercalary month. Since Vṛitra is periodically destroyed by Indra, the god of the seventh intercalary month, and since he is made one with Indra himself, it is also clear that Vṛitra is the first half of the broken eighth intercalary month. The reference to cold and fever in the passage seems to indicate the rainy season. The passage itself runs as follows:—

त्वष्टा हतपुत्रो वींद्रं सोममाहरत्. तस्मिन्निंद्र उपहवमैच्छत्. तं नोपाह्वयत पुत्रं मेऽवधीरिति. स यज्ञवेशसं कृत्वा प्रासहा सोममपिबत्. तस्य यदत्यशिष्यत तत् त्वष्टाहवनीयमुपप्रावर्तयत् स्वाहेंद्रशत्रुर्वर्धस्वेति. यदवर्तयत् तद्वृत्रस्य वृत्रत्वं. यदब्रवीत् स्वाहेंद्रशत्रुर्वर्धस्वेति तस्मादस्य इन्द्रः शत्रुरभवत्. स संभवन्नग्नीषोमावभिसमभवत् स इषुमात्रमिषुमात्रं विष्वङ्ङवर्धत. स इमाल्लोकानावृणोत्. तद्वृत्रस्य वृत्रत्वं. तस्मादिन्द्रोऽबिभेत्. स प्रजापतिमुपधावत् शत्रुर्मेऽजनीति. तस्मै वज्रं सिक्त्वा प्रायच्छदेतेन जहीति. तेनाभ्यायत तावब्रूतामग्नीषोमौ मा प्रहारावसंतः स्व इति. मम वै युवं स्थ इत्यब्रवीत् मामभ्येतमिति. तौ भागधेयमैच्छेताम्. ताभ्यामेतमग्नीषोमीयमेकादशकपालं पूर्णमासे प्रायच्छत्. तावब्रूतामभिसंदष्टौ वै स्वो न शक्नुव एतुमिति. स इन्द्र आत्मनः शीतरूरावजनयत्. तच्छीतरूर्वोर्जन्म. य एवं शीतरूर्वोर्जन्म वेद नैनं शीतरूरौ हतः. ताभ्यामेनमभ्यनयत्. तस्माज्जञ्जभ्यमानादग्नीषोमौ निरक्रामताम्. प्राणापानौ वा एनं तदजहिताम्. प्राणो वै दक्षोऽपानः क्रतुः. तस्माज्जञ्जभ्यमानो ब्रूयान्मयि दक्षक्रतू इति.. प्राणापानावेवात्मन्धत्ते. सर्वमायुरेति. स देवता वृत्रान्निर्हूय वार्त्रघ्नं हविः पूर्णमासे निरवपेत्. घ्नन्ति वा एनं पूर्णमासा आ अमावास्यायामाप्याययंति तस्माद्वार्त्रघ्नी पूर्णमासेऽनूच्येते वृधन्वती अमावास्यायाम्. तत् संस्थाप्य वार्त्रघ्नं हविर्वज्रमादाय पुनरभ्यायत तेऽब्रूतां द्यावापृथिवी माप्रहारावयोर्वै श्रित इति. तेऽब्रूतां वरं वृणावहै नक्षत्रविहिताहमसानीत्यसावब्रवीच्चित्रविहिताहमिति. इयं तस्मान्नक्षत्रविहितासौ चित्रविहितेयं य एवं द्यावापृथिव्योर्वरं वेदैनं वरो गच्छति. स आभ्यामेव प्रसूत इंद्रो वृत्रमहन्. ते देवा वृत्रं हत्वाग्निषोमावब्रुवन् हव्यं नो वहतमिति. तावब्रूतामपतेजसौ वै त्यौ वृत्रे वै त्ययोस्तेज इति. तेऽब्रुवन् क इदमाच्छैतीति. गौरित्यब्रुवन् गौर्वाव सर्वस्य मित्रमिति. साब्रवीत् वरं वृणै मय्येव सतोभयेन भुनजाथ्वा इति तद्गौराहरत्. तस्माद्गवि सतोभयेन भुञ्जते

एतद्वा अग्नेस्तेजो यद्घृतेमतत् सोमस्य यत्पयः य एवमग्नीषोमयो-स्तेजो वेद तेजस्व्येव भवति. ब्रह्मवादिनो वदंति किन्दैवत्यं पौर्णमासमिति. प्राजापत्यमिति ब्रूयात्. तेनेन्द्रं ज्येष्ठं पुत्रं निर-वासयदिति. तस्माज्ज्येष्ठ पुत्रं धनेन निरवसाययंति. T.S.II. 5.2.

"Tvashṭri whose son was killed (*by Indra*) began to perform a Soma sacrifice without inviting Indra to it. But Indra wanted to be invited to it. But he did not invite Indra, because the latter killed his son. But Indra drank the Soma by force after obstructing the sacrifice. Tvashṭri poured (*prāvartayat*) into the fire what Soma here remained, and said (*addressing the fire*): 'Grow with Indra as thy enemy.' Vṛitra [the demon that rose from the fire in consequence of the above libation] is so called, because the act of pouring down Soma into the fire is from the root *Vṛit*. Since he said: 'Grow with Indra as thy enemy,' Indra became his enemy. While coming out of the fire, he (Vṛitra) became Agni and Soma. By the measure of an arrow, he grew on all sides and pervaded these three worlds. Because he pervaded them, he is called Vṛitra, 'pervader.' Indra became afraid of him, and going to Prajāpati, said: 'there has arisen an enemy to me.' Having sprinkled the thunderbolt with water, he gave it to him to kill the demon. Indra advanced with the thunderbolt. Then Agni and Soma said: 'Do not kill; we are within (*him*). Indra said: 'You are for me; and so, come to me.' They asked for a share (*in the sacrifice*). Indra promised to them a cake on eleven pot-sherds, to be offered to them every full-moon. They said: 'We are bitten (*by his teeth*), and cannot come out (*of his mouth*.) Then Indra created out of his own body cold and fever. This is how cold and fever came into existence. Whoever knows this origin of cold and fever, will not be attacked by cold and fever. Indra transferred cold and fever to them (*or* to Vṛitra). When he (Vṛitra) began to shiver, Agni and Soma came out: it is *prāṇa* (air inhaled) and *apāna* (air exhaled) that left him. Prāṇa is *Daksha* and Apāna is *Kratu*. Hence the sacrificer should begin to shiver and say: 'Daksha and Kratu are within me.' Thereby he will have Prāṇa and Apāna in himself, and live the whole length of life. Having released the gods from Vṛitra, Indra offered an oblation at the full-moon on account of his slaying Vritra; for they kill him at full-moon, and revive him at new-moon. Hence a Ṛik-verse about the slaying of Vritra is recited at full-moon, while another about his revival is sung on the occasion of new-moon. Having offered an oblation for slaying Vritra, Indra again faced Vritra with his thunderbolt. Then the Sky and the Earth said: 'Do not kill him, for he is lying upon us.' And they said again: 'We request a gift (*if he is to be killed*); I shall like to be decked with stars—so said the Sky; and I shall like to be variously formed,—so said the Earth.' Hence the Sky is decked with stars, while the Earth is variously formed. Whoever knows this gift of the Sky and the Earth will have the same gift. Having been born out of these two (the Sky and the Earth), Indra killed Vṛitra. Having killed Vṛitra, the gods asked Agni and Soma to carry their oblations. They said: 'We have lost our energy; for it is in Vṛitra. The gods inquired among themselves, saying 'who can secure that energy?' Some replied: 'The cow (*can do that*); for the cow is the friend of all.' The cow said: 'I shall request a gift: you live upon the two things that exist only in me.' The cow secured that energy. Hence they live upon the two things that exist in the cow alone. What is called *ghī* is the energy of Agni, and what is called milk is that of Soma. Whoever knows thus the energy of Agni and Soma will be energetic. The Brahmavādins debate: 'Of what deity is the full-moon?' One should reply: 'Prajāpati.' Hence Prajāpati gave to Indra, his eldest son, a firm footing. Hence men give to their eldest son a firm footing by bestowing upon him a large portion of wealth."

The following passage of the Taittirīya Saṁhita (VI, 5, 1) seems to furnish additional evidence about Vritra being a half month:—

इन्द्रो वृत्राय वज्रमुदयच्छत्. स वृत्रो वज्रादुद्यतादबिभेत्. सोऽब्रवीन्मा मे प्रहारास्ति वा इदं मयि वीर्यं तत्ते प्रदास्यामीति. तस्मा उक्थ्यं प्रायच्छत्. तस्मै द्वितीयमुदयच्छत्. सोऽब्रवीन्मा मे प्रहारास्ति वा इदं मयि वीर्यं तत्ते प्रदास्यामीति तस्मा उक्थ्यमेव प्रायच्छत्. तस्मै तृतीयमुदयच्छत्. तं विष्णुरन्वतिष्ठत जहीति. सोऽब्रवीन्मा मे प्रहारास्ति वा इदं मयि वीर्यं तत्ते प्रदास्यामीति. तस्मा उक्थ्यमेव प्रायच्छत् तं निर्मायं भूतमहन् यज्ञो हि तस्य मायाऽऽसीत्.

"Indra raised the thunderbolt against Vritra. Then Vritra became afraid of this raised thunderbolt; he said: 'Do not kill me; there is some power in me; that I shall give you.' So saying he gave Ukthya (Fifteen) to Indra. Indra raised weapon against him for a second time. He said: 'Do not kill me; there is some power in me; that I shall give you.' So saying he gave the latter the same Ukthya (Fifteen.) Then Indra raised the weapon against him. for a third time; then Vishṇu followed Indra, saying 'kill him.' He said: 'Do not kill me; there is some power in me; I shall give you that.' So saying he gave the same Ukthya to Indra.. Indra then killed this guileless demon. It was, verily, the sacrifice which was his guile."

We are told in the above passage that while breathing out, Vṛitra gave *Ukthya* to Indra. Ukthya is a word used in the Vedic literature in the sense of 'fifteen.'[12] The word Vajra, the weapon of Indra, is also used in the same sense.[13] Accordingly the wielding of Vajra or 'fifteen' by Indra as well as the gift of fifteen by Vṛitra to Indra clearly mean the growth of fifteen days over and above the seventh intercalary month.

Contemporary religious records also furnish evidence that the Ādityas are the Gods of intercalary months. It is known that the Adityas are the sons of Aditi. Aditi in the Ṛigveda (X.100. 1, 94) is requested to protect the poets from *Amhas*, 'sin.' She and her sons also are requested to release the poets from guilt or sin (R. V. I. 24; II. 27; VII. 93; I. 162; VII. 87).[14] I have shown in the first chapter how the word *Amhaspatya* is used in the sense of an intercalary month and an intercalary month alone. There is no doubt that this word is philologically identical with the Zend word *Ameshaspenta*. The number of Ameshaspentas is also seven Prof. Macdonell says (*Vedic Mythology*, P. 44). "It is here to be noted that the two groups have not a single name in common, even Mithra not being an Ameshaspenta; that

12 See *Tai, Sam*. VII. 2, 5, 17.

13 See *Ibid*. VII. 3, 6, 15; 4, 7, 25.

14 See Macdonell's *Vedic Mythology*, p. 121.

the belief in the Adityas being seven in number is not distinctly characteristic and old; and that though the identity of Adityas and Ameshaspentas has been generally accepted since Roth's essay, it is rejected by some distinguished Avestan scholars."

Whatever might be the reason of the Avestan scholars for rejecting the identity, this much is clear, that the words Amhaspatya and Ameshaspenta are identical; and that when the former word is invariably used in the sense of an intercalary month in the Yajurveda, there is no doubt that the forgotton meaning of the latter word must also be the same and that when the Ameshaspentas are seven, the number of Amhaspatyas must also be and is, as we have already seen, seven. As regards the difference in the names of the Ameshaspentas and of the Adityas, it does not appear to be of much importance, for the seven Amhaspatyas or intercalary months are found variously named both in the Ṛigveda and the Atharvaveda.

The following are some of the passages of the Atharvaveda (VIII. 9) in which the seven Adityas or the Gods of intercalary months are called in various ways:

षडाहुश्शीतान्षडु मास उष्णानृतुं नो ब्रूत यतमोऽतिरिक्तः ।
सप्त सुपर्णाः कवयो निषेदुः सप्त छंदांस्यनु सप्त दीक्षाः ॥ 17

सप्त होमाः समिधो ह सप्त मधूनि सप्त ऋतवो ह सप्त ।
सप्ताज्यानि परिभूतमायताः सप्त गृध्रा इति शुश्रुमा वयम् ॥ 18

अष्ट जाता भूता प्रथमज ऋतस्याष्टेंद्र ऋत्विजो दैव्या ये ।
अष्टयोनिरदितिरष्टपुत्राष्टमीं रात्रिमभि हव्यमेति ॥ 21

अष्टेंद्रस्य षड्यमस्य ऋषीणां सप्त सप्तधा ।
अपो मनुष्यानोषधीस्तां उ पंचानु सेचिरे ॥ 23

"Six they call the cold, and six the hot months.

Tell ye us the season, which one is in excess; seven eagles, poets, sat down; seven metres after seven consecrations."17

Seven are the offerings, the fuels seven, the seasons seven; seven sacrificial butters went about the existing thing; they are such as have seven heavenly birds, so have we heard." 18.

"Eight are born the beings first-born of Ṛita; eight, O Indra! are the priests who are of the gods; Aditi has eight wombs, 'eight sons; the oblation goes unto the eighth night." 21.

"Among the seers, eight are with Indra, and six are in pairs; they are seven-fold and seven; waters, men, and herbs,—over these the five (*years*) have showered." 23.

In verse 17 the poet clearly mentions the intercalary months (Atirikta Ṛitu) and numbers them in various names as seven. The expression 'seven seasons,' when taken with the expression 'the excessive season,' leaves no doubt that they are intercalary months and seven in number. In verses 21 and 23 the poet refers to the story of Aditi, and seems to hesitate to count her sons as eight, though that was the number fixed at first. In the following passages of the Atharvaveda (IX, 9. and R. V. I. 164) the seven months are called seven horses and seven sisters:—

सप्त युंजंति रथमेकचक्रमेको अश्वो वहति सप्तनामा ।
त्रिनाभि चक्रमजरमनर्वं यत्रेमा विश्वा भुवनानि तस्थुः ॥ 2

इमं रथमधि ये सप्त तस्थुः सप्तचक्रं सप्त वहंत्यश्वाः ।
सप्त स्वसारो अभि संनवंत यत्र गवां निहिता सप्त नामा ॥ 3

द्वादशारं न हि तज्जराय वर्वर्ति चक्रं परिद्यामृतस्य ।
आ पुत्रा अग्ने मिथुनासो अत्र सप्त शतानि विंशतिश्च तस्थुः ॥ 13

सनेमि चक्रमजरं वि वावृत उत्तानायां दश युक्ता वहंति ।
सूर्यस्य चक्षू रजसैत्यावृतं यस्मिन्नातस्थुर्भुवनानि विश्वा ॥ 14

साकंजानां सप्तथमाहुरेकजं षडिद्यमा ऋषयो देवजा इति ।
तेषामिष्टानि विहितानि धामशः स्थात्रे रेजंते विकृतानि रूपशः॥ 16

"Seven harness an one-wheeled chariot; one horse, having seven names, draws it. Of three naves is the wheel, unassailed, whereon stand all those existences. 2

"The seven that stand on this chariot, seven horses draw it, seven-wheeled; seven sisters shout at it together; where are set down the seven names of the kine?"

"The twelve-spoked wheel,—for that is not to be worn out,—revolves greatly about the sky of Ṛita; there, O Agni!, stood the sons, paired, seven hundred and twenty." 13.

"The unwasting wheel, with rim, rolls about; ten paired ones draw upon the upper side (*uttāna*); the sun's eye goes surrounded with the welkin in which stood all existences." 14.

"Of those born together the seventh they call the sole-born (single-born); six, they say, are twins, god-born seers; the sacrifices of them, distributed according to their respective stations and modified in form, move to the one permanent (*sthatre*)." 16.

The seven hundred and twenty sons, spoken of in verse 13, are evidently the 720 days and nights of the civil year; and the ten twins on the upper side of the chariot, referred to in the next verse, must necessarily be the 10 days and nights above the 360 days of the year.

This shows that the poets were well acquainted with the real length of the solar year. It is the seven Ādityas or the gods of the intercalary months, that are referred to in verse 16. The expression that the seventh was single-born clearly shows the break in the eighth intercalary month, as will be pointed out.

In the following verses of the Atharvaveda (X. 8.) the mention of the number 'one thousand' in connection with seven swans seems to furnish additional evidence that the seven Ādityas, eagles, or swans, as they are variously called, are the seven intercalary months.

द्वादश प्रधयश्चक्रमेकं त्रीणि नभ्यानि क उ तच्चिकेत ।
तत्राहतास्त्रीणि शतानि शंकवष्षष्टिश्च खीला अविचाचला ये ॥ 4
इदं सवितर्विजानीहि षड्यमा एक एकजः ।
तस्मिन्हापित्वमिच्छंते य एषामेक एकजः ॥ 5
एकचक्रं वर्तते एकनेमि सहस्राक्षरं प्रपुरो निपश्च ॥
अर्धेन विश्वं भुवनं जजान यदस्यार्धं क्व तद्बभूव ॥ 7
सहस्राह्ण्यं वियतावस्य पक्षौ हरेर्हंसस्य पततस्स्वर्गम् ।
स देवान्सर्वानुरस्युपदद्य संपश्यन्याति भुवनानि विश्वा ॥ 18

" Twelve fellies, one wheel, three naves,—who understands that? Therein are inserted three hundred and sixty pins, pegs that are immovable."[15] 4.

" This, O Savitṛi!, do thou distinguish; six are twins, one is sole-born; they seek participation in him who of them is the sole sole-born." 5.

"One-wheeled it rolls, one-rimmed, thousand-syllabled forth in front, down behind; with a half it has generated all existence; what its other half is,—what has become of that?" 7

---

15 Comp. R.V.I. 164, 48.

" By a thousand days are the wings expanded of him, of the yellow swan flying to heaven ; he, putting all the gods in his breast, goes, viewing together all existence."[16] 18

In verse 4, the Sāvana year of 360 days is discribed: and in verse 5, the three pairs of intercalary months together with the single seventh month are referred to. In verse 7, the cycle of 20 years is described as containing a thousand syllables, *i.e.*, days. The question about the other half seems to refer to the loss of fifteen days in the eighth intercalary month. In verse 18, the last cycle of five years with $7\frac{1}{2}$ intercalary months seems to be described as a special period or great year, each wing or half of which is measured by a thousand days. The yellow Swan is the seventh intercalary month. Now, if we expand the wings by putting 1,000 on each, its duration becoms equal to 2,000 days. In 2,000 days there are $\frac{2{,}000}{29d,\ 12h,\ 45m,} = \frac{2{,}000 \times 32}{945} = \frac{12{,}800}{189} = 67$ lunations and 22 days, taking a lunation to be equal to 29 days, 12 hours, and 45 minutes.[17] It is clear, therefore, that by the expressions 'thousand-syllabled chariot," and 'a wing of thosand days' duration,' the poet refers to the last cycle in the greater cycle of 20 years, in-as-much-as that cycle is approximately equal to five lunar years and seven and a half lunations. It is also to be noted that five lunar years are $= 5 \times 354 = 1{,}770$ days and twenty-times 12 extra days $= 20 + 12 = 240$ days. Putting

---

16 Comp. A.V. XIII. 2,38.

17 But the Vedic estimate of the synodic lunar month, as shown by the Jyotisha Vedāṇga, was 1830 days divided by 62 lunations=29 days, 12 hours, 23.2258......Second.

these together, we have 1,770+240=2,010 days, which is greater by 10 days than the duration of 2,000 days, as described in verse 18. We shall see that the same cycle of five years with seven and a half intercalary months is also termed Purusha, 'man' or Sapta-purusha, 'seven men.' Hence it is probable that the rising up of the thousand-headed, thousand-eyed, and thousand-legged Purusha by 10 *aṅgulās* or days above the earth, described in the Purushasūkta, refers to the same cycle of 2,010 days, which was made equal to 2,000 days. It is probable that the use of *aṅgulas* to mark days was a common practice among the Vedic poets, as among the Arabians. Regarding the use of fingers by an Arabian prophet to mark days, this is what Alberūnī says[18] :—

"—'We are illiterate people, we do not write, nor do we reckon the month thus and thus and thus,' each time showing his ten fingers, meaning a complete month or thirty days. Then he (the prophet) repeated his words by saying 'And thus and thus and thus,' and at the third time he held back one thumb, meaning an incomplete month of twenty-nine days."

In the following verses of the Atharvaveda (XII, 3, 16; and XIII, 2, 24) the same intercalary months are described as seven sacrifices and seven yellow steeds :—

सप्तमेधान्पशवः पर्यगृह्णन् य एषां ज्योतिष्मानुत यश्चकर्ष ।
त्रयस्त्रिंशद्देवतास्तान्सचंते स नः स्वर्गमभिनषे लोकम् ॥
सप्त त्वा हरितो वहंति देव सूर्य शोचिष्केशं विचक्षणम् ।
अयुक्त सप्त शुंध्यवः सूरो रथस्य नप्त्यः ताभिर्याति स्वयुक्तिभिः ॥

"Seven sacrifices the cattle obtained; of which some were full of light, and others were pining; to them the

18 *Chronology of Ancient Nations*, P. 78; 1879.

three and thirty attach themselves; do thou conduct us unto the heavenly world."

"Seven yellow steeds, O heavenly sun, draw in the chariot thee, the flame-haired, the out-looking: the sun hath yoked the seven neat daughters to the chariot; with them who are self-yoked, he goeth."

The only point to be considered in this is the number 33. Here, again, the allusion seems to be the same thousand days by which each wing of the heavenly swan was said to be expanded; for 1,000 is equal to $\frac{1,000}{30} = 33$ months and 10 days or half of 67 lunations and 20 days.

In the following verse of the Atharvaveda (X, 8, 7 and 13; and XII. 4,22) the poets speak of the same cycle as one of eight wheels or eight intercalary months:

अष्टाचक्रं वर्तते एकनेमी सहस्राक्षरं प्रपुरो निपश्चा ।
अर्धेन विश्वं भुवनं जजान यदस्यार्धं कतमः स केतुः ॥

"The eight-wheeled (*choriot*) rolls, having one rim, thousand-syllabled, forth in front, down behind; with a half it has generated all existence; what its other half is,—which sign is that?"

In the following passage of the Atharvaveda (IX, 10, 17) the poet counts the intercalary months neither as eight nor as seven, but exactly as seven and a half and calls them embryos:—

सप्तार्धगर्भाभुवनस्य रेतः विष्णोस्तिष्ठंति प्रदिशा विधर्मणि ।
ते धीतिभिर्मनसा ते विपश्चितः परिभुवः परिभवंति विश्वतः ॥

"Seven and half, embryos, the seed of existence, stand in front in Vishṇu's distribution; they, by thoughts, by mind, they, inspired, surround on all sides the surrounders."

In the following verses of the Atharvaveda (X, 3, 8-10), the poet mentions the thirteenth month, and refers to the seven intercalary months as seven eagles and seven suns, making Kaśyapa the head of them :—

अहोरात्रैर्विमितं त्रिंशदंगं त्रयोदशं मासं यो निर्मिमीते ।
तस्य देवस्य क्रुद्धस्य एतदागः...........॥
कृष्णं नियानं हरयस्सुपर्णा अपो वसाना दिवमुत्पतंति ।
त आववृत्रन्सदनादृतस्य तस्य देवस्य क्रुद्धस्य एतदागः ॥
यत्ते चंद्रं कश्यप रोचनावद्यत्संहितं पुष्कलं चित्रभानु ।
यस्मिन्सूर्या अर्पितास्सप्त साकं तस्य देवस्य क्रुद्धस्य एतदागः ॥

" He who measures out the thirteenth month, fabricated of days and nights, having thirty members,— against that god, angered, is this offence.

The yellow eagles, clothing the recurring Krishṇa[1], eclipse in waters, fly up to the sky ; they have come hither from the seat of Rita ; against that god, angered, is this offence.

" What of thee, O Kaśyapa, is bright, full of shining what that is combined, splendid, of wondrous light, in which seven suns are set together ; against that god, angered, is this offence."

In the following verses of the Atharvaveda (XIX. 53, 1 and 2) the Poet describes the same seven intercalary months as time in the form of a thousand-eyed horse with seven reins, and also as seven wheels :—

कालो अश्वो वहति सप्तरश्मिः सहस्राक्षः अजरो भूरिरेताः ।
तमारोहंति कवयो विपश्चितः तस्य चक्रा भुवनानि विश्वा ।
सप्त चक्रा वहति काल एव सप्तास्य नाभीरमृतं न्वक्षः ।
स इमा विश्वा भुवनान्यर्वाङ्कालः स हीयते प्रथमो नु देवः ।

1 " Iyānah Krishṇah daśabhih śahasraih " Tai. Ar. I, 6.

"Time drives a horse with seven reins, thousand-eyed, possessing much seed; him the inspired poets mount; his wheels are all beings.

"Seven wheels doth this Time drive; seven are his naves; immortality forsooth his axle; he, Time, including all these beings, goes on as first god."

The meaning of a thousand eyes is the same as that of a thousand syllables, or a thousand days expanding a wing of the heavenly swan, explained above.

Regarding Aditi's eight sons who became seven Adityas, the Tait. Ar. (I. 13) says as follows:—

अदितिर्जातमदितिर्जानित्वम् । अष्टौ पुत्रासो अदितेर्ये जातास्तन्वः परि । देवानुपप्रैत्सप्तभिः परा मार्ताण्डमास्यत् ॥ सप्तभिः पुत्रैरदितिः उपप्रैत्पूर्व्यं युगम् । प्रजायै मृत्यवे तत्परा मार्ताण्डमाभरदिति ॥

ताननुक्रमिष्यामः–मित्रश्च वरुणश्च धाता चार्यमा च अंशश्च भगश्च इन्द्रश्च विवस्वांश्चेत्येते . . . .

गर्भः प्राजापत्यः । अथ पुरुषस्सप्तपुरुषः ।

"Aditi is past and Aditi is future; of the eight sons of Aditi, who were born from her body, she approached the gods with seven sons and cast out Mārtāṇḍa; with seven sons Aditi approached the gods in the former *Yuga* (cycle of 20 years); she brought thither Mārtāṇḍa again for birth and death. We enumerate them: Mitra and Varuṇa, Dhātā and Aryamān, Aṁśa and Bhaga and Indra and Vivasvān,—these are they." . . . . .

"The seed belongs to Prajāpati, Father Time; and the Purusha (*born thereof*) is sevenfold."

The *Śatapatha Brāhmaṇa* identifies the seven logs and tongues of Agni with purushā and also with Indra. The passages in which this identification is made are thus translated by Prof. Eggeling :—

" He offers with Vag. S. XVII. 79, ' thine, O Agni, are seven logs,'—logs mean vital airs, for the vital airs do kindle him ;—' seven tongues,'—this he says with regard to those seven persons which they made into one person ;—' Seven Rishis,'—for seven Rishis they indeed were ;—' seven beloved seats,'—this he says with regard to the metres ' sevenfold the seven priests worship thee,'—for in sevenfold way the seven priests indeed worship him ;—' the seven homes,'—he thereby means the seven layers of the altar ; ....' seven,' he says each time,—of seven layers the fire-altar consists, and of seven seasons the year, and Agni is the year."[23]

" This same vital air in the midst doutless is Indra. He, by his power, kindled those other vital airs from the midst ; and in as much as he kindled, he is the kindler (Indra) : the kindler indeed,—him they call ' Indra ' mystically (esoterically), for the gods love the mystic. They (the vital airs) being kindled, seven separate persons (Purusha)[24]."

I presume that I have made it clear that the various expressions, such as the eight sons of Aditi, the seven or eight Adityas, seven eagles or swans, seven butters, seven logs of fire, seven tongues of Agni, seven Vāyus, seven cattle, seven breaths, seven Agnis, seven Purushas, seven horses, seven sisters, seven priests, seven seers and seven and a half embryos, are all of the same meaning,

---

23 Sat. Bra. IX. 2, 3, 44-45. 24 I bid VI. 1, 1, 2.

viz., the seven and a half intercalary months occurring in the cycle of twenty luni-sidereal years, and that the act of ġetting rid of the intercalary months is described as a recurring conflit between Vṛitra, the demon of the intercalary months and eclipses, and Indra, the god of the seventh intercalary month. That this conflict was a periodic and recurring phenomenon, is so well known to all Vedic scholars that it needs no proof. The expression that Indra killed Vṛitra three times, securing thereby three *ukthyas* or ' fifteens,' evidently signifies the cycle of sixty years, which consists of three cycles of 20 years each or twelve cycles of 5 years each. Since Indra is said to be the slayer of Vṛitra, Śambara, Bala and other demons, it is clear that those demons represent the same evil nature or side of the same intercalary months. The expression that Indra found out Śambara and killed him in the fortieth year (*R. V.* II. 12, 11) proves the same fact.

CHAPTER III

MANVANTARA CYCLE.

IN the first and second Chapters treating of the Vedic Sacrificial Calender and the Seven Adityas, it has been clearly pointed out how the units of '1000' and 'seven' are closely associated with each other. In the later Siddhānta works and also in the Amara Kośa, the same unit of '1000' is connected not with 'seven,' but with 'fourteen' which is double the number of 'seven'; but in doing so, the unit of 1000 is interpreted as a thousand Yugas and the unit of fourteen as fourteen Manvantaras of seventy-one or seventy-two Yugas each. Furthermore in making 1000 Yugas equal to fourteen times seventy-one or seventy-two Yugas, the word Yuga is given a wide meaning of 43,20,000 years and the whole is treated as an era of Brahmakalpa of 1000 fourfold Yugas split into fourteen Manvantaras of seventy-one or seventy-two fourfold divine Yugas each. Thus a simple arithmetical problem of 1000 days with one intercalary month, and 7000 days with seven intercalary months or 14,000 days with fourteen intercalary months became in the time of the Siddhāntas and the Amara Kośa a huge era of Brahmakalpa and Manvantaras made up of a huge number of years. It cannot at all be denied that this huge era of Brahmakalpa and Manvantaras has for its basis the same two simple units, the unit of 1000 and the unit of seven. This can be easily perceived if the era of Brahmakalpa and Manvantaras described in the Purāṇas and the Siddhānta works is analysed and traced to its simple source. Let us then see what these works say of this era.

In the Sankalpa or declaration of good intention which every man belonging to the twice-born class makes before undertaking any religious rite, the era of Brahmakalpa coupled with the sixty years' cycle is stated as follows—

"On the auspicious day and hour, in the second Parārdha of First Brahma, which is called the Kalpa of the White Boar, in the period of Vaivasvata Manu, in the Kaliyuga, in the country of Jambūdvīpa, in Bhāratavarsha, in the county of Bharata, in the luni-solar cycle of the sixty years which begins with Prabhava and ends with Kshaya or Akshaya and which is current, as ordained by Lord Vishṇu, in the year (name), of the cycle, in the Southern or the Northern Ayana, as the case may be, in the white or dark half, on the Tithi, I (name) begin to perform the rite (name) with the object of pleasing the Almighty."

That this Sankalpa in terms of the era of Brahmakalp is not a later compilation is evedent from the references made to it in the Mahābhāshya of Patanjali, in the Purāṇas, Māhātmyas, and all astronomical works.[1]

पुरा कल्पे एतदासीत्–षोडशमाषाः कार्षापणम्.

Regarding this era of Brahmakalpa, the Padmapurāṇa says as follows—

In terms of his own measurement, he is said to have a hundred years. It is called a Para; and its half is

---

1 Patanjali, I. 1, 50; Vishṇu Purāṇa, I. 1, 12; Setumāhātmya, XXII. 10; XXVI. 96; Brahmsidhānta, Introduction; Parāsaramādhavīya Vivāhaprakaraṇa.

called a Parārdha. Fifteen Nimeshas are said to make a Kāshṭha; Thirty Kāshṭhas make a Kalā; thirty Kalās make a Muhūrta; The same number (30) of Muhūrtas make a day and night for men: the number (30) of days and nights make a month; a month consists of two wings (fortnights); six of them (months) make an Ayana, and a year is made of two Ayanas: southern and northern. The southern Ayana is a night and the northern a day of the gods.

Of twelve thousand divine years (43,20,000) the four Yugas, Krita, Tretā, Dvāpara, and Kali are made; Know from me, O! King, their divisions; four, three, two, and one of a thousand divine years are there in the four Yugas in order, as settled by the ancients; an early Sandhyā or twilight period of the same number of hundreds is also affixed to them in order; a Sandhyāmśa or part of twilight period of the same length there is at the close of each Yuga. The above periods with Sandhyā and Sandhyāmśa constitute, O! King, a fourfold Yuga of Krita, Tretā, Dvāpara and Kali. One thousand of these makes a day-time to Brahma. O! King, there are fourteen Manus in a day of Brahma; listen to me about the measure of their (Manus') time; the seven sages, the gods, Indra, Manu and his sons are all created simultaneously and likewise they are put an end to. Seventy-one of the fourfold Yugas with a little more make a Manvantara, constituting the period of a Manu and the gods. To eight-hundred thousand of divine years, likewise fifty-two thousand, and thirty Koṭis they amount to. Likewise sixty-seven Niyutas, O! wise king, and twenty thousand;—no more than this.—This is the period of a Manu called Manvantara in terms of human years. Fourteen such Manvantaras make a day time to Brahma; then the three worlds are burnt up; gods residing in the world called Mahah feel the heat and come up to the world called Janah. O! King, learned in the knowledge of Brahma, hear me of that time when all the three worlds are submerged under water; the Lord swollen with the worlds in his belly sleeps on the snake-bed. He is meditated upon by the Yogis in the Jana-world. Then he creates a night of the same duration. Thus a year of Brahma is made up of 360 such days and nights. His life extends to a hundred of such years. In his period one Parārdha has elapsed; at the end of that Parārdha one Kalpa called Padma has ended; to the second Parārdha which is now current the name, 'Varāha,' is given.

The above scheme may be cast in a tabular form as follows:—

15 Nimeshas	A Kāshṭha
30 Kāshṭhas	A Kalā
30 Kalās	A Muhūrta
30 Muhūrtas	A day and night
30 Days and nights	A month of light and dark halves or wings.
6 Months	An Ayana, Southern or Northern called a night and day to Gods respectively.
12,000 divine years or 43,20,000 human years	A fourfold Yuga
A Kṛita Yuga	4,800 Divine Years
A Tretā Yuga	3,600 Divine Years
A Dvāpara Yuga	2,400 Divine Years
A Kali Yuga	1,200 Divine Years

Thus the fourfold Yuga consists of 43,20,000 human years. 71 Fourfold Yugas or 30,67,20,000 years make a Manvantara. A day or night Kalpa of Brahma is equal to 14 Manvantaras, or 4,29,40,80,000 human years.

A day and a night Kalpa put together comes to 8,58,81,60,000 years. The Padmakalpa that has already passed is equal to the above number of years.

In the present Varāhakalpa a Day-Kalpa equal to 4,29,40,80,000 years and 6 Manvantaras equal to 1,84,03,20,000 years have also passed. The Vaivasvata Manu is now current.

The Amarakośa corroborates the above view. It says :

A month is a day and night to the Pitṛis.

A human year is a day and night to the Gods.

Two thousand divine Yugas make a day and night to Brahma.

They (the day and the night of Brahman) make two Kalpa periods to men.

Seventy-one divine Yugas make a Manvantara. According to the Sūrya Siddhānta a fourfold Mahāyuga is equal to 43,20,000 human years, made up of a Krita period of 17,28,000 years, a Tretā period of 12,96,000 years, a Dvāpara period of 864,000 years and a Kali period of 43,20,000 years. A Manvantara is equal to 71 Mahāyugas or 30,67,20,000 years. A day and night Kalpa of Brahma is equal to 8,63,00,00,000 years made of 28 Manu periods and 28 Manusandhi periods and one Krita period.

If the epithet 'divine' to Yuga is left out of consideration in the above calculations, the scheme may be recast in simple words as follows :—

71 Yuga years make a Manvantara.

14 Manvantaras or 1,000 Yuga years make a day or night Kalpa of Brahma.

If 1,000 is divided by 14, the quotient obtained is 71 and 6/14. This fraction is referred to as "a little more" (sādhika) in the passage of the Padma Purāṇa quoted above. In the Amara Kośa and also in the Purāṇas this fraction seems to have been omitted and only the whole number 71 is taken as a fourteenth part of 1,000 In the astronomical works, however, a fourteenth part of 1,000 is put as 72 and a Manvantara is stated to be equal to 72 Yugas. All are, however, agreed in saying that a lunar month makes a day to the Pitris, that a human year of 360 days is a day and night to the Gods, and that two thousand Yuga years make a day and night to Brahma. According to Amara these two thousand Yuga-years make two Kalpas, a day Kalpa of 1,000 Yuga-years and a night Kalpa of 1,000 Yuga-years.

Now the question that crops up is what is the meaning of a Yuga. It is used in four different senses:

(i) A fourfold Yuga of 12,000 years made up of a Krita Yuga of 4,800 years, a Tretā Yuga of 3,600 years, a Dvāpara Yuga of 2,400 years and a Kali Yuga of 1,200 years;

(ii) A cycle of five human years, as expounded in the Vedāngajyautisha;

(iii) A single human year; and

(iv) A Parva or half a lunation, known as a white or dark half of a lunar month. In the Setumāhātmya which is said to form part of the Skānda Purāṇa, one of the eighteen Purāṇas, the word 'Yuga' is used in the sense of a year of 360 or 366 days as follows:—

In olden times a saint called Gālava devoted to Vishṇu performed penance on the shore of the South Indian ocean, not far from Hālāsya, quite close to the village called Fullagrāma, near the tank Kshīrapushkariṇī, and also on the bank of the tank called Dharmapushkariṇī for *ten-thousand Yugas* extolling the eternal Brahma; kind to all, observing fast, truthful, with full control over his senses, regarding all creatures as identical with himself, and with no desire for all worldly pleasures. For some years he lived on ripe leaves of trees and plants, for some years on water, and for some years on air. Thus he performed his penance for *five thousand years*. Then for the remaining *five thousand years* he continued his penance taking no food whatever, and without breathing in the air. In the rainy season he allowed himself to be drenched, in winter he kept standing in water, and in summer he sat in the midst of 'five fires', always being devoted to Vishṇu, meditating on the meaning of the 'Eight syllabled Mantra' and thinking of God Janārdana. Thus Gālava, the saint, performed his penance; thus there passed away ten thousand years when at last Vishṇu was pleased with him."

Here the equation of *ten thousand Yugas* with the sum of 'five thousand years' and the remaining 'five thousand years' clearly proves that Yuga is used in the sense of a year. Its use in the sense of a Parva of 15 Tithis is frequently met with in the Sūryapragnapti and other Jaina astronomical works. It says (P. 167) that in the Parvarāśi number, that which is divisible by four is called krita; that number which, when divided by four, leaves a remainder 3 is called Tretā; those numbers which, when divided by four, leave remainders 2 or 1 are called Dvāpara and Kali respectively. The verses run as follows—

चउहिं हियम्मि पव्वे एक्के सेसम्मि होइ कलिओगो ।
वेसु य दावरजुम्मो तिसु तेया चउसु कडजुम्मो ॥
कलिओगे तेणवई पक्खेवो दावरम्मि वावट्ठी ।
तेऊए एक्कतीसा कडजुम्मे णत्थि पक्खेवो ॥
सेसद्धे तीसगुणे वावट्ठीभायियम्मि जं लद्धम् ।
जाणे तइसु मुहुत्तेसु अहोरत्तस्स तं पव्व ॥

छाया

चतुर्भिर्हृते पर्वणि एकस्मिन् शेषे भवति कल्योजः ।
द्वयोः द्वापरयुग्मं त्रिषु त्रेता चतुर्षु कृतयुग्मम् ।
कल्योजसि त्रिनवतिः प्रक्षेपो द्वापरे द्वाषष्टिः ।
त्रेताया मेकात्रिंशत् कृतयुग्मे नास्ति प्रक्षेपः ।
शेषार्धे त्रिंशद्गुणे द्वाषष्ट्या भाजिते यल्लब्धम् ।
जानीहि तेषु मुहूर्तेषु अहोरात्रस्य तत्पर्व ॥

When the Parva-number is divided by four, and when there is a remainder of one, it is Kalyojah; when the remainder is two, it is Dvāparayugma; when it is three, it is Tretā; and when it is four, that is, when there is no remainder, it is Kritayugma. In the case of Kalyojah, ninty-three is added to the parva-rāśi; in the case of Dvāpara' it is sixty-two; in Tryojah, it is thirty-one: and in the case of Kritayugma, there is no *prakshepa* or addition. When half of the remainder is multiplied by thirty and is divided by sixty-two, know that the quotient indicates the time in Muhūrtas of the expiration of the parva.

Commenting on the above Malayagiri says:

Parva means parva-rāsi-number. When this number, divided by four, leaves a remainder, one, it is called Kalyojan; when the remainder is two, it is Dvāpara-yugma; when it is three, it is called Tretaujah; and when there is no remainder, it is called Kṛtāyugma. in the case of Kalyojorāsi, 93 is added to the remainder, making it 94; in the case of Dvāpara it is 62; in the case of Tretaujah, it is 31; and in the case of Kṛitayugma, there is no addition or *prakshepa.* The resulting numbers after the addition are divided by 124. Then what the rule says is this:

The remainder is divided by 124 and is multiplied by 30; and the whole is reduced to half, as both the numerator and denominator is divisible by two. Then the product shows the number of *muhūrtas* and *muhūrta-bhāgas.* The disciples are thus told to consider the expiration of the parva at the close of the *muhūrtas* and *muhūrtabhāgas*, just arrived at, on the last day. The following examples make the meaning clear:

Suppose one asks at the close of how many *muhūrtas* on the last day does the first Parva expire? Then we take 1, as the number of *Parvas* is one; as it is Kalyojah, we add 93 to it; it becomes 94; we divide this by 124; it is not divisible, as the number is less; then reducing it by 2 and multiplying it by 30, we have $47 \times 30/62$; it is equal to $1410/62 = 22$ *muhūrtas* and 23 parts of 31 divisions of a *muhūrta.* This means that the first parva·

was completed after 22 *muhūrtas* and 23 parts of 31 divisions of a *muhūrta* on the last day of the Parva. Likewise, the second Parva was completed after $64 \times 30/124 = 32 \times 30/62$; it is equal to $960/62 = 15$ *muhūrtas* and 15 parts of 31 divisions of a *muhūrta*; similarly the third Parva attains completion at the close of $34 \times 30/124 = 17 \times 30/62 = 510/62 = 8$ *muhūrtas* and 7 parts of 31 divisions of a *muhūrta*. In the case of the 4th Parva no addition is made. It attained completion at the close of $4 \times 30/124 = 30/31$, or 30 parts of 31 divisions of a *muhūrta*. If the time of the 124th Parva's completion is asked, then $124/124 = 1$; thus there is no fraction. It means, therefore, the Parva in question was completed at the close of the day, that is at the close of 30 *muhūrtas* of the day.

Here the Prakshepa or addition of 93, 62, 32 to Kali, Dvāpara and Tretā is merely the fractional Parva-Raśi numbers corresponding to those Parvas. As a yuga of 5 years consists of $5 \times 366 = 1830$ days distributed in 124 Parvas, a Parva is equal to $1830/124 = 14$ days and 94 parts of 124 divisions of a day; this is called Kaliparva or Kaliyuga. Similarly the second Parva is $1830 \times 2/124 = 29$ days and 64 parts of 124 divisions of a day. Likewise the third Parva is $1830 \times 3/124 = 5490/124 = 44$ days and 34 of 124 divisions of a day. The fourth Parva is $1830 \times 4/124 = 59$ days and 4 of 124 divisions of a day. The 124th Parva is $1830 \times 124/124 = 1830$th day complete. Evidently the Vedic poets gave the name Kali to that new moon or full moon which attained its completion at midnight; Dvāpara to that which attained completion just after sunset; Tretā to that which attained completion at midday; and Krita to that which attained completion just at sunrise. From the above it is clear that the first Parva closes at or after three quarters or Pādas of the day; the second after two quarters or Pādas of the day; the third Parva after one quarter or Pāda of the day; and the fourth Kṛita Parva just at sunrise. According to the Vedāngajyautisha (Verses 12-14) a pāda is equal 31 Kalās which is exactly one-fourth of 124 Kalās of a Parva day. We are told that the Parva day closing at one, two, or three quarters of the day should be rejected (dyu heyam Parva chetpāde), and that Parva day which closes at sunrise or within one-fourth part of a day should be observed. In other words Krita Parva was observed and Tretā, Dvāpara and Kali Parvas were rejected and the Parva rite was performed on the previous day. This view is supported by what is stated in the Aitareya Brāhmaṇa (5-2-15). The passage is as follows—

कलिश्शयानो भवति सञ्जिहानस्तु द्वापरः ।
उत्तिष्ठन् त्रेता भवति चरन्सम्पद्यते कृतम् ॥

Kali is lying; Dvāpara is retiring; Tretā is standing up; and Krita is moving. As Kali Parva closes at midnight, it is said to be lying fast asleep; as Dvāpara closes at after sunset, it is said to be retiring from work abroad; as Tretā closes at midday, it is said to be standing up and as Krita closes at or after sunrise it is said to be moving for work. Even now rites in honour of gods are performed on those days on which the Tithi closes before midday and ancestral rites (Pitṛikarma) are performed on the day on which the Tithi of the death of the ancestor closes at or after midday.

There can, therefore, be no doubt that the words Kṛita, Tretā and Dvāpara used in the Krishṇa Yaj.

(4, 3, 3) mean the Parvas of those names, as pointed above. Dr. Keith has, however, taken them to mean the throws of dice.

We are also told in the passage of the Satapatha Brāhmaṇa (XIII, 3, 2, 2) that the sacrificer of the Vedic times attained his firm foundation by 'Chatushṭoma' and by Kṛita among dice. The information so far gathered by a number of Oriental scholars regarding the game of dice is summarised by Prof. Eggeling in a note on his translation of the Satapatha Brāhmaṇa (Vol. IV, P. 107) as follows:—

"The allusions to the game of dice in the early literature are not sufficiently definite to enable us to form a clear idea as to the manner in which the game was played. Sāyaṇa, on Taitt. S. 1, 8, 16, remarks that the dice here used consisted either of gold *cowries* (shells) or of gold (dice shaped like) Vibhītaka nuts. That the brown fruit of the Vibhītaka tree (Terminalia Bellerica)—being about the size of a nutmeg, nearly round, with five slightly flattened sides—was commonly used for this purpose in early times, we know from the Ṛigveda; but we do not know in what manner the dice were marked in those days. According to the commentators, the game is played with five dice, four of which are called Kṛita, whilst the fifth is called Kali; and if all the dice fall uniformally (ekarūpa)—*i.e.* with the marked sides either upwards or downwards—then the player wins, and in that case the Kali is said to overrule the other dice. In this case the Kali would seem to represent the king. Kāty. Sr. S. XV, 7, 18-19, however, admits of another mode of playing, by which the Kali represents the Sajāta tribesman, whilst the king and those that come after him (in the enumeration in paragraphs 15-20) play the Kṛita, &c., To understand this mode we have probably to turn to Chānd. Up. (IV, I, 4) where it is said of the saint Raikva, that everything good fell to him, just as the lower dice (or casts) submit to the conquering Kṛita. Here the commentators assign the names Kṛita, Tretā, Dvāpara, and Kali to different sides of the dice, marked respectively with 4, 3, 2, and I marks. (anka).—In Taitt. Br. I. 7, 10, the game at dice at the Rājasūya, is referred to as follows:—With 'this king has overcome the regions' he hands (to the king) five dice; for these are all the dice: he hereby renders him invincible. They engage to play for a dish of rice (ödana), for that is (a symbol of) the chief: he thus makes him obtain every prosperity. He addresses them (with the epithets of) 'far-famed, most prosperous, true king.' The commentary and the Sūtras then supply the following explanation:—'The keeper of the dice (akshāvāpa) having (marked off and) raised the gambling ground (by means of the wooden sword), and sprinkled it, throws down more than a hundred or more than a thousand gold dice. From them he takes five dice and hands them to the king: these, as representing the five regions, are taken to include all those dice.' These explanations, so far from clearing up the doubtful points, seem rather to add to them. It may be noted, however, that in the well-known hymn Ṛik. S. X, 34, in which the gambler's state of mind is pictured in a very expressive language, the dice of the game are apparently spoken of as Tripanchāśavrata, or 'the troop of fifty three' (or thrice five, according to Ludwig's rather improbable conjecture)."

The above remark of Prof. Eggeling leaves no doubt about the prevalence of the game of dice from the earliest down to our own times. But the manner in which the game was played in ancient times cannot certainly be the same as it is now. Now-a-days two rectangular pieces of ivory or horn marked on each of the four sides with 6, 4, 2, and 1, are thrown on the ground; and each player leads his eight dice-men through 54, 60, or 72 out of 96 houses marked on a piece of cloth, by taking one or two dice-men at a time through as many houses as the number on the surface of the two thrown pieces. If the number on the thrown pieces is such as leads one or two dice-men of a player to the same house as is occupied by one or two of his opponent's dice-men, then the latter takes back his dice-men to the first house again and begins his play afresh with the defeated pieces. But in ancient times the method of play was quite different. The game of dice formed part of the sacrificial ceremony connected with the establishment of the sacred fire. Taking a cow belonging to the sacrificer, a number of players used to go along the streets of a town or village, and making the cow the stake, they used to play at dice in different batches with those who deposited grain as their stake. Each player used to throw on the ground a hundred or more *cowries* (shells); and when the number of the *cowries* thus cast and fallen with their face upwards or downwards, as agreed upon, was exactly divisible by four, then the sacrificer was declared to have won: but if otherwise, he was defeated. With the grain thus won, four Brāhmans used to be fed on the day of sacrifice. That this was the nature of the custom that prevailed in those days, is clear from the following Sūtra of Āpastamba and Agniswāmi's commentary thereon.

कृतेन यजमानो विजिनाति ॥ Āp. Śr. S. V, 20, I.

"The sacrificer wins by Kṛita."

The commentary runs as follows—

कृतत्रेताद्वापरकलिनामानो द्यूतप्रकारा ये कृतमायानामिति श्रुतावया इत्याख्यायंते । न्युप्तेष्वक्षेपु चतुष्कशो विभज्यमानेषु यत्र सर्वे भागास्समा भवंति तत्कृतं नाम । अथ यत्रांततस्त्रयोवशिष्यंते सा त्रेता । यत्र द्वौ स द्वापरः यत्रैकस्सकलिः । तथा च श्रुतिः 'ये वै चत्वारस्स्तोमाः कृतं तत् । अथ ये पंच कलिस्सः' इति । तत्राक्षाणां शतत्वात् कृतेन प्रकारेण यजमानो विजिनाति विजयते.

"Kṛita, Tretā, Dvāpara, and Kali are the names of the different forms of the game of dice as referred to in 'Kṛita among the forms of dice-play,' (Krishṇa Yajus IV, 3, 3). When the number of dice thrown can be arranged into groups of four each and all the groups consist of the same number, it is called Kṛita; and when there is a remainder of three at last, it is Tretā; when two, it is Dvāpara; when one, it is Kali. Accordingly the Veda (Taittirīya Brāhmaṇa I, 7, 10) says:—'That is Kṛita which consists of groups of four each; and that is Kali which consists of five.—As here there are a hundred dice it is by the form of Kṛita play that the sacrificer wins."[1]

It is, therefore, clear from the above passages that the word, 'Kṛita,' whether used in connection with Parvas or dice, was a name signifying a set of four or multiples of four; that the word 'Tretā' was a name given to three or any number which, when divided by four, left a remainder three; that the word Dvāpara was a name

---

1 Compare Mahidhara's Commentary on *Vajasaneyi Samhita*, X, 2, 8, and X, 30.

given to the number two or any number which, when divided by four, left a remainder two; and that Kali signified one or any other number which, when divided by four, left a remainder one.

The recurrence of Kṛitayugma, the same as Kṛita-yuga, either in the dark half or light half of the month is also alluded to in the Bhagavatī Sūtra. Surprised, perhaps, at the thought that though born on the auspicious day of the Kṛitayugma, some persons commit sin and are thereby destined to hell, Gautama addresses Mahāvīra as follows:—

कण्हपक्खियरासिजुम्मणेरयियाणं भंते कनु उववज्जंति? सुक्ख-पक्खियरासिजुम्मकयजुम्मणेरयियाणं भंते कनु उववज्जंति ॥[1]

"Why are persons destined to hell born on the day of Kṛitayugma in the dark half of the month, O, lord? And why are they born on the day of Kṛitayugma in the light half of the month, O, lord?"

Mahāvīra's answer to this question is not of so much importance to us as the question of Gautama referring to the occurrence of Kṛitayugma in the dark or light half of the month. It need not be explained that what Gautama meant by a Kṛitayugma in the light or dark half of the month was the close of the Parva of that name at sunrise or before midday, in the light half or dark half of the month.

---

1 P. 1931 *Bhagavati Sutra.*

"The date of the collection, or perhaps more correctly, the composition of the Jaina Canon would fall somewhere about the end of the fourth or the beginning of the third century B.C." P. 43, Preface to Translation of the *Achārānga Sutra of the Jainas by* Hermann Jacobi.

Whether the game of dice was also played in connection with the laying of Apānabhṛit bricks on the altar reciting the Yajurmantra (4, 3, 3) which contains Kṛita and other Parva words, is a question which cannot be definitely answered. As the names of seasons, number of months and sets of intercalary days are given in that passage, I am inclined to think that Kṛita and other words in the passage signify Parvas rather than games of dice. The word 'Trivṛit' means a set of nine days as well as a set of nine verses. Eighteen lunar months differ from eighteen solar months by 18 days which are divided into two sets of nine days each; and one set of the nine days is inserted before, the central day and another after the central day of the year. Likewise in the course of 3½ years the difference between the two years amounts to 42 days, of which 30 days are divided into two sets of 15 days each and inserted before and after the central day of the year, as before, leaving the remaining 12 days to be disposed of later. Likewise in the course of six years the difference amounts to 72 days, in which 68 days are divided into sets of 17 days each; and of these, two sets of 17 days are inserted before the central day and the other two sets after the central day of the year. The remaining 4 days are carried forward to be adjusted in the 9th year. In the 9th year the difference of 108 days between the lunar and the solar years are split into sets of 21 days each, half of them being inserted before and the other half after the central day. Here also the number of days that remain are carried forward to be treated of in the 12th or the 13th year. This seems to the meaning of the passage. It is as follows:—

(1) "The East of the quarters; the spring of the seasons; Agni the deity; holy power the wealth; the Trivrit the Stoma, and it forming the path of the fifteenfold Stoma; the eighteen month old calf the strength; the Kṛita of the throws of dice; the east wind the wind; Sanaga the Rishi.

(2) The South of the quarters; the summer of the seasons; Indra the deity; the kingly power the wealth; the fifteenfold the Stoma, and it forming the path of the seventeenfold Stoma; the two year old calf the strength; the Tretā of the throws; the South wind the wind; Sanātana the Rishi.

(3) The West of the quarters; the rains of the seasons; the All-gods the deity; the peasants the wealth; the seventeenfold the Stoma, and it forming the path of the twenty-one-fold Stoma; the three-year old the strength; the Dvāpara of the throws; the West wind, the wind; Ahabuna the Rishi.

(4) The North of the quarters; the autumn of the seasons; Mitra and Varuṇa the deity; prosperity the Stoma, and it forming the path of the twenty-seven-fold Stoma; the four-year-old the strength; the Āskanda of the throws; the North wind, the wind; Pratna the Rishi.

(5) The zenith of the quarters; the winter and the cool season of the seasons; Bṛihaspati the deity; radiance the wealth; the twenty sevenfold the Stoma, and it forming the path of the thirty three-fold; the draught ox the strength; the Abhibhū of the throws; the wind all through the wind; Suparṇa the Rishi."

The question that is under consideration is not, however, the meaning of the passage of the Kṛishṇa

Yajurveda quoted above. The main question under our consideration is whether the words, Kṛita Yuga, Tretāyuga, Dvāparyuga, and Kaliyuga are anywhere used in the sense of a Parva of fifteen Tithis. I think that the consideration of the quotations made from the Sūryapragnapti and the Vedāngajyautisha leaves us in no doubt as to the use of those words originally in the sense of a Parva of 15 Tithīs terminating at sunrise, at midday, at evening or at midnight. It follows, therefore, that the word Yuga once meant a Parva of fifteen days also. Accordingly the word Yuga in the phrase of 'Yugasahasra' of a Brahmakalpa may be taken in any of the four senses. It remains now to be decided which of these four meanings is historically sound. According to the Padma Purāṇa passage quoted above there had passed away three thousand Yugas equivalent to three Brahmakalpas and six Manvantarās equivalent to four hundred and twenty six Yugas, together with twenty eight Yugas in the seventh Manvantara, in short three thousand and four hundred and fifty four years. Taking the Yuga in each of the four senses, we have—

(i) 3454 Yugas $=3454 \times 43{,}20{,}000$ years
$=14{,}92{,}12{,}80{,}000$ Years.

(ii) 3454 Yugas $=3454 \times 5$ Years $=17{,}270$ Years.

(iii) 3454 Yugas $=3454 \times 1$ Year $=3{,}454$ Years.

(iv) 3454 Yugas $=3454 \times 1/24$ Years $=143\frac{22}{24}$ Years.

References to Brahmakalpa are found in the Sukla Yajurveda 20, 4. It is as follows:—

"In the beginning rose Hiraṇyagarbha, born, only Lord of all created beings."

"He fixed and holdeth up this earth and heaven."

Reciting this verse, the sacrificer puts down a lotus leaf on the altar ground, a piece of gold on the lotus leaf and a golden figure of man on the lotus to represent Hiraṇyagarbha Brahma, or Prajāpati, the God of time. Similarly the Tait. Āra. (1, 25) Speaks of a lotus leaf, a piece of gold and a golden figure of man being laid down in the pit of an altar ground. In Kṛishṇa Yajurveda (5, 2, 7) the sacrificer is required to lay down a gold disc in the pit of the altar ground and a golden figure of man on the disc to represent the creation of Brahma, the Creator. The great fire altar described in the fourth Kāṇḍa of the Kṛishṇa Yajurveda seems to symbolise the closing the Padma Kalpa of Brahma.

The names of the fourteen Manūs mentioned in the Purāṇas are—

1. Svāyambhuva	8. Sūryasāvarṇi
2. Svārochisha	9. Dharmasāvarṇi
3. Uttama	10. Dakshasāvarṇi
4. Tāmasa	11. Rudrasāvarṇi
5. Raivata	12. Brahmasāvarṇi
6. Chākshusha	13. Indrasāvarṇi
7. Vaivasvata	14. Vedasāvarṇi

These and other references to a Kalpa of 1000 years, Parvas or days divided into 14 periods of 71 or 72 years, Parvas or days, called Manvantarās or reigns of Manus establish that something like an era of Manvantaras, started synchronously with the eras of the cycles of sixty and of thirtythree years in B. C. 3101 was current in India till A.D. 353.

What, however, is to be borne in mind is the division of 1,000 into 14 Manvantarās of 71 or 72 Yugas or Parvas each. Now let us see how this era is connected with the cycles of 20 and 33 years.

The Vedic Cycle of Thirty Three Years :—

The Vedic Poets seem to have been observing a grand cycle of 33 lunar years made up of six smaller cycles for adjusting the lunar year of 354 days with the solar year of 365 days. The Vedic terms in which the cycles are spoken of in the Vedas are so enigmatic as to defy interpretation. Still a comparative study of all the parallel passages found in the various schools of the Yajurveda will not fail to unravel the astronomical and religious mysticism of the Vedic poets. The passages taken for comparative study are Krishna Yajus. 4, 3, 3; 4, 3, 5; Maitra. S. 3, 11, 11; Śukla Yajus. 14, 10; 18, 27-27; 21-12-28; 24, 12-13; 28, 24-34. The passage of the Krishna Yajus runs as follows:—

(1) "The East of the quarters; Agni the diety; holy power the wealth; the Trivrit the Stoma, and it forming the path of the fifteenfold Stoma; the eighteen-month old calf the strength; the Krita of the throws of dice; the east wind the wind; Sanaga the Rishi."

(2) "The South of the quarters; the summer of the seasons; Indra the deity; the kingly power the wealth; the fifteenfold the Stoma, and it forming the path of the seventeenfold Stoma; the two-year-old the strength; the Tretā of throws; the South wind the wind; Sanātana the Rishi."

(3) "The West of the quarters; the rains of the seasons; the All-gods the deity; the peasants the wealth; the seventeenfold the Stoma, and it forming the path of the twenty one-fold Stoma; the three-year-old the

strength; the Dvāpara of throws; the West wind the wind; Ahabuna the Rishi."

(4) "The North of the quarters; the autumn of the seasons; Mitra and Varuṇa the deity; prosperity the wealth; the twenty-one-fold the Stoma, and it forming the path of twenty-seven-fold Stoma; the four-year-old the strength; the Āskanda of throws; the North wind the wind; Pratna the Rishi."

(5) "The zenith of the quarters; the winter and the cool season of the seasons; Brihaspati the deity; the radiance the wealth; the twenty seven-fold the Stoma, and it forming a path of the thirty-three-fold; the draught-ox the strength; the Abhibhū of throws; the wind all through the wind; Suparṇa the Rishi."

There are six minor cycles clearly stated in the above passage. The first is a cycle of 18 months (lunar), which falls short of 18 solar months by 18 days. According to the Laṭyāyana Srauta Sūtra (4, 6, 12, 10 and the Śatapatha Brāhmaṇa) 9 days are inserted before the central day of the year and 9 days after the central day. It is these 9 days which are here called 'Trivrit Stoma.'

The second is the cycle of 3½ years with 42 intercalary days. Of these 30 days are disposed of by adding 15 days before the central day of the year and 15 days after it. The remaining 12 days are reserved for the seventeen-fold Stoma. This is the meaning of the word 'Saptadaśavartani', forming the path of the seventeenfold Stoma.

The third cycle is a cycle of 6 years with 12×6=72 intercalary days. Of these two sets of seventeen days are added before the central day and 2 sets after the central day. The remaining 4 days are reserved to be disposed of in the next cycle of 9 years. The intercalary days of the next cycle are $12 \times 9 = 108$, which form 5 sets of 21 days each nearly. As in other cycles they are disposed of by adding about 52 days both before and after the central day of the cycle, 3 days being reserved for the next cycle. The next is the cycle of 13 years ($1\frac{1}{2}+2+2\frac{1}{2}+3+4=13$) with 156 intercalary days of which 5 sets of 27 days each are disposed of as before and 21 days are reserved for adjustment in the next cycle of 17 years. ($1\frac{1}{2}+2+2\frac{1}{2}+4+3+4=17$).

The number of intercalary days in this cycle are $12 \times 17 = 204$ days; which are disposed of by adding 3 sets of 33 days both before and after the central day of the cyclic year.

The adjustment of the lunar year with the solar year by adding to the lunar year one or two sets of 9, 15, 17, 21, 27 and 33 days reserving at the same time a few days is evidently to equalise the actual and theoretical difference between the two kinds of years. By the repetition of the same process for another 16 years, the cycle of 33 years is completed bringing the solar and lunar years to the same starting point. Each year is called a god and the cycle of 33 years is called a set of 33 gods. If even after 33 years the two years were not found to come to the same starting point, the cycle was continued for another year and somehow or other equalized at the close of the 34th year. Accordingly Prajāpati, the god of the cyclic year is sometimes called '33'. (Tait. S. IV. 6, 9). Seventeen which is half of thirtyfour is also another important number frequently used to signify Prajāpati, the Year-God, just as the Vedānga-jyautisha calls Prajapati the lord of the five years' cycle.

The necessity for the division of the cycle of 33 years into sub-cycles is not merely the convenience of adjusting the two years at easy intervals, but the recurrence of certain astronomical and natural phenomena at those intervals. These phenomena are specified in the grand 'Apri' hymn of the White Yajus (28, 24-34) which runs as follows:—

"Let the priest worship Agni, Indra, kindled, excellent strength-giver, lending him mighty power, Gāyatri metre, a cow aged eighteen months and vital vigour. Let him enjoy the butter, Hotar, worship. (24.)

Let the priest serve with sacrifices him who breaks forth, Tanūnapāth, the germ which Aditi conceived, pure Indra who bestoweth strength, bringing him mighty power, the Ushnih metre, an ox two years old, and of vital vigour. Let him enjoy the butter, Hotar, worship. (25.)

Let the priest sacrifice to Soma, Indra, adorable, adored, best Vritra-slayer, strength-giver, mighty, to be abored with viands, bringing him mighty power, Anushṭup metre, a cow of thirty months, and vital vigour. Let him enjoy the butter, Holar, worship. (26.)

Let the priest worship Indra, strength-bestower, immortal, with fair grass, allied with Pūshan, seated on sacred grass, dear, everlasting, bringing him mighty power, Brihati metre, a steer of three years' age and vital vigour. Let him enjoy the butter, Hotar, worship. (27.)

Let the priest worship the wide-opening Portals, easy to pass, divine, law-strengthening, golden, Indra, the Brahman priest, the strength-bestower, bringing him mighty power, the Pankti metre, a bullock four years old and of vital vigour. Let him enjoy the butter, Hotar, worship. (28.)

Let the priest worship lofty Night and Morning, well decked, of varied hue, lovely to look on, Indra, the Universal, strength-bestower, bringing him mighty power, the Trishṭup metre, a bullock, four years old and of vital vigour. Let him enjoy the butter, Hotar, worship. (29.)

Let the priest worship both celestial Hotars, the gods' best glory, sages famed for wisdom, the two companions, Indra strength-giver, bringing him mighty power, Jagati metre, an ox that draws the wain and vital vigour. Let him enjoy the butter, Hotar, worship. (30)

Let the priest sacrifice to the three well decorated goddesses, gold-decked, lofty, Bhāratis, Indra, their lord who gives strength, bringing him mighty power, Virāj, the metre, and a cow in milk. Let him enjoy the butter, Hotar, worship. (31)

Let the priest worship Tvashṭar, the prolific, strengthener of growth, maintaining varied growth and form, Indra who gives vital force, bringing him mighty power, Dvipād metre and an ox full grown. Let him enjoy the butter, Hotar, worship. (32)

Let the priest worship the forest lord Vanaspati, the Immolator, hundred-powered, praiseworthy, golden leafed, who wears the girdle, loved, the gracious lord, Indra, who gives the strength of life, giving him mighty power, Kakup metre, a barren, calf-slipping cow, and vital vigour. Let him enjoy the butter, Hotar, worship. (33).

Let the priest offer worship to the Svāhākṛiti goddesses, to Agni House-hold lord apart, to Varuṇa the leech, and sage, mighty Indra who bestoweth strength, bringing him mighty power, Atichhandas metre, a strong bull in his prime, and vital vigour. Let them enjoy butter, Hotar, sacrifice. (34)

Here in the six verses from 24 to 29 the number of years given is 17 and in verses from 30 to 34 no years are given, but only sacrificial victims are named. In S. Yajus. 21, 12-17, however, the number of years given comes to 19. The verses run as follows:—

"Kindled is Agni with the brand, yea, kindled well, the excellent; the metre Gāyatri, the steer of eighteen months, give power and life. (12)

Tanūnapāth whose acts are pure, our bodie's guard Sarasvati, Ushṇihā metre and the steer of two years age give power and life. (13)

Agni with offerings, meet for praise, and Soma the immortal god, Anushṭup metre, and the steer of thirty months, give power and life. (14)

Agni with goodly grass spread out, deathless with Pūshan at his side, Bṛihati metre and a steer of three years age, give power and life. (15)

The Doors divine, the mighty regions, Brahma, god Bṛihaspati, the metre Pankti, here a bull in his fourth year, give power and life. (16)

The two young Dawns of lovely form, the deathless universal gods, the Trishṭup metre, here a bull in his sixth year, give power and life. (17)

The two celestial Hotars, both physicians, Indra's close-knit friends, the metre Jagatī, an ox who draws the wain, give power and life. (18)

The three Iḍā, Sarasvatī and Bhāratī, the Marut folk, Virāj the metre, here a cow in milk, a bull, give power and life. (19)

Tvashṭar, the wondrous, full of seed, Indrāgnī furtherers of weal, Dvipāda metre, a cow and a vigorous bull give power and life, (20)

Our slaughterer Vanaspati, Savitar who promoteth wealth, the metre Kakup, here a cow who casts her calf, give power and life. (21)

Here the number of years given is $1\frac{1}{2}+2+2\frac{1}{2}+3+4+6=19$. The remaining four bulls for which no age is given may be taken to signify 14 years, making the total 33 years.

While in the above passages of the S. Yajurveda age-grades are given in terms of cows and bulls, the M. S. II. 7, 20 and the Tait. Brāh. II 6, 17, substitute Tryavi (three sheep), and Panchāvi (five sheep) for 'a calf of 18 months', and 'a bull of 30 months; respectively. Solar eclipses are called Avis, sheep.

These thirty three years are described as thirty three gods, made up of 11 on the earth, 11 in the atmosphere and 11 in the sky and the total number of such gods is stated to be 3,349 in the Tait. Brāhmaṇa (III. 11, 3, 25), while in the Śukla Yajurveda (XXXIII, 7) only 3,339 gods are stated to have worshipped Agni.

There are two more points deserving consideration in the Śukla Yajurveda passage (28, 24-34) quoted above. They are—

(*a*) the Samiddha and other nine gods of the Apri Verses and

(*b*) the metres assigned to the months of the Samiddha and other gods.

In verse 24 a cycle of 18 months is mentioned as the time of kindling the fire. No season is mentioned. The mention of the Gāyatrī metre, however, indicates the gnoman's shadow of 24 Angulas. It is probably the rainy season, the time being the full moon or the new moon of the month of Śrāvaṇa. The fire or the sun of this season is called Samiddha, well kindled.

In verse 25 the fire or the sun is called Tanūnapat, son, born of himself; the Ushnih-metre of 28 syllables indicates the time of the full-moon or the new-moon of the month of Bhādrapada. It is two years with 30 intercalary days after the time of the Samiddha or Kindled fire.

In verse 26 the name of the fire or the sun is 'Narāśamsa,' praise of men. It is more clearly stated in verse 3 of this hymn than here. The shadow measure is 32 Angulas and the month is Aśvayuja, the time of autumnal equinox. The time is 30 months after the time of the Tanūnapāth fire or six years from the beginning. During the rainy season no Vedic hymns are chanted on account of frequent thunders and lightnings when silence is enjoined. In the month of Aśvayuja the fury of the rains is at an end and the priests are glad to resume their Vedic recitation. Hence, the autumn months are called time of speech. In verses 6-8, Book 3 of the White Yajus the same time is spoken of, as the time of the arrival of the spotted cow or bull before the arrival of the Mother and the Father, (the solstices) mounting up to heaven (the equinoctial heaven). He is clearly spoken of as 'ruling thirty stations,' the 30 Muhūrtas of the equinoctial day. It is then that 'speech or song is bestowed on the Bird, the sun or the fire.

In verse 27 the time of the grass or 'Barhis' is mentioned. The shadow measure is 36 Angulas and the month is Kārtika. It is in this month that the sacrificial grass is ripe for cutting. It is a cycle of 9 years.

In verse 28 the month that is indicated is 'Mārgaśīrsha' with its shadow measure of 40 angulas; when, we are told, the portals of the heaven are opened. The month being that of the season of winter solstice, the heavenly doors are said readily opened for the entrance of the departed souls of the good. The time is four years after the previous cycle of 9 years; that is, a cycle of thirteen years, when the number of accumulated intercalary months comes to be five.

In verse 29 we are told that the shadow measure is 44 angulas. Accordingly it is the month of Pushya, the sixth from Śrāvaṇa. The cycle that is devised is that of 17 years or 16½ years when the number of intercalary months amounts to six. If no adjustment is made, the beginning of the unadjusted lunar year falls back by six months and the solstitial periods are reversed, that is, the winter solstice takes the place of summer solstice. The Nakta or the night period from summer solstice to winter solstice becomes the Ushasa period of six months from winter solstice to summer solstice. The priests would be obliged to speak of Ushasā-nakta instead of speaking of Naktoshasā as usual.

In verse 30 the shadow me asure is said to be 48 Angulas in the mouth of Māgha when gods are called upon by the Hotris to assemble in the sacrificial hall. The number of the years composing the cycle is not, however, mentioned.

In verse 31, the metre mentioned is Virāj of 30 syllables which indicates a fall of about 18 Angulas from 48 Angulas of the Hotri month mentioned before. Accordingly it is the beginning of the season of the vernal equinox which is a time of speech, Sarasvatī. Here also the number of the years composing the cycle is not mentioned. It is probably seven or eight years after the last cycle of 17 years, for the vernal equinox comes three months after the winter solstice and because there are three intercalary months in 8 years. It is also the time when the sun has come down from the sky of winter solstice to the Earth or the spring season which is here called Iḍā. It is also the time of nurishment or 'Bhāratī.' The Purushasūkta says (White Yajus. 31, 22) "Beauty and Fortune are thy (Vishṇu's) wives; each side of thee are day and night. The constellations are thy form; the Asvins are thy open jaws."

The Maitrāyaṇīya Samhitā (3, 2, 1) says explaining the verse "Naktoshasā," that the day and night are requested to return again with Ūrja, (Punarūrjānivartasva); bearing upon this, it says (4, 9, 14) that others substitute the verse "Udutyam . . . . chitram . . . ." for the verse "Punarrūrja nivartasva." In saying this the Maitrāyaṇīya Samhitā seems to mean that while some sacrificers pray for the return of the quinoctial days (naktoshasā) in the month of Āśvina or Kārtika when the sun passes through the constellations Aśvini, Bharaṇi and Krittikā, others pray for the return of the day and night with the constellation 'Chitrā' and recite the verse "Udutyam-Chitram."

In 3, 42, the same Samhitā says that Kaṇa Srāvayasa found out that the day and night are associated with the constellation Chitrā. In 4, 3, 10 and 44, 7 the Samhitā says that Vishuvān is Varuṇa and is associated with Chitrā. In 3, 8 and also in 3, 1, 6 and 3, 28 it says that the day is Mitra and that the night is Varuṇa ; that the day is termed as 'Satya' and that the night is called 'Ṛita.' In this connection it is to be noted that of the two equinoctial days one happens in what is called the dark half of the year, that is, the six motnhs from the summer solstice to winter solstice and that the other occurs in what is called the day-part of the year, that is, the six months from the winter solstice to the summer solstice. Hence, it is clear that the equinox in Āśvina or Kārtika was called night, Ṛita and Varuṇa and that the other equinox in Chaitra or Vaiśākha was called day, Satya and Mitra. In 3, 29 the M. Samhita further says that the Aśvinis have ceased to be physicians of the gods, because they have become promulgators of what is not klipta, established, and that the sacrifice to the Aśvins (on the day of vernral equinox) is now obsolete (Utsanna). This is evidently a reference to the precession of equinoxes. The poets were certainly unaware of the law of precession of equinoxes. Yet it cannot be denied that they could observe the change that has gradually taken place in the locus of equinoctial colure. They could certainly observe that the occurrence of day and night of equal length deviated from its original place in the constellations from Aśvin to Krittikā and that it began to take place in a constellation that preceeds the constellation that marked its occurrence before. Hence they seem to have called the Aśvins as 'Asatya', not the vernal equinox, while others persisted in asserting that they were 'na-asatyas', not-untrue."

The term 'Maitrāyaṇīya, itself suggests that it is the Samhitā of those who once observed the autumnal equinox in the Mitra or Anurādha constellation. The Atharvaveda (19, 7) refers in distinct terms to the occurrence of the equinox in the 'Chitras.' "The Chitras are together in the sky and the serpents in the earth."

In verse 32 god Tvashṭri is the deity and the metre is Dvipād of 10 syllables, indicating a shadow of about 10 Angulas. Hence the month must necessarily be that of summer drawing near to the summer solstice. Next in verse 33 the metre is Kakup of 8 syllables, less than that of the previous month by about 2 syllables or Angulas. Hence it seems to close the summer nearly after the vernal equinox. The interval between the vernal equinox and the close of the summer is about 3 months. As three intercalary months occur in the course of 7½ or eight years, it follows that the interval from the sacrificial time of the three goddesses to the end of the sacrificial time of Vanaspati or the Forest-lord is about 8 years. Accordingly the period of Tvashṭri and Vanaspati including the periods of Hotri and three goddesses preceding comes to 16 years, which with the seventeen years to the end of the period of Naktoshasā makes 33 years.

Thirty-three lunar years are equal to $33 \times 354 =$ 11,682 days. Since thirty-two solar years of 365 days each are equal to $32 \times 365 = 11,680$ days, it follows that

33 lunar years are equal to 32 solar years. The thirty-third lunar year is an intercalary year and is here divided into ten periods called—

1. Kindled Fire;
2. Tanūnapāth;
3. Narāśamsa;
4. Barhis or Grass;
5. Divine Doors;
6. Day and Night;
7. The Divine Hotris;
8. The three goddesses Iḍā, Sarasvatī and Bhāratī;
9. God Tvashṭri; and
10. Vanaspati.

In interpreting the Āpri verses as meaning the above idea, I have given no place to imagination or etymological hair-splitting. I have simply pointed out what the Apri verses can possibly mean as they are. These verses are found in the White Yajurveda in five places; 4, 10; 18, 26; 21, 1-39; 24, 12-13; 28,24-33. A careful consideration of all these passages will leave no doubt that the various age-grades mentioned in them are minor cycles making up a big cycle of 33 lunar years.

Metres, when applied to months, seem to signify the measure of the shadow cast by Purusha, a gnoman, in those months; for the shadow measures of the several months given in the Sūryapragnapti, Kālalokaprakāśa and other Jaina astronomical works are identical with the Vedic metrical measure of those months. This is what the Kālalokaprakāśa says on the shadow measures of the months:—

"In order to know the shadow-measure of months and days in a Yuga or cycle of 5 years, a simple astronomical rule (Karaṇa) will be explained:—

The gnoman's shadow decreases in the Dakshiṇāyana and increases in the Uttarāyaṇa. A Śanku or gnoman is a wedge-shaped cube or the body of a man and is called Purusha. Its shadow is called Paurushī. The Nandichūrṇa says that Purusha means a gnoman or the body of a man. Paurushi is the shadow cast by anything equal to its own height; it is called Paurushi. This is so on the first day of the Dakshiṇāyana. Thereafter, the Paurushī increases and will be twice as much on the first day of the Uttarāyaṇa. Thereafter it decreases till it becomes equal to the height of the body casting it. In the Dakshiṇāyana it goes on decreasing at the rate of 1/183 part every day and increases at the same rate in the other Ayana. For example, a gnoman of 24 angulas in height casts a shadow of 24 angulas in length in the beginning of the Dakshiṇāyana; and then it goes on increasing, as stated above, at the rate of 1/183 of its height a day. Now if the increase in 183 days is 24 Angulas, then in one day it will be 24/183, *i.e.*, 8/61. Then in 30½ days of a solar month, the increase will be $8/61 \times 61/2$ *i.e.*, 4 Angulas. Accordingly the ancients say that in every month of the two Ayanas the increase or decrease is 4 Angulas. Here the month is solar. Hence in three months the increase will be $4 \times 3 = 12$; in six months $6 \times 4 = 24$ Angulas, which is equal to the height of a gnoman. On the first day of the Uttarāyaṇa, it will be 48 Angulas, that is, twice the height of a gnoman. Hence on the Dakshiṇāyana it measures two Pādas which is equal to the height of the knee-joint from the foot. A Pāda or Pada is 12 Angulas. Two Padas is a constant in the decrease in the Dakshiṇāyana; and the increase in the other Ayana is 4 Padas and it is constant.

On the first day of the dark half of the month of Śrāvaṇa in the first year of a cycle of 5 years the shadow measure is two Pādas; and it will go on increasing till on the seventh day of the dark half of the month of Māgha it measures 4 Pādas. Then increase commences on the 13th day of the dark half of Śrāvaṇa in the second year. The decrease begins on the 4th of the white half of Māgha. There-after the increase begins on the 10th of the white half of Śrāvaṇa; and the decrease begins on the first day of the dark half of Māgha. In the 4th year, the increase commences on the 7th of the white half of Śrāvaṇa and the decrease begins on the 13th of the dark half of Māgha. In the 5th year it begins to increase on the 4th of the white half of Śrāvaṇa and to decrease on the 10th day of the white half of Māgha. The rate of increase or decrease per day is $\frac{8}{61}$ of an Angula; and for seven days it is $8/61 \times 7 = \frac{56}{61}$ = an Angula—1/5 of an Angula; the same per 15 days is $8/61 \times 15 = \frac{120}{61}$ = two Angulas—2/5.

As to the statement made in some works that the increase or derease will be an Angula in 7 days; and two Angulas in a Paksha, this is only a rough measure for popular use. For, the commentator on the Uttarā dhyayana Sūtra says that "an Angula by seven days should be taken to mean an Angula by seven days and a half." This comes to an Angula minus 8/61 of an Angula. In thirty days it amounts to $240/61 = 3\frac{57}{61} = 4 - \frac{4}{61}$. Hence, 4 Angulas should be lessened by 4/61 of an Angula. Accordingly if 4/61 of an Angula relating to half a day is added to 30 days, it will be $4 - 4/61 + 4/61 = 4$ Angulas for $30\frac{1}{2}$ days of a solar month. Accordingly an increase or decrease of one Pāda or 12 Angulas in three months there will be. In one Ayana or 6 months, there will be an increase or decrease by two pādas. In the Jyotishkaraṇda 4/31 of an Angula is stated to be the rate of increse or decrease. This should, however, be taken to relate to Tithis, but not to days. What is stated here in this work holds good with reference to days. For 31 Tithis are equal to 30½ days. This has already been shown above. There is an increase or decrease of 4 angulas in 30½ days or 31 Tithis. For 4×31 is equal to 124. This divided by 31 gives 4. The excess of a part of a Tithi month by month (solar) should be inferred. Thus in 61 solar days there are 62 Tithis. In each Ayana of 183 days there are 186 Tithis. The increase of 4/31 of an Angula a day can be verified arithmetically. If in 186 Tithis there is an increase or decrease of 24 Angulas, then in one Tithi it will be 24/186 equal to 4/31 of an Angula.

A rule or Karana to find out the Paurushi measure on a Tithi of a Parvạ will now be given :—

Take the number of Parvas that have elapsed and multiply it by 15 to obtain the number of Tithis elapsed. Add to this the number of Tithis that are there up to the Tithi in question. Divide the sum by 186. The quotient, if it is an unevan number, denotes Dakshināyana. If it is an even number, it is Uttarāyana. If there is a remainder, multiply it by four and divide it by 31, thus making 31 parts of an Angula. This corresponds to the daily Paurushi measure derived from the constant of two Pādas in the Dakshiṇāyana or from the constant of four Pādas in the Uttarāyaṇa.

Say what the Paurushi measure will be on the 5th Tithi of the 85th Parva of a cycle of five years? Here

84×15=1,260 Tithis. Add 5 to this. Hence, the number of Tithis is 1,265. Divide this by 186. 1,265/186 =6 and 149/186. Hence, it means that six Ayanas are passed and that the Dakshināyana is current. Now 149×4=596. This divided by 31 will give 596/31 which equals 19 and $\frac{7}{31}$. And 7 multiplied by 8 will be 56 Yavas. This divided by 31 will be 56/31 which equals 1 and 25/31 of a Yava. As this relates to the Dakshināyana, this 7 angulas, 1 Yava, and 25/31 of a Yava should be added to the constant of increase, that is, to two Pādas. Hence the answer is 3 Pādas, 7 Angulas, 1 Yava and 25/31 of a Yava on the 86th Parva of the cycle.

To take another example,—What is the Paurushi on the 5th Tithi after 96 Parvas have elapsed?

Here 96×15=1,440. By adding 5 Tithis it will be 1,445 Tithis. Now, 1,445/186=7 and 143/186. Hence, it means that 7 Ayanas are passed and that the current Ayana is the Uttarāyaṇa. Now 143/186×4= 572/186. Now 572 divided by 31 gives 18 and 14/31. Hesce the Paurushi measure is one Pāda, 6 Angulas and 14/31×8 which equals 3 and 19/31. As this is Uttarāyaṇa, this should be deducted from the constant of 4 Pādas. Hence 4 Pādas—one Pada, 6 Angulas and 3 and 19/31 of a Yava=2 padas, 4 Yavas and 12/31 of a Yava.

How to find a Tithi that has elapsed?—

Now the Tithi's measure is 4/31 of an Angula. Hence 4 Angulas are equal to 4/31×31. Hence 31 Tithis have elapsed. Now if 4 Pādas are less by 8 Angulas, say how many Tithis in the Uttarāyaṇa have elapsed? If 4/31 of Angula is less, then it indicates one

Tithi. Hence 8 Angulas will give 8×31/4, which is equal to 62. Hence 62 Tithis have elapsed. When we say that Paurushi measure is less by a Pāda, it should be taken to mean 6 Angulas in the morning in the month of Jyeshṭha, Āshāḍha and Śrāvaṇa. In the 3 months of Bhādrapada, Āsvayuja and Kārtika, it will be 8 Angulas. In the three months of Mārgaśira, Pushya and Māgha, it will be 10 Angulas. And in the remaining three months 8 Angulas should be added.

If in the month of Pushya there is a shadow of a human body measuring 9 Pādas, then by every month it goes on decreasing by one Pāda. Hence in the month of Ashāḍha it will be 3 Pādas. In the same way the increase should be ascertained. If in the month of Pushya there is a shadow of one Vitasti in length, that is, 12 Angulas, then it means the forenoon of the day. Then it decreases at the rate of 2 Angulas per month; then in Ashāḍha at midday there will be no shadow; that is, it vanishes like the friendship of a wicked man. If in Pushya the shadow of a human body is 6 Pādas, then it will decrease by one Pāda per month till it vanishes in the month of Ashāḍha. Whatever is said here regarding a Paurushi shadow less by one Pāda, all that relates to the full-moon-day of the month in question."

From the above it is clear that if there is a shadow of two Pādas or 24 Angulas by the end of Ashāḍha when the Dakshināyana happens, the measure of the shadow on the Uttarāyaṇa day in the month of Pushya or the beginning of the month of Māgha will be 4 Pādas or 48 Angulas. The monthly increase or decrease is at the rate of 4 Angulas. This is purely Vedic and holds good in the strides of Vishṇu. Vishṇu whose dwarf incarnation is Vāmana seems to mean both the sun and a gnoman (Purusha).

This is how the strides of Vishṇu are described in the Krishṇa Yajurveda (4, 2, 1)—

"Thou art the step of Vishṇu, overcomimg hostility, mount the Gāyatrī metre, step along the earth; excluded is he whom we hate. Thou art the step of Vishṇu, overcoming imprecations, mount the Trishṭubh metre, step along the atmosphere; excluded is he whom we hate. Thou art the step of Vishṇu, overcomer of the enemy, mount the Jagatī metre, step along the sky; excluded is he whom we hate. Thou art the step of Vishṇu, overcomer of the foe, mount the Anusṭubhā metre, step along the quarters; excluded is he whom we hate."

The White Yajurveda (2, 25) speakes of the strides of Vishṇu as follows—

"By Jagati metre in the sky strode Vishnu; therefrom excluded is the man who hates us and whom we hate. By Trishṭubh metre in the air strode Vishnu. Therefrom excluded is he who hates us and whom we hate. By Gāyatrī upon the earth strode Vishṇu. Therefrom excluded is he who hates us and whom we hate." In the same way in Book 12, 5, the strides of Vishṇu are spoken of:—

"Thou art the rival-slaying stride of Vishnu. Mount the Gāyatrī metre; stride along the earth. Thou art the foe-destroying stride of Vishnu. Mount the Trishṭubh metre. Stride along the mid-air. Thou art the traitor-slaying stride of Vishnu. Mount the Jagatī metre; stride along the sky. Thou art the foeman-slaying stride of

Vishnu. Mount the Anushṭubh metre. Stride along the quarters."

"Let me not with my foot offend thee, Vishnu; Agni, may I approach thy shade abounding in store of riches. Thou art Vishnu's mansion." (White Yajus, 2, 8. Compare Krishna Yajus 4, 2, 1.)

"Purusha hath a thousand heads, a thousand eyes, a thousand feet. Pervading the earth on every side, he fills a space of ten Angulas." (Sukla Yajus. 31, 1.)

From these passages and specially from White Yajus 2, 8, it is easy to perceive that by Vishnu's strides are meant the varying lengths of a gnoman's shadow measuring 24, 10 or 8 Angulas on the day of the Dakshināyana and 48 Angulas on the Uttarāyana in the month of Pushya. The expression "May I approach thy shadow" in White Yajus 2, 8, leaves no doubt that it is the shadow that is meant. A shadow of ten Angulas on earth, that is, on the day of the Dakshināyana is clearly a gnoman's shadow measured in another latitude, as clearly stated in the Kālalokaprakāśa quoted above.

Vishnu's Incarnations:—

It follows, therefore, that corresponding to the gradual increase of the gnoman's shadow from two Pādas or 24 Angulas at the commencement of the Dakshināyana to four Pādas or 48 Angulas on the day of the Uttarāyana, the metrical measure of the day of kindling of fire rises from Gāyatrī of 24 Angulas to 28 on the day of Tanūnapāth fire, to 32 on the day of Narāśamsa fire, to 36 on the day of Barhis, to 40 on the day of Divine Doors, to 44 on the day of Day and Night, and to 48 on the day of the two Divine Hotri Priests. It follows therefore that the seven occasions from that of Kindled Fire to that of the Divine Hotris are the seven months from Śrāvaṇa to Māgha of the same year or of different cyclic years characterized by their respective shadow-measures from Gāyatri to Jagatī. Now it is easy to perceive that the seven measures are really the seven footprints of Vishnu and that among them prominance is given to the three footprints, Gāyatrī in the spring season, Trishṭubh in the windy or autumnal season and Jagatī in the winter season. It is also easy to perceive that the Purusha's or gnoman's shadow in the beginning of the rainy season at the close of the spring and summer is dwarfish and that it rises to its full height in winter. It, therefore, follows that the dwarf incarnation of Vishṇu is no other than the dwarfish shadow of the gnoman called Purusha on the day of summer solstice. If this is so, as I think it is, it follows that the other nine incarnations of Vishnu are no more than forms of shadow cast by the gnoman on other occasions.

The fish incarnation of Vishnu is no other than the fish figure formed by the intersection of the two circles drawn round the gnoman with radius equal to the length of the shadow cast by the gnoman in the forenoon and afternoon in the spring season. The fish-figure thus formed is indeed called 'Matsya', a fish, in the Śulba-sūtras of Āpastamba and other writers of the Śrautasūtras. The same fish figure takes the form of a tortoise, if it is rather bulged out than elongated. The same takes the form or is imagined to take the form of a boar at the close of the real spring season and at the commencement of the rainy season; for the raising of the earth from the waters means the demarkation of the spring from the rainy season. Since the earth is called Hiranyakasipu, its slaying by Narasimha means the shortening of the spring to give room to the rainy season arriving with its thunder like a lion's roar. The description of the arrival of a lioness in Śukla Yajus. 5, 12 implies the arrival of the rainy season at the close of the spring and summer; for the lioness is said there to have won Brāhmans and Nobles. Readers of the Vedas need not be told that in the Vedas Brāhman and Kshattra are frequently used to mean the spring and the burning summer. In the Vedic passage "Brahma drimha, Kshatram drimha" the words really mean the spring and the summer. The passage means—"make the spring firm and make the summer firm." It appears that originally the word Nārasimha meant a waterlion instead of a man-lion. The term "Nāra" in Nārāyaṇa is taken to mean water and Vishnu is said to have his abode in water or milk of the ocean. Vishnu is the god of both the summer solstice and the winter solstice. In summer solstice he is Nārāyana, Nārasimha, Vāmana, Indra or Upendra. As a god of winter solstice he is Vāsudeva or god of Vasu, wealth or the constellation of Dhanishṭhā. In winter solstice he keeps his height, while in summer solstice he becomes dwarf. Likewise the other three incarnations of Vishnu. Jāmadagnya Rāma is the god of summer solstice. Jāmadagnya Rāma is the son of Jamadagni, kindled fire. He is said to have made a gift of the earth to Brāhmans, exterminated the Kshatriyas twenty one times and reclaimed the earth from the ocean. This means that as a god of summer solstice, he drew a line of demarkation between the spring and summer, the seasons of the Brāhman and Kshatra, and also a line of demarkation between the spring and summer on the one hand and the rainy season called the ocean on the other hand. This means that by proper adjustment of the lunar year, the spring season of two months from Chaitra to the end of Vaiśākha and the summer of two months from Jyeshṭha to the end of Ashāḍha were given their proper places in the year. The reclaiming of the earth from the ocean means that by proper adjustment of the lunar year with the solar year, the four months of spring and summer were kept apart from the rainy season. The story of his slaying the Kshatriyas by moving round the earth twenty one times means that by adjusting the lunar with the solar year on twentyone intercalary periods, the spring, the season of the Brāhman, and the summer, the season of the Kshatriyas, were kept in their proper places in the year. If, for example, the lunar year is not adjusted with the solar year for five years, the beginning of the year on the first of Śrāvaṇa in the rainy season falls back by two months and the year begins not in the rainy season, but in the beginning of the summer season. If it is not adjusted for another five years, the beginning of the year falls back by four months and it begins on the first day of the spring season. Similarly, failure to adjust the lunar year with the solar year for about 25 years brings back the beginning of the year by ten months. That is, the rainy season whith its plough, Sita, comes to arrive in the Asvina month, the time of vernal equinox. Hence Rāma, the god of summer solstice, has to cross the ocean and go to Lanka on the equator, behead the ten heads or recurring ten heads of Rāvaṇa, and recover

Sīta, the plough, after purifying her in the heat of early summer, and give her the legitimate place in the rainy season. The breaking of Rudra's bow by Rāma to win Sītā seems to refer to the shifting of the solstitial position from its usual place in the sky. According to the Tait. Ar. 1, 5, one end of the bow lay on the earth and the other on the sky. Here the earth means the end of the spring and summer at the close of the month of Ashāḍha and the sky means the beginning of the winter solstice in the month of Māgha. In other words, the semicircular path of the sun from the rainy season to the winter season changed its semicircular form owing to the commencement of the rainy season earlier by a month or so than before. The cutting of the five hundred and thousand heads of Ahi and Mahi Rāvaṇas by Sīta, the wife of Rāma, as narrated in the Ānanda Rāmāyana, seems to refer to the adjustment of five hundred and thousand intercalary months in the course of about 2,500 years.

It is now necessary to have a clear notion of what the Vedic poets meant by "World." The two worlds are the two Ayanas, the Dakshiṇāyana and the Uttarāyaṇa. The Sata. Br. (12, 7, 3, 7) says "There are two Vedis (altar-grounds), two worlds in truth there are, they say, the world of the gods, and the world of the fathers, one of the Vedis is in the North, and the other in the South, for the world of the Gods is in the North, and the world of the fathers is in the South. By the Northern one he secures the world of the gods, by the Southern one the world of the fathers."

The following passage of the Tai. Br. (2, 10, 4, 3) makes the meaning of the worlds still clearer:—

दिवं मे यच्छ अन्तरिक्षं मे यच्छ पृथिवीं मे यच्छ पृथिवीं मे यच्छ अन्तरिक्षं मे यच्छ दिवं मे यच्छ अह्ना प्रसारय रात्र्या समच रात्र्या प्रसारय अह्ना समच.

"Give me the sky, give me the air, and give me the earth, give me the earth, give me the air, and give me sky, expand by the day and decrease by night, expand by night and decrease by day."

While in other passage the lengths of Vishnu's feet on the two Ayanas are stated, in this passage variation in the length of days and nights in the three worlds, that is, the three seasons of four months each, is clearly stated. From the Uttarāyaṇa to the Dakshiṇāyana the day increases in length, and the night decreases. From the Dakhiṇāyana to the Uttarāyaṇa, the night increases and the day decreases. Accordingly the earth means the two months from Śrāvana to the end of Bhādrapada. The air means, the two months from Āsvayuja to the end of Kārtika; the sky means the two months from Mārgaśira to the end of Pushya; again in the decending order the sky means the two months from Māgha to the end of Phālguna; the air means the two months from Chaitra to the end of Vaiśākha; and the earth means the two months from Jyeshṭha to the end of Ashāḍha. In other words, the earth here means the four months of the rains from Jyeshṭha to the end of Bhādrapada; the air means the four months from Asvayuja to the end of Pushya; the sky means the four months from the end of Pushya or from the beginning of Māgha to the end of Vaiśākha. In the same way it should be noted that the six worlds are the six months of any one Ayana; and seven worlds are the six months of any one Ayana and the intercalary month at the end of Pushya or at the end of Ashāḍha. As the characteristics of months and the seasons varied, the vedic poets assigned each month to a different sun, six months to six suns, and seven months to seven suns.

The one contrivance which the Vedic poets seem to have made use of to ascertain the Nāḍikas of day and night was the recitation of the Vedic Hymns. As in the Pancha-Siddhāntika of Varāhamihira (P. 82) a hundred Bṛihatī verses or 3,600 syllables are taken to measure a Nādikā or 24 minutes. The recitation of 6,000 Bṛihatī verses is taken to measure a day and night. According to the Nidāna Sūtra of the Sāmáveda (IX, 6) the shortest day measured only 12 Muhūrtas or 24 Nādikās. It says " In the Agnishṭoma sacrifice the twelve Stotras or sets of verses chanted leave three Muhūrtas unmeasured." According to the Sūryapragnapti of the Jainas the longest day measured 18 Muhūrtas, a phenomena occurring only in latitude 40°. The gnoman's shadow-measure of 48 angulas in winter solstice is equally a feature of the same latitude.

The splitting up of the cycle of 33 years into ten minor cycles seems to have been resorted to not merely for the sake of easy adjustment and verification, but also to make them to correspond to the various sacrificial periods of every ordinary year. The poets seem to have been aware of the defective nature of their cycle of 33 years and therefore frequently convert it into 34 years and say that Prajāpathi, the year-god, is 34. Seventeen, which is half of 34, is also called Prajāpati. Vishṇu is also identified with Prajāpati and is spoken of as follows :—

चतुर्भिश्च चतुर्भिश्च द्वाभ्यां पञ्चभिरेव च ।
हूयते च पुनर्द्वाभ्यां स मे विष्णुः प्रसीदतु ॥

" He who is called upon by four and four, by two, by five, and who is called upon again by two (on the whole 17 times)—may that Vishṇu be gracious to me." What these five calls are in the course of 17 years (4+4+2+5+2=17) has been made clear in the Apri verses explained above.

The Era of the cycle of thirty-three years :—

This era is merely a chain of the Cycle of Thirty-three lunar years during which the lunar year comes back to its original starting point or the new year's day after revolving through all the twelve months of the year like the Moharam of the Mahamaḍans. In the second year, it falls back by 11 days, in the third year by 22 days, in the fourth year by 33 days and so on, and thus falling back by 363 days in 33 years it makes almost an additional year. In other words 33 lunar years are equal to 32 solar years; 33×354=11,682=32×365=11,680 days. Each of the 33 years is called a god and the total number of the synodic gods is counted and recorded in the Śukla Yajurveda (33, 7) as 3,339.

The cycle of sixty years is a twelve-fold cycle of five years of the Vedāngajyautisha, a correct and reliable edition of which was published by the author in A.D. 1935. The five years are named in the Vedas as " Samvatsara, Parivatsara, Anuvatsara, Idvatsara and Iḍāvatsara. (Sukla Yaj. 27, 45 and Krishna Yaj. 5, 7, 2). The sixty years are called Prabhva, Vibhava and so on up to Akshaya, the sixtieth year. The Tait. Araṇyaka mentions Prabhava and Akshaya and compares them to a big river made up of small rivers and says that though the intercalary year is perceptible, the intercalary month (Adhisatva) is not so perceptible, but only inferrible. (Tait. Ar. 1. 1). The Atharvaveda (12, 3, 41) speaks of the cycle as "sixty years." Aryabhaṭa, the celebrated astronomer of India, says in his "Aryabhaṭīya" (A.D. 497) that when he was 23 years old, the cycle of sixty years completed its sixtieth revolution. This means that by A.D. 497, there had elapsed 3,600 years in the Era and that it began in B.C. 3,100. Evidently, the Brahmakalpa Era stopped short of this Era by 174 years.

These two Eras, the Era of Brahmakalpa and the Era of the cycle of Sixty years are now blended in the Samakalpa or declaration of the date and the rite about to be performed and recited by every Brāhman day after day throughout India—

"In the Kalpa of Brahma, in the second half of the Svetavarāha Kalpa, in the Manvantara called Vaivasvata, in Kaliyuga, in the year (named) in the series of sixty years beginning with Prabhava, in the Northern or Southern Ayana, (as the case may be), in the season, (named), in the month, in the half (bright or dark) and on the Tithi, one is going to perform . . . . . ."

Thus, these three Eras began in B.C. 3,100 and were current till about the second or the third century A.D. Only the Eras of Brhmakalpa and the Cycle of sixty years are blended and continued in the name of Era of Kaliyuga even to the present day.

Having thus shown how the cycles of Brahmakalpa, and Manvantarās as well as the cycles of 20 and 33 years are treated like Eras in the Purāṇas and Siddhāntas, I now proceed to point out the primary significance of these cyclic numbers in the next chapter.

CHAPTER IV.

THE CYCLE OF ECLIPSES.

AS already pointed out in the second chapter, the Vedic poets knew how to determine a lunation before hand. At first they took a lunation to be equal to 29½ days and adjusted the lunar with the solar year by intercalating one month in every 30 months, 8 months in 8×30=240 months or 20 years, and 12 months in 12×30=360 months=30 years. According to Kātyāyana Śrauta Sūtra (P. 326) the followers of the Dākshāyaṇīya system of sacrifice performed only full moon and new moon sacrifices and regained the season in the course of 30 years.

The Sataptha Brāhmṇa (XI. 1, 2, 10) describes the Dākshāyaṇīya sacrifice is some detail, as follows—

"One ought to perform full and new moon sacrifices during fifteen years. In these fifteen years there are 360 full moons and new moons and there are in a year 360 nights. It is the nights he thus gains. If he perfoms it for another fifteen years, the year itself is gained."

The meaning of the above passages is this—

The diference between 30 lunar 30 solar years is 12×30=360 days. Thus by performing full and new moon sacrifices for 30 years, the sacrifice gains 360 days. In other words the lunar year falls behind the solar year by about 12 days per year and thus falling back by 360 days in the course of 30 years, it comes back to its original starting point agreeing with the starting point of solar or sidereal year.

As already shown, the eight intercalary months of 20 years' cycle are described as the eight sons of Aditi. But this cycle of 20 years with 8 intercalary months is found to be wide of the mark.

It was replaced by another Chakra or cycle with seven and a half intercalary months. In the cycle Aditi is said to have brought forth eight children. In course of time the eighth was not fully developed and was half born. It was called Mārtāṇḍa and kept aside, and only seven and half sons called seven sages, of whom three were born in pairs making six and the seventh was born single, were retained. This is evidently a cycle of nearly 20 years withe $7\frac{1}{2}$ instead of 8 intercalary months. This is the cycle which is described in the Vedāngajyautisha. This was also found to be defective and corrected by making the number of children to be only seven. This cycle is called a wing of 1,000 days of Hamsa who is said in the Atharvaveda (10, 8, 8) to fly to the heavens with his wings stretched with 1,000 days each and having all the gods clinging to his breast. A. V. 10, 8, 7 says that one Chakra or cycle is of 1,000 syllables. In the Purushasūkta it is called a cycle of 1,000 eyes, heads and feet. Tait. Ar. says that the time of the three eclipses called Paṭara, Viklidha and Pinga is the time of the cycle of 1,000 days. It also says that the Earth contains 1,000 and the sky also contains 1,000. In Tait. Ar. 1, 6, 3 the same cycle is called a Drapsa and that Krishṇa of the Drapsa is said to come once in 20,000 days put on its two wings. Now 1,000 days are nearly equal to 33·86 lunations, or in round numbers 34. Hence 20,000 days are equal to $20{,}000 \times 34/12 = 56$ years and eight months. The Vedic Poets say that each eclipse comes in three colours, once in black colour, once in red colour and finally in white colour. Accordingly dividing 56 years and 8 months by 3, we have 18 years, 10 months and 10 days as the period of recurrence of an eclipse of the same colour. It must be noted that the number of intercalary months is only seven in 19 years. Likewise let us take the case of Purusha or Rudra who is said to have 3,000 heads, eyes and feet. Doubling the number, we have 6,000 days, which is equal to $6{,}000 \times 34/12 = 17$ years. Prajāpati is frequently said to be 17. They say that the Vashaṭkāra formula which consists of the words "Āśrāvaya, Astu Śraushaṭ, Yaja, Yeyājamahe, and Vaushaṭ" contains 17 syllables and that *seventeen* is *Prajāpati*. From this it follows that the Vedic poets had three cycles in order, that is, 17 years, 19 years and 20 years, to enable them to rectify their errors.

In his "Indian Ephemeris," Vol. 1, P. 136, the late L. D. Swamikannu Pillai says regarding the cycles of eclipses as follows—

"The Metonic cycle can be applied to eclipses when the sun is very near a node at one of the eclipses compared. For example, on the 87th day of A.D. 1, (Table IV. L. first line) the distance of the sun from node was 3·41°. Nineteen years later the distance must have decreased by 7·57°, that is the distance then was 4·16°, at which distance an eclipse was still probable. We find from Table IV-L that on the 87th completed day of A.D. 20 there was a solar eclipse at which the sun's distance from node was 4·16°.

(*a*) 358 synodical months (Sūryasiddhānta)
=10571·9504682 dys.

(*b*) 30½ joint revolutions of sun and node (Ephemeris)
=10571·909390 days.

(*c*) 29 Indian sidereal years (Sūryasiddhānta)
=10592·503924 days.

Difference between (*a*) and (*c*)=20·6 days.

Difference between (*a*) and (*b*)= ·041 days.
or ·4258.

This cycle of 29 years, less 20·6 days is the foundation of the next two cycles."

(*a*) 716 synodical months (Sūryasiddhānta)=
21143·9009364 days.

(*b*) 61 joint revolutions of sun and node (Ephemeris)
=21143·818780 days.

(*c*) 58 Indian sidereal years (Sūryasiddhānta)=
21185·007848 days.

Difference between (*a*) and (*c*)=41·107 days.

Difference between (*a*) and (*b*)=+·082156 day
or — ·08516°.

This cycle of 58 years less 41·1 days is the one most frequently used in Table IV—L.

That the cycle of 1,000 days and multiple of 1,000 days is well adapted to predict eclipses occurring in a cycle of 18 years and 11 days with opportunities to correct errors is clearly stated in the following extract from "Handbook of Astronomy" by Chambers, Vol. 2, Pages 464-467—

1. "That the 2,300 years' cycle is the primary natural cycle for the correction of the Metonic cycle, and that it may be considered as simply an expanded Metonic

cycle. Before Meton earned Athenian honours by discovering the celebrated 19 years cycle which bears his name, the prophet Daniel had embodied in his revealed chronology the grand cycle naturally adapted to correct the error of the Metonic cycle. The use of the 19 year cycle or 'golden number' is indispensable in a calendar adjusted to the months and years of nature. The slight error of that cycle (2h, 4m, 4s,) accumulates to nearly an entire day (22h, 44m) in eleven cycles or 209 years, and to nearly 11 days (10d, 10h) in 11 × 209 cycles or 2299 years, which error is balanced by 11 days epact in the 2300th year. The epact of any single year is 11 days, and the 11 days' excess of the lunar months above 2299 solar years, in 121 Metonic cycles, becomes the epact of the 2300th year. The 2300-year cycle which is 33 times more exact than the Metonic and is the primary cycle adapted to its correction is thus the natural secular basis for a calendar correctly adjusted to solar and lunar revolutions.

2. That the accumulated epacts of 2300 years or the difference between 2300 lunar years and the same number of solar make up exactly 70 lunar years and 7 months, a measure harmonious with the 7 months epact in the Metonic cycle and with the prevailing septiform character of the prophetic times. The rate of epact is remarkable as being *one month in 1000 days.* (992d); 12 months or a lunar year in 12,000 days (the 33 years' Messianic cycle) and 70 times this in 2,300 years.

3. That 2,300 years is not only a cycle of the solar and synodical month, as De Cheseaux discovered, but is also a cycle of the anomalistic lunar month. The

anomalistic month is 27·5546 days and 30,487 anomalistic months=840,057·09 d; 28,447 synodical months= 840,056·67 d; 2,300 solar years=850,057·09 d. Lunar cycles are contracted or enlarged in their various returns according to the moon's anomaly. The maximum difference is 9h, 47m. acceleration or retardation of new moon from this cause; together 19½ hours. The 2,300 years being a cycle of the anomalistic month is free from this important error. The addition of this element to the other cyclical features in 2,300 years, makes it an unrivalled lunar cycle.

4. That the number 2,520 of which 1,260 is an evident fraction is the least common multiple of the first ten numbers. Twelve hundred and sixty years are 3½ times (each time being a period of 360 years) and are simply half the week of (7/2) times or 2520 years. The arithmetical property of 2520 just mentioned gives rise to a large number of divisors and adapts it to the position which it holds as the measure of the many-jointed vertebral column, so to speak, of the prophetic times.

5. That the addition of 75 years to 1,260 years and thus to the whole "seven times" or 2,520 years which is made in the prophecies of Daniel is also made by natural solar and lunar revolutions. If any 2,520 lunar years and 2,520 solar years are taken as commencing at the same point, the former will terminate 75 solar years before the latter.

6. That the error of the 30 year cycle (the month of the prophetic times) is self-corrected in the "42 months" of prophecy or 1,260 years (30×42=1,260). Thirty years is a cycle both in its lunar and solar forms adapting the prophetic times to both reckonings. 2,520

lunar years are a chain of 84 cycles of the latter; the one chain being 75 years longer than the other. Thirty lunar years constitute a very exact cycle of the month and day and indeed form the basis of the Mahamadan Calendar. If the lunar month be reckoned as 29d 12h 44m. (which is simply to omit 3 seconds), the cycle is absolutely exact. Thirty solar years on the other hand form a cycle of the solar year and the lunar month with an error of 1d, 10h, 2m This error accumulates to a month in 360 years and therefore to 2 months in 1260 years which latter is a cycle of the month and year, a hundred times more accurate than the 30 years' cycle whose error it corrects.

7. That 2,300 years is exactly 40 days of the "Annus Magnus" of the revolution of the solar perigee with reference to the equinoxes. The stellar direction of the earth's elliptic orbit has a slow forward revolution completed in 109,800 years while the equinoxes sweep back in a contrary direction accomplishing their revolution in 25,870 years. The compound result is an advance of the solar perigee with reference to the equinoxes of one day in 57½ years or 40 days in 2,300 years and 365 days or a complete revolution in nearly 21,000 years."

The cycle of thirty years observed by the Dākshāyanīya school of sacrificers has already been pointed out. The error of this cycle is also selfcorrected in 1260 years which is 42 times the cycle of 30 years. This idea is implied by 42 suns and 84 suns referred to in the Sūryapragnapti (XIX Prābhrita, P. 271) and also by the number of suns from 7 to 1,000 mentioned in the Tait. Araṇyaka. The passage of the Sūryapragnapti is as follows:—

(१) ता कति णं चन्दिमसूरिया सव्वलोयं ओभासन्ति उज्जोविंति तवेंति पभासेंति य हि तेत्ति वदेज्जा? तत्थ खलु इमाओ दुवालसपडिवत्तिओ पण्णत्ताओ. तत्थेगे एवमाहंसु. ता एके चन्दे एगे सूरे सव्वलोयं ओभासति उज्जोपत्ति तवेत्ति पभासेत्ति, एगे एवमाहंसु. एगे पुण एवमाहंसु. ता तिण्णि चन्दा तिण्णि सूरा सव्वलोयं ओभासंति.[4]

(२) एगे एवमाहंसु—ताआउट्ठं चन्दा ताआउट्ठं सूरा सव्वलोयं ओभासंति उज्जोवेंति तवेंति पगासंति.

(३) एगे एवमाहंसु—एतेणं अभिलावेणं णेतव्वम् । सत्त चिंदा सत्तसूरा, दस चन्दा दस सूरा भारस चन्दा भारस सूरा बातालीसं चन्दा बातालीसं सूरा बावत्तरि चन्दा बावत्तरि सूरा बातालीसं चन्दसतं बातालीसं सूरसतं बावत्तरि चन्दसयं बावत्तरि सूरसयं बातालीसं चन्दसहस्सं बातालीसं सूरसहस्सं बावत्तरं चन्दसहस्सं बावत्तरं सूरसहस्सं सव्वलोयं ओभासेंति उज्जोवेंति तवेंति पगासंति.

"How many moons and how many suns light, heat and illuminate the whole world? What should be said about this? About this, there are twelve opinions: About this some say—

(1) That one moon and one sun light the whole world, heat it and illuminate it. Some others are of opinion—

(2) That three moons and three suns light, heat and illuminate the whole world. With the same repetition of words other opinions should be understood. Others say—

(3) that $3\frac{1}{2}$ moons and $3\frac{1}{2}$ suns light, heat and illuminate the whole world. Others says—

(4) that seven moons and seven suns light, heat and illuminate the whole world. Some others says—

---

4 Other three words as before.

(5) that ten moons and ten suns,—

(6) twelve moons and twelve suns,—

(7) forty-two moons and forty-two suns,—

(8) seventy-two moons and seventy-two suns,—

(9) forty-two-hundred moons and forty-two-hundred suns,—

(10) seventy-two-hundred moons and seventy-two-hundred suns,—

(11) forty-two-thousand moons and forty-two-thousand suns,—

(12) seventy-two-thousand moons and seventy-two-thousand suns,—

light, heat and illuminate the whole world."

From the above extract from Chamber's Astronomy and also from the reference made in the Sūryapragnapti and the Tait. Āraṇyaka to 42, 82, or 7 to 1000 suns, and also from the mention made in the Tait. Āraṇyaka of the reappearances of Krishṇa once in 20,000 days, one can clearly perceive that the numbers 7 and 1,000 and their multiples are all connected with the 18 or 19 years' cycle of eclipses. "Seven suns" and "Prathamajas" are names of one cycle of eclipses in terms of seven intercalary months with seven different suns presiding over those months. Similarly "fourteen suns" or "fourteen moons" is a name of two cycles of eclipses equivalent to 36 or 38 years. Likewise twenty-one suns or cows is a name of three cycles of eclipses equivalent to 54 or 57 years, which is made to terminate in 60 years with opportunities for corrections and adjustment of the cycles.

We are not, however, left to guess or infer from the evidence of these and other cyclic numbers alone that there was a sacrificial cycle of eclipses in observance during the Vedic times. There are other records to show that these numbers are associated with cycles of eclipses.

In order to show that this is the real significance of the numbers, it is, however, necessary to know the exact nature of the conception which the Vedic poets had entertained about eclipses and the name or names which they gave to eclipses, lunar and solar. Speaking of a solar eclipse a Vedic poet says (Black Yajus. 2. 1. 2) as follows—

"Svarbhānu, the Asura, pierced the sun with darkness: the gods desired an atonement for him: the first darkness of his which they struck off became a black sheep; the second a bright-coloured one; the third a white one. What they cut off from the upper-part of the bone became a barren ewe."

From this we can understand that the first solar eclipse which the poets observed was of a black colour and that when that recurred in the second cycle it was of reddish colour; and that when it recurred for a third time, that is after 54 years and 33 days, it was dimmly white; and that on the fortyfourth day after 72 years there was no trace of any eclipse and that the sun was in his usual bright splendour. Another thing to be noted in this connection is that one of the many names given to solar eclipses is "Avi, Sheep." Again in B. Yj. 2, 1, 4 the poet speaks of a solar eclipse as follows:—

"Yonder sun did not shine; the gods desired an atonement for him; for him they offered this offering of ten bulls; verily thereby they restored his brilliance; For him who desires splendour he should offer this offering of ten bulls; verily he has recourse to yonder sun with his own share; verily he bestows on him splendour; he becomes resplendant. He should offer in the spring in the morning three with spots on the forehead; in the summer at midday; in the autumn in the afternoon three with white tails. Three are the brilliances of the sun, in the spring in the morning; in the summer at midday; in the autumn in the afternoon. Verily he wins whatever brilliances there are. They are offered in the course of the year. The year is the giver of splendour."

Evidently the offering of bulls in sets of three each in each of three seasons of the year is to make an atonement for the eclipses which occur in sets of three in the three seasons in a cycle of 18 years. In the three seasons of the cycle sometimes the solar eclipses amount to eleven, sometimes to ten and at other times to nine. The variation in number is due to other causes which it is unnecessary to discuss here. All that is to be noted in this connection is that bull or cow is another name given to a solar eclipse in addition to " Avi " or sheep which, as already pointed out, is given to it.

Similarly 'Goat' is the name given to a lunar eclipse. In Bl. Yaj. 2, 1, 1 a lunar eclipse is alluded to as follows:—

" He who desires cattle should offer one of a triplet to Soma and Pūshan; the she-goat has two teats; two are born together; the third for strength and growth; Verily he has recourse to Soma and Pūshan with their own share; verily they produce cattle for him; Soma is the depositor of seed; Pūshan the producer of cattle."

Again in Bl. Yaj 6, 5, 10 the poets connect goats with Soma cups and sheep with Aditya cups. In the same passage it is said in defence of the drawing of Soma cups twice or thrice at a time as contrasted with the drawing of one Aditya cup at a time that "the female goat gives birth to two or three (at a time), but sheep are more numerous," alluding of course to the recurrence of more solar eclipses than lunar eclipses in a cycle, though lunar eclipses recur twice or thrice close to each other in the cycle of eclipses (*vide* A.D. 1901 and 1906 of "Chart of the Cycle.") In the most famous verse, "Ajāmekām, Tait. Ar. 10, 10, which the Sānkhyas take as referring to Prakṛiti and Purusha, and the Vedāntins to Māya and two kinds of Jīvas, one addicted to, another aloof from, worldly pleasures, lunar and solar eclipses seem to be really referred to. The verses run as follows:—

अजामेकां लोहितशुक्लकृष्णां बह्वीं प्रजां जनयन्तीं सरूपाम् ।
अजो ह्येको जुषमाणोऽनुशेते जहात्येनां भुक्तभोगामजोऽन्यः ॥
हंसः शुचिषद्वसुरन्तरिक्षसद्धोता वेदिषदतिथिर्दुरोणसत् ।
नृषद्वरसदृतसद्व्योमसदब्जा गोजा ऋतजा अद्रिजा ऋतं बृहत् ॥

One he-goat, desirous of enjoyment, lies near the she-goat who produces numerous offsprings of similar nature and possessed of red, white and black colours; another abandons her after complete enjoyment.

He is called Hamsa while in "Śuchi"; Vasu while in the atmosphere; Hota while in the altar (earth); Atithi while in Duroṇa (his own house); Abjāh (flower of water) while among the people ; Gojāh (born of a cow) while among the nobles (royal people) ; Ritajāh (born of the Rita) while in the place of Ṛita ; Adrijāh (born of the mountain) while in the sky ; he is the Brihat Ṛita."

What are really referred to here, as in the whole of the Vedic literature, by the name Aja (masculine and feminine) are really the sun and the moon. The female is the sun, while the male goat is the moon. On the occasion of full moon, both look to each other lovingly as two lovers, but without union or without disfiguration after lunar eclipse ; on the occasion of the new moon, however, the sun (male) enjoys the moon (female) and abandons her.

In Tait. Ar. 3, 14 we are told that Prāṇa (the sun) abandons the just-born (the crescent), but never leaves out the decrepit one (the new moon) ; the sun not only sports with the moon when the latter is decrepit and lying, but also she never leaves him on the days close to his decrepitude ; Also Prāṇa (the sun) kindles the fire and is the father of the offspring, and as sacrifice he is Soma.

In the next Anuvāka of the Tait. Ar. a solar eclipse is described as the slaughter of a bull and the sun is asked to leave the winter solstice (Devayāna) and to go on his way to the other heaven (the Pitṛiyāna, the summer solstice) and the poet says that he has sacrificed for the death of one of the Prathamajās, a name given to the seven intercalary months of the eighteen years' cycle of eclipses.

In Bla. Yaj. 3, 4, 3 Vāyu is addressed to come with his thousand teams and a barren goat (barren on account of absence of a lunar eclipse at the commencement of a fourth cycle of eclipses) is asked to become Manu and bring the divine host, as follows—

(*a*) "O! Vāyu, Drinker of the Pure, Come to us;

A *thousand* are thy teams, O! Thou that hast all choice boons; For thee the sweet drink hath been prepared, whereof, O! God, Thou hast the first drink."

(*b*) Thou, the fourth, art the barren, the eager one;

Since once in thought the embryo hath entered the womb do thou, the barren, go eagerly to the gods, be the desire of the sacrificer fulfilled.

(*c*) Thou art the goat, resting on wealth, sit on the earth, mount aloft on the atmosphere, in the sky be thy great radiance.

(*d*) Stretching the thread of the atmosphere do thou pursue the light; guard the paths of light made by prayer; Weave ye without a flaw the work of the singers; *Become Manu*, produce thou the *host divine*. Thou art the offering of mind."

It has already been pointed out how Krishṇa is expected to come back once in 20,000 days or 56 years and 8 months. Now 56 years and 8 months are equivalent to three cycles of eclipses. On the occasion of the prayer made by the poet in Bl. Yaj. 3, 4, 2 the expected lunar eclipse of black colour did not recur and the cycle counted in sets of thousand days proved a failure. In a cycle there are 71 or 72 eclipses forming a *Manu*. A Manvantara is a period of 71 or 72 Yugas or lunations with eclipses. In the present case the eclipse did not occur, as expected on the basis of the cycle of eclipses. This conclusively proves that the counting of time in sets of thousand days found in the Vedas, the references to

Manu, the sacrifice of *goats*, *sheep*, *cows* or *bulls*, and *horses* are all connected with the cycle of eclipses and that whenever a sacrifice is made, it may be taken for granted, it indicates the recurrence of an eclipse, solar or lunar, corresponding to the animal immolated.

The description of Purusha becoming the hymns of the Ṛigveda, the Sāmans of the Sāmaveda and the sacrificial formulas of the Yajurveda in the Purusha Sūkta is evidently a reference to the counting of the cycles of eclipses in terms of the hymns of the Ṛigveda, the Sāmans of the Sāmaveda and Yajus formulas of the Yajurveda. For example the Vashaṭkāra formula of 17 syllables indicates the expiration of 17 years in a cycle and Āpri hymn, 33 years. The number 31 counted in terms of eight Vasus, eleven Rudras and twelve Adityas seems to refer to the occurrence of 31 eclipses in the course of half a cycle of eclipses. The word Rudra is a name given to a solar eclipse with two lunar eclipses, preceding or following, or one preceding and another following the solar. Rudra is Phālāksha and Tryambaka, *i.e.*, a god with three eyes. of which one is fire, that is, a solar eclipse capable of reducing Kāma, the moon, to ashes. When eclipses are evidently deaths to gods, such as the moon and the sun, they must surely be death to men. Hence the Vedic poets thought that some atonement for the sin committed by them was necessary and they instituted the sacrifices to avoid Mṛityu, death, and performed it on the occasions of eclipses.

In the Purushasūkta the number 1,000 in the head of Purusha is evidently a reference to the cycle of eclipses. The horses and animals are those of cyclic sacrifices.

The goats and sheep signify lunar and solar eclipses. The transformation of Purusha's face into a Brāhman means the beginning of the year in the spring season; for the spring is taken as the symbol of Brāhman. It may also mean the sun, as he is represented as a learned god (Suryo vipaschit manasā punātu),an Achārya to the way-ward moon and to men, his offspring. The latter seems to be the meaning of the word Brāhmaṇa in the Sūkta, because the words, " Prabhrājamāna, Vṛishabha, Lohitāksha found in the Sāvitri hymn recited day after day at noon by every Brāhman mean solar eclipses ; and Brahma means the sun, and also the teacher in the word 'Brahmopadeśa' made to a Brāhman in his Upanayana rite. Likewise the two arms of Purusha, the time-god, represent the moon, for the moon is also called the king in the Vedas and the arm is spoken of as Rajanya. The two thighs represent the Maruts or the seven winds. The earthly representatives of the heavenly gods are of course the Brāhmans, the kings and the trading people. The Śūdras are the earthly representatives of the labourers of the gods. His mind is the moon. Accordingly the word "Manas" in the Vedas means the moon. His nostrils represent seven winds each; his eye the sun; his mouth Agni and sometimes Indra also. His navel represents seven winds of the sky and his feet the earth (the summer solstice).

Usually there are 7 butter sacrifices (Pākayagnyas) in the spring and 7 cooked rice offerings in the autumn. They seem to represent so many solar and lunar eclipses in those two seasons. In the head of Purusha there are seven or nine Prāṇas representing 9 lunar eclipses with three solar eclipses signified by Agni and Indra in his mouth and his head (the winter solstice?) The nine lunar with three solar eclipses in the face of Purusha become three Rudras, each Rudra being made by one solar eclipse and two lunar eclipses. His two arms represent 14 lunar eclipses together. The seven Paridhis are seven circles of the celestial sphere, for Paridhi means an 'upasuryaka or Maṇḍala.'[1] The twenty one kindling sticks represent 21 solar eclipses with seven lunar; and the earth a solar eclipse (on the summer solstice?). On the whole there are 30 lunar eclipses and 32 solar eclipses in all.

According to expert astronomers there can be no more than 42 or 43 solar and 30 lunar eclipses in a cycle. Of these about six or eight may become invisible in one region of the globe and become visible in other latitudes. When such things happen, the poets wonder and pray to Varuṇa to maintain his Ṛita or law. The Purusha hymn will again be considered in detail in connection with Chakras later on.

As the poets say that their Uttarāyaṇa from winter solstice to summer solstice consisted of the seven self-born months represented by the seven Ādityas or the seven intercalary months of the nineteen years of the cycle of eclipses, the other half of the year must necessarily consist only of 5 months. Accordingly their longest day and the shortest night must be as 7/12 and 5/12 respectively. In other words their day must be 14 hours long and their latitude 40° to which a day of 14 hours corresponds.

The Rudrasūkta furnishes abundant proof to say that Rudra signifies an eclipse. His 1,000 eyes, 1,000 arrows and 1,000 bows refer to the cycle of 1,000 days

---

1 See Digram II.

which forms part of an eclipse cycle. His braided hair, his black neck, white throat, his dusky, brown, and ruddy colours signify various partial lunar eclipses. His cord or Yajnyopavīta and his unstrung bow represent the appearance of two bodies on the same line so as to eclipse and to be eclipsed. The followers of the Tantra cult say that "Bindudvayāntāre daṇḍah śivarūpo Maṇiprabhah,"[2] a line between two circular bodies is Śiva of gem-like hue. His drawing the bow in various forms symbolises the various forms of eclipses. His visibility even to cow-herds and women carrying water from tanks outside the villages implies the striking nature of Rudra eclipses occuring in the course of 30 days which all look at with wonder and fear. His creepingaway signifies the slow disappearance of eclipses. As some lunar eclipses appear like smoke and dust in far off wilderness, he is called a hunter and pursuer of cattle running or driven away by thieves. As lunar and solar eclipses that have occured one after another in the same region, he is called the lord of cattle of plants, for the moon is said to have been recovering from his periodical consumption by eating the plants. Hence Rudra is a great physician. As eclipses are compared to kidnapping of men and women by burglars and thieves, he, being a sharer of eclipses, is the lord of thieves. Mention of the word, 'Manu' in the Sūkta is a clear reference to eclipse cycles. His appearance ten times in each of the five regions signifies to four sets of Rudras or eleven Tryambakas and six Tryambakas more in the upper region.

2 See the author's Theory of the Origin of the Devanāgari Alphabet, I.A. 1905-6.

Both in the Mait. S. (2, 9, 1) and the Tait. Ar. (10, 1) Rudra is praised as 'Skanda,' the son of six mothers, as the 'elephant-faced god' born of two mothers, as 'Brahma' seated on a lotus, as 'Keśava' (possessed of braided hair) and 'Nārāyaṇa,' as 'King Soma,' as 'Vainateya,' as a 'Trader,' and as 'Fire.' He is also described as a ferry sailing on water, and as an assembly of ministers and as a chief of an assembly. Now like ordinary lunations, eclipses also are regarded as the main cause of fecundity in the world. Also eclipses are spoken of as the offspring of eclipses that preceded them, solar eclipses as daughters and lunar as sons. A glance over the Chart of all the eclipses from 1,840 A.D. to 1919 A.D. appended here, will show the probability of the lunar eclipse of April 1902 being regarded as the son of the six eclipses of the year of 1901.—Likewise, it seems likely that a partial lunar eclipse in the rainy season was looked upon as the son of the two preceding solar or lunar eclipses, for the crescent looks like one tusk of an elephant. It is also possible that a lunar eclipse at midnight was looked upon as a ferry sailing on a river or as a merchant sailing on, or two friends sailing in a ferry. If the sun or moon set eclipsed, the next day in the morning or in the evening the sun or moon might be spoken of as a deserter of his friend who sailed with him. A series of three or more eclipses in a year might possibly be spoken of as an assembly of ministers in consultation.

There are also many other things connected with eclipses. The seven musical notes, the drum, the circular bronze pieces to produce rhythm, and the art of writing also seem to have some connection with eclipses. These are, however, questions too wide to be discussed here. Hence leaving them out, I next take up the list of eclipses given in the Tait. Āraṇyaka.

In the Tait Āraṇyaka the poet describes the year with an intercalary month as Adhisamvatsara and begins it with the rainy season. He describes also the signs and characteristics by which its arrival can be easily recognized. As the Araṇyaka furnishes additional evidence about the cycle of eclipses, some of its passages bearing on the subject are quoted with translation and notes, wherever necessary, and without original passages, where unnecessary. The Araṇyaka begins as follows:—

आपमापामपस्सर्वा अस्मादस्मादितोऽमुतः ।
अग्निर्वायुश्च सूर्यश्च सह सञ्चस्करर्धिया ॥
वाय्वश्वा रश्मिपतयः मरीच्यात्मानो अद्रुहः ।
देवीर्भुवनसूवरीः पुत्रवत्त्वाय मे सुत ।
महानाम्निर्महामानाः महसो महसस्स्वः ॥
देवीः पर्जन्यसूवरीः पुत्रवत्त्वाय मे सुत ।
अपाश्न्युष्णिमपा रक्षः अपाश्न्युष्णिमपा रघम् ।
अपाघ्रामप चावर्तिं अप देवीरितो हित
वज्रं देवीरजीतांश्च भुवनं देवसूवरीः ।
आदित्यानदितिं देवीं योनिनोर्ध्वमुदीषत ।
शिवा नः शन्तमा भवन्तु दिव्या आप ओषधयः
सुमृडीका सरस्वति मा ते व्योम सन्दृशि ।

"I have obtained and obtained all waters from this and that side; may Agni, the sun, and the wind make the waters prosperous. (1)

"O! Waters, whose steeds are the *seven winds*, whose lords are the rays of the sun, whose body is formed of shining rays, who are not malicious to any one, and who are the mothers of all beings, allow me to *have sons.* (2)

"O! Waters, who are of pleasing names, who are worthy of worship, who are of shining form, who are productive of *food*, and who are the mothers of the raining clouds, allow me to *have sons.* (3)

"O! Waters, take away the excessive heat and fever, take away the demon, take away the bad smell, and take away our poverty. (4)

"O! Waters, hold up the thunderbolt, hold up life and all beings; O! mothers of gods, hold up the Adityas as well as the goddess Aditi together with her womb (*bringing forth the Ādityas or intercalary months*). (5)

" May the heavenly waters and herbs be auspicious to us, and may they bring happiness to us; O! water, thou art the bestower of comforts; I have not seen thy abode in the sky." (6)

In the next passages the poet proceeds to define the cycle and its characteristics:—

स्मृतिः प्रत्यक्षमैतिह्यमनुमानश्चतुष्टयं ।
एतैरादित्यमण्डलं सर्वैरेव विधास्यते ॥ 7

सूर्यो मरीचिमादत्ते सर्वस्माद्भुवनादधि ।
तस्याः पाकविशेषेण स्मृतं कालविशेषणं ॥ 8

नदीव प्रभावात्काचिदक्षय्यात्स्यंदते यथा ।
तां नद्योऽभिसमायंति सोरुस्सती न निवर्तते ॥ 9

एवं नानासमुत्थानाः कालास्संवत्सरं श्रिताः ।
अणुशश्च महशश्च सर्वे समवयंत्रि तं ॥ 10

स तैस्सर्वैस्समाविष्टः उरुस्सन्न निवर्तते ।
अधिसंवत्सरं विद्यात् तदेवलक्षणे ॥ 11

अणुभिश्च महद्भिश्च समारूढः प्रदृश्यते ।
संवत्सरः प्रत्यक्षेण नाधिसत्वः प्रदृश्यते ॥ 12

पटरो विक्लिधः पिंगः एतद्वरुणलक्षणं ।
यत्रैतदुपदृश्यते सहस्रं तत्र नीयते ॥ 13

एकं हि शिरः नाना मुखे कृत्स्नं तदृतुलक्षणं ।
उभयतस्सप्तेंद्रियाणि जल्पितं त्वेव दिह्यते ॥ 14

शुक्लकृष्णे संवत्सरस्य दक्षिणवामयोः पार्श्वयोः ।

तस्यैषा भवति :—

शुक्रं तेऽन्यद्यजतं तेऽन्यत् ।
विषुरूपे अहनी द्यौरिवासि ॥ 15

विश्वा हि माया अवसि स्वधावः ।
भद्रा ते पूषन्निच रातिरस्त्विति ॥ 16

नात्र भुवना न पूषा न पशवः नादित्यः ।
संवत्सर एव प्रत्यक्षेण प्रियतमं विद्यात् ॥ 17

एतद्वै संवत्सरस्य प्रियतमं रूपं योऽस्य महानर्थे उत्पत्स्यमानो भवति ।
इदं पुण्यं कुरुष्वेति तमाहरणं दद्यात् ॥ 18

साकंजानां सप्तथमाहुरेकजं षड्यमा ऋषयो देवजा इति ।
तेषामिष्टानि विहितानि धामशः स्थात्रे रेजन्ते विकृतानि रूपशः इति ।
को नु मर्या अमिथितः सखा सखायमब्रवीत् जहाको अस्मदीषत इति ।
यस्तित्याज सखिविदं सखायं न तस्य वाच्यपि भागो अस्ति ।
यदीं शृणोत्यलकं शृणोति । न हि प्रवेद सुकृतस्य पन्थामिति ॥

"Remembrance of past experience in records, seeing with the eyes, tales heard from others, and inference as the fourth,—with all these (*four kinds of evidence*), the circle of the (*seven or eight*) Adityas is laid up. (7)

" The Sun takes up the water from the whole world; by means of the peculiar and ripe form of the waters (*i.e.*, raining clouds) the characteristics of the time are remembered.

" Just as a river flows from an imperishable source, and just as other streamlets join her, and just as she, growing in volume, never returns, so the moments of various birth from Prabhava, the first year, to Akshaya, the last of the sixty years, are merged in the year, by small bits and big periods; they all form the year; being formed of them it grows in length and never returns.

" One should understand this as a year with intercalation of 7 months (Adhisamvatsara), and that by means of the characteristics (to be spoken of); formed of small and big bits of time, the ordinary year is visible to the eye; but not so the swollen thing (*i.e.*, the year in which intercalation is to be made.") (12)

The poet has defined the year as being formed of a number of small and big moments; and has pointed out the difficulty of seeing the intercalated year. Now he is going to describe those characteristics by which its arrival can be inferred :—

" Paṭara, Viklidha and Pinga—these (eclipses) are the characteristics of Varuṇa. When this is seen, then there is inserted *1000* (The cycle of 1000 days.).

One is the head; but faces are many; this is in brief the sum total of the characteristics of the seasons. Seven organs of sense on both sides; speech alone is to be added.

On the right and left sides of the year there are the white and dark portions. The following is said of it :—

"O! Year, thy white part is one thing; and thy worshipable part is another; of such different forms are thy days and nights; thou art like the sky; O! Food-productive year, various illusions thou keepest; O! Nurse of the world, may thy yield be good.

There is no world here; neither god Pūshan; nor cattle; nor Āditya; the year alone should with direct perception be regarded as the loveliest; this is what makes the year dear to all on account of its great yield in store for us for the purpose of performing a meritorious rite; one should return this gift.

They say that of the twins born together, the seventh is single born; the six alone are born twins, sages born from the gods; these have their stations formed for them; they vary in their nature for the good of the world.

O! Men, tell me whether there is a friend who, though not hurt and deserted by his friend says concerning the other that "deserting me, this friend runs away."

He who deserts a friend that knows the obligations of friendship is not worthy of being mentioned even by name; if one hears of him or hears of 'Im' sound, one hears of a false friend; for the deserter does not know the path of virtue.[1]

ऋतुर्ऋतुना नुद्यमानः विननादाभिधावः ।
षष्टिश्च त्रिंशका वल्गाः शुक्लकृष्णौ च षाष्टिकौ ॥

सारागवस्त्रैर्जरदक्षः वसन्तो वसुभिस्सह ।
संवत्सरस्य सवितुः प्रैषकृत्प्रथमः स्मृतः ॥

---

1 This seems to refer to one of three successive or Tryambaka eclipses which becomes invisible. The symbol of 'ī' letter in Brāhmī alphabet is three dots like a Tryambaka.

अमूनादयतेत्यन्यान् अमूंश्च परिरक्षतः ।
एता वाचः प्रयुज्यन्ते यत्रैतदुपदृश्यते ॥
एतदेव विजानीयात् प्रमाणं कालविपर्यये ।
विशेषणं तु वक्ष्यामः ऋतूनां तन्निबोधत ॥
शुक्लवासा रुद्रगणः ग्रीष्मेणावर्तते सह ।
निजहन्पृथिवीं सर्वां ज्योतिषाऽप्रतिख्येन सः ।
विश्वरूपाणि वासांसि आदित्यानां निबोधत ।
संवत्सरीणं कर्मफलं वर्षाभिर्ददतां सह ।
अदुःखो दुःखचक्षुरिव तद्मापीत इव दृश्यते ।
शीतेनाव्यथयन्निव रुरुदक्ष इव दृश्यते ।
ह्लादयते ज्वलतश्चैव शाम्यतश्चास्य चक्षुषी ।
या वै प्रजा भ्रंश्यन्ते संवत्सरात्ता भ्रंश्यन्ते ॥
याः प्रतितिष्ठन्ति संवत्सरे ताः प्रतितिष्ठन्ति ।
वर्षाभ्य इत्यर्थः इति.

One season driven on by another coming behind it makes a noise: They are sixty-fold and thirty-fold; the white and black parts are also sixty-fold.[1]

The spring with the Vasus (Vasu-eclipses) clothed in reddish white garment is the first messenger, it is said, of the productive year. When the spring comes, these words are usually spoken and heard of:—" Feed these persons and protect those others."

One should know this characteristic of time: We are going to describe the distinct characteristics of the seasons: know them from us.

---

1 This may mean thirty revolutions of sixty years. The white and black parts are the Northern and Southern Ayanas of the year. If this is so, it means that 1800 years were passed in the era of the cycle of 60 years.

16. The group of the Rudras (Rudra eclipses) wearing white garments comes always with the summer, striking the earth with its dazzling hot splendour; the garments of the Ādityas who bring in to us the results of our work in the year are of various colours.

17. Though not unhappy, he is unhappy in having sore-eyes; it appears brown; its cold does not cause grief; it has plenty of deer; it pleases all with its shining plenty; its eyes are also cured; those who fall from work in this season fall from the year; those who stand in the year do really stand in the *rainy season.*

अक्षिदुःखोत्थितस्येव विप्रसन्ने कनीनिके।
आङ्क्ते चाद्रणं नास्ति ऋभूणां तन्निबोधत।
कनकाभानि वासांसि अहतानि निबोधत।
अन्नमश्नीत मृज्मीत अहं वो जीवनप्रदः।
एता वाचः प्रयुज्यन्ते शरद्यत्रोपदृश्यते।
अभिधून्वन्तोऽभिघ्नन्त इव वातवन्तो मरुद्गणाः।
अमुतो जेतुमिषुमुखमिव सन्नद्धास्सह ददृशे ह।
अपध्वस्तैर्वस्तिवर्णैरिव विशिखासः कपर्दिनः
अक्रुद्धस्य योत्स्यमानस्य क्रुद्धस्येव लोहिनी।
हेमतश्चक्षुषी विद्यात् अक्ष्णयोः क्षिपणोरिव।
दुर्भिक्षं देवलोकेषु मनूनामुदकं गृहे।
एताः वाचः वदन्तीः वैद्युतो यान्ति शैशिरीः॥

ता अग्निः पवमाना अन्वैक्षत। इह जीविकामपरिपश्यन्।
तस्यैषा भवति।

इहेह वः स्वतपसः मरुतः सूर्यत्वचः शर्म सुप्रथा आवृणे। इति.

18. The pupils of the eyes are bright like those of a person who has recovered from sore-eyes; he applies a

pigment to the eyes ; there is nothing in this season to cause sore-eyes ; this is said of the Ribhus (autumnal eclipses).

19. The garments are as yellow as gold and fresh ; the following words are usually spoken now: " eat plenty of food ; cleanse your edible things ; I am the bestower of food upon you ". These words are used when there arrives the *autumn*.

20. The groups of the windy *Maruts* shake and strike all things ; they appear, as it were, ready prepared for war to conquer others in battle with arrows ; they bear twisted hair in varying form and put on their dirty coats which are like sheep's bladder. They are not angry, though their eyes are red like those of an angry person.

21. There is draught in the world of gods ; water is plentyfull in the houses of men ; speaking the following words they spend the *winter-nights* with lightnings : Agni saw the plight of the people in not finding water on the way ; and requested the Maruts with the words: "O! Maruts, Ye are powerful and shine like the sun ; I request that ye may bestow happiness on these people."

अतिताम्राणि वासांसि अष्टिवज्रशतघ्नि च ।
विश्वेदेवा विप्रहरंति अग्निजिह्वा असश्चत ॥ 22

नैव देवो न मर्त्यः न राजा वरुणो विभुः ।
नाग्निर्नैन्द्रो न पवमानो मातृक च न विद्यते ॥

दिव्यस्यैका धनुरार्तिः पृथिव्यामपराश्रिता ।
तस्येंद्रो वज्रिरूपेण धनुर्ज्यामच्छिनत्स्वयम् ।
तदिन्द्रधनुरित्यज्यं अभ्रवर्णेषु चक्षते ।
एतदेव शंयोर्बार्हस्पत्यस्य एतदिन्द्रस्य धनुः ॥ 25

रुद्रस्येव धनुरार्त्निः शिर उत्पिपेश
अत्यूर्ध्वाक्षेऽतिरश्चात् शिशिरः प्रदृश्यते ।
नैव रूपं न वासांसि न चक्षुः प्रतिदृश्यते ॥

अन्योन्यं तु न हिंस्रातः सतस्तद्देवलक्षणम् ।
लोहितोऽक्षिण शारशीर्ष्णि सूर्यस्योदयनं प्रति ॥

त्वं करोषि न्यंजलिकां त्वं करोषि निजानुकाम् ।
निजानुका न्यंजलिका अमी वाचमुपासतामिति ॥ 29

तस्मै सर्वे ऋतवो नमन्ते मर्यादाकरत्वात् ।
प्रपुरोधां ब्राह्मण अश्नोति य एवं वेद ॥ 30

स खलु संवत्सर एतैस्सेनानीभिस्सह
इन्द्राय सर्वान्कामानभिवहति ।

22. Very red are the garments; there are the weapons called Ashṭi, Vajra and Śataghni; the All-gods strike their enemy and make a loud noise with their fiery tongues: "There is no god here; nor man; not even the powerful king Varuṇa; no Fire, no Indra; no Pavamāna; there is none like ourselves; there is one end of the bow in the sky; and the other end is on the earth; Indra disguised as Vamris (white ants) cut it into two pieces; this is the string-less bow of Indra; so say the people looking at the colours of the rainbow; this is the bow of Śamyu, the son of Bṛihaspati; this is the bow of Rudra; the broken end of the bow, when cut into two, carried off Rudra's own head: it became Pravargya; hence whoever performs a Pravargya sacrifice restores Rudra's head: Him Rudra does not injure; whoever knows this prospers.

23. There is seen the *winter*, in which man raises his eyes; no regard for beauty; nor for fine garments;

not even the eyes are turned to the mirror ; in this season men (even enemies) do not hurt each other ; this is the characteristic of *winter* (dewy season) running on its course ; eyes are red ; and the head has the colour of a parrot at this time when the sun turns on his *northward journey* ; " thou, doest fold thy hands in reverence ; thou bendest thy knees "—these words are obeyed even by young men ; to him all the seasons bow down ; because he fixes the functions of all men ; the Brāhman takes up his leadership as a chief priest ; whoever knows this prospers, the year with its troops of seasons (eclipses) brings in all desires to Indra ;

The Cycle of Eclipses :—

In the last chapter the cycle of 1,000 days borne on each of its two wings by the heavenly bird called Hamsa has been clearly explained. In that recurring cycle Aditi is said to give birth to 7½ sons casting out the other half. Evidently it is a cycle of 20 years made up of 4 cycles of 5 years each. Now the Tait. Aranyaka introduces us to the same cycle in terms of days. It is called Drapsa. It is a cycle of 20,000 days.

स द्रप्सः । तस्यैषा भवति । अवद्रप्सोंऽशुमतीमतिष्ठत् । इयानः कृष्णो दशभिस्सहस्रैः । प्रावर्तमिन्द्रश्शच्या धमन्तं । उपस्तुहि तं नृम्णामथद्रामिति । एतयैवेन्द्रः सालावृक्या सह असुरान्परिवृश्चति । पृथिव्यंशुमती तामन्ववस्थितस्संवत्सरः दिवं च । नैवं विदुषा आचार्यान्तेवासिनौ अन्योन्यस्मै द्रुह्याताम् । यो द्रुह्यति भ्रश्यते स्वर्गाल्लोकात् इत्यृतुमण्डलानि । सूर्यमण्डलान्याख्यायिकाः अतऊर्ध्वं सनिर्वचनाः.

It is the Drapsa; the following is said about it, The Drapsa has taken its stand on the one shining Amśumatı. This Krishṇa has come by ten thousand (days). Thou art Indra. O! Thou, with thy might, melt the recurring, fuming and ever-remembered black one. Indra slays the Asura with the Sālāvrikī. Anśumatī is the earth. The year has spread itself from the earth as far as the sky. The teacher and the student who know this should not hate each other. Whoever hates so will fall down from heaven. Thus are explained the circles of seasons. Now the Āditya-maṇdalās and their stories will be explained."

Now it has been shown in the last chapter that a cycle of two thousand days is a cycle of 67½ lunations. Hence 20,000 days are equal to 675 lunations or 56 years and 4 months. As already pointed out, the Vedic poets distinguished the recurrence of eclipses in three different colours, such as black, red and white in course of three cycles and the reappearance of the eclipse in its original colour in the fourth cycle. Hence if we divide 56 years and 4 months by 3, we get 18 years and nine months for one cycle of eclipses.

While describing the same *Drapsa*, the Atharva-veda (XVIII, 18, 28, 29) calls it 'hundred-streamed,' and says as follows—

द्रप्सश्चस्कन्द पृथिवीमनुद्यामिमं च योनिमनु यश्च पूर्वः ।
समानं योनिमनुसञ्चरन्तं द्रप्सं जुहोम्यनुसप्त होत्राः ॥ 28

शतधारं वायुमर्कं स्वर्विदं नृचक्षसस्ते अभिचक्षते रयिम् ।
ये पृणंति प्रच यच्छन्ति सर्वदा ते दुहते दक्षिणां सप्तमातरम् ॥ 29

"The Drop leaped toward the earth, the sky, and toward both the sources, and the one that was of old. To the drop that goes about toward the same source, do I make oblations after the seven priests.

"A hundred-streamed Vāyu (wind), a heaven-finding sun, do those men beholders look upon; whoso bestow and present always, they milk a sacrificial gift having seven mothers."

It is to be noted how the author of the Araṇyaka connects the 'drop' with Indra, the god of the seventh intercalary month, and ten thousand days, while the Atharvaveda combines it with seven priests and mothers (*i.e.*, seven intercalary months) and one hundred streams. Now let us revert to the poet of the Araṇyaka and hear what he says about the seven suns.

अरोगो भ्राजः पटरः पतङ्गः स्वर्णरो ज्योतिष्मान्विभासः ।
ते अस्मै सर्वे दिवमापतंति ऊर्जं दुहाना अनपस्फुरंत इति ॥
कश्यपोऽष्टमः स महामेरुं न जहाति ।

तस्यैषा भवति : —

यत्ते शिल्पं कश्यप रोचनावत् इंद्रियावत् पुष्कलं चित्रभानु ।
यस्मिन्सूर्या अर्पिताः सप्त साकम् ॥

तस्मिन्राजानमधिविश्रयेममिति । ते अस्मै सर्वे कश्यपाज्ज्योतिर्लभन्ते । तान्सोमः कश्यपादधिनिर्धमति भस्त्राकर्मकृदिवैवम् । प्राणो जीवानि इंद्रियजीवानि । सप्तशीर्षण्याः प्राणाः सूर्या इत्याचार्याः । अपश्यमहमेतान्सप्तसूर्यानिति पञ्चकर्णो वात्स्यायनः सप्तकर्णश्च प्लाक्षिः॥

आनुश्रविक एव नौ कश्यप इति उभौ वेदयिते । न हि शेकुमिव महामेरुं गंतुमु । अपश्यमहमेतत्सूर्यमण्डलं परिवर्तमानं-गार्ग्यः प्राणत्रातः । गच्छतं महामेरुम् । एकं चाजहतम् ।

भ्राजपटरपतङ्गा निहने तिष्ठन्नातपंति । तस्मादिह तप्तितमाः ।
अमुत्रेतरे तस्मादिहातप्तितमाः । तेषामेषा भवति.
सप्त सूर्या दिवमनुप्रविष्टाः तानन्वेति पथिभिर्दक्षिणावान् ।
ते अस्मै सर्वे घृतमातपंति ऊर्जं दुहाना अनपस्फुरन्तः इति ॥

सप्तर्त्विजस्सूर्या इत्याचार्याः । तेषामेषा भवति । सप्तादशा नाना-सूर्याः सप्त होतार ऋत्विजः । देवा आदित्या ये सप्त तेभिस्सोमाभि-रक्षण इति,

तदप्याम्नायः । दिग्भ्राजः ऋतून्करोति । एतयैवावृता आसहस्र-सूर्यताया इति वैशंपायनः । तस्यैषा भवति.

यद्द्याव इन्द्र ते शतं शतं भूमीः उतस्युः । न त्वा वज्रिन् सहस्रं सूर्याः अनु न जातमष्ट रोदसी इति ।

नाना लिङ्गात्वादृतूनां नानासूर्यत्वमिति अष्टौ तु व्यवसिता इति.

"Aroga (one without disease), Bhrāja (shining), Paṭara (one covered with gray layers), Patanga (flying), Svarṇara (golden), Jyotishman (one with mass of light), and Vibhāsa (one with splendour),—these illumine the heavens for him, milking strength (for the sacrificer) and never losing their splendour. Kaśyapa is the eighth; he never leaves the mountain called the great Meru. (Pole). The following is said of him :—

"What contrivance of thine, O! Kaśyapa, is that which is full of shining, vigorous, splendid, and of wondrous light, and in which the seven suns are set together."– "

"In him (Kaśyapa) may we seek a king; they all (the suns) obtain light from him; the moon blows them out from Kaśyapa, just like a goldsmith blowing his bellows (over the fire with gold)."

"(The seven suns are) the vital breaths; they are the forces of life; they are the principles of the vital breaths; the seven vital breaths in the head are the suns,—so say the Teachers; Panchakarṇa, the son of Vātsyāyana says:

"I have seen the seven suns;" so also Saptakarṇa, the son of Plākshi."

"We have only heard of Kaśyapa; thus both of them tell each other; we cannot go to the great Meru. Gārgya Prāṇatrāta says: "I have seen the circle of the (seven) suns which are moving around; go to the great Meru and also to the one (Sun) who never leaves it."—"

"Bhrāja, Paṭara, and Patanga shine, standing below; hence they are productive of heat to this world; of them, the following is said"—

"The seven suns have entered into the heavenly world; whoever has paid sacrificial fees will follow them; they all illumine the *ghi* for him, milking strength and causing no heat.

"—'The seven sacrificial priests are the suns, '—so say the Teachers; of them, the following is said: 'The seven regions with many suns, the seven Hotri Priests, and the shining Adityas who are also seven; by means of them the moon is maintained (*i.e.*, the lunar year is prevented from rotating further)."

"—'Accordingly there is the saying: (the sun called) Digbhrāja (illuminator of the regions) makes the seasons; in this way the suns are multiplied up to a thousand,'—so says Vaiśampāyana. About this, the following is said:—"

"O! Indra, if the number of both heaven and earth comes to a hundred, even then, O! Wielder of the thunderbolt, no thousand suns will follow thee, born as thou art between those two worlds. For the seasons are of different signs, the suns are many; but it is settled that they are eight."

What are called heaven and earth in this and other passages are the two limits between which the seven intercalary months are inserted, so as to bring both the lunar and solar years to the same starting point.

34. " But that they are only eight, is once for all fixed. There are eight spaheres of the suns above. This is said of them—

35. The Varied troops of the gods went up, the troops that are the eyes of Mitra, Varuṇa and Agni; these troops have pervaded the three worlds, the sky, the earth and the air. The sun is the life and soul of the moveable and immoveable worlds."

After mentioning the names of seven suns with seven solar eclipses, viz., "Aroga, Bhrāja, Paṭara, Patanga, Svarṇara, jyotishmat and Vibhāsa which belong to the air, or sky, the Araṇyaka names the eight solar eclipses in the eight quarters of the compass colled eight Vasus as:—

1. Agni,
2. Jātavedas,
3. Sahojas,
4. Achirāprabha,
5. Vaiśvānara,
6. Naryāpas,
7. पंक्तिराधस्
   and
8. Visarpin, (located on earth.)

Then 11 lunar and solar eclipses occurring in the middle region are of the same names and they are:—

1. Prabhrājamāna,[9]
2. Vyavadāta,[10]
3. Vāsuki,
4. Vaidyuta,[11]
5. Rājata,[12]
6. Parusha,[13]
7. Śyāma,[14]
8. Kapila,[15]
9. Atilohita,
10. Ūrdhva,
11. Avapatanta.

Then the eleven solar eclipses, (of which the first seven are also the names of the seven lunar eclipses occurring on each of the two sides of Purusha or the celestial circle of 7000 days) are named as follows—

1. Svāna,[1]
2. Bhrāṭ,[2]
3. Anghāri,
4. Bambhāri,
5. Hasta,
6. Suhasta,
7. Kriśānu,[3]
8. Viśvāvasu,[4]
9. Mūrdhanvān,
10. Sūryavarchas,
11. Kṛiti.

These are Nīlakanṭhas, (Garagirah), Blacknecks, swallowers of poison.

Then the seven lunar eclipses of the atmospheric region are named as follows:—

1. Varāhu,[5]
2. Svatapas,[6]
3. Vidyunmahas,[7]
4. Dhūpi,
5. Svāpi,[8]
6. Gṛihamedhas,[9]
7. Asimividviṭ. (These in the sky also).

1 and 2 frequently mentioned in the Rigveda. 3 Rigveda I. 57, 5. 6 Rigveda III. 96, 1. 7 Rigveda I. 40, 3. 9 Rigveda VII. 63, 3. 10 Rigveda X. 117, 7. 11 Rigveda V, 105, 10. 12 Rigveda X. 13, 5. 13 Rigveda X. 53, 1. 14 Rigveda X. 128, 3. 15 Rigveda X. 27, 16

1 Rigveda I. 104, 1. 2 X. 125, 2. 3 IV. 27, 3. 4 X. 139, 5. 5 I. 121, 11. 6 I. 64, 7. 7 V. 54, 3. 8 IV. 41, 7.

Then three eclipses (solar?), one on the summer solstice called Sarama, one in the middle of the year called Agni, and one on the winter solstice called Bṛihaspati are mentioned.

It is probable that the two solstices are also named as 'Ahas' day and 'Rātri', night, and also as the two Aśvins.

As the poets believed that the earth and the sky (the summer and winter solstices) do not move from their places as fast as the sun and moon, they seem to have given them two asses to drag their chariot. The asses are idle and do not move fast from their places even in the face of danger. Then the Tait. Ar. (1, 10) describes in obscure and yet vivid terms a complete solar eclipse as follows—

सवितारं वितन्वन्तं अनुबध्नाति शम्बरः ।
आपपूरु षम्बरश्चैव सविताऽरेपसो भवत् ।
त्यं सुतृप्तं विदित्वैव बहुसोमगिरं वशी ।
अन्वेति तुग्रो वक्रियां तं । अयसूयान्त्सोमतृप्सुषु ।
स सङ्ग्रामः तमोद्यः अत्योतः ।

वाचो गाः पिपाति तत् । स तद्गोभिः स्तवाऽत्येत्यन्ये । रक्षसाऽनन्विताश्च ये ।

....

अमी य ऋक्षा निहितास उच्चा । नक्तं ददृश्रे कुह चिद्दिवेयु । अदब्धानि वरुणस्य व्रतानि । विचाकशच्चन्द्रमा नक्षत्रमेति इति ।

....

अपागूहत सविता तृभीन् सर्वान्दिवो अन्धसः ।
नक्तं तान्यभवन्दृशे । अस्थ्यस्था सम्भविष्यामः ।

"Just as a niggardly person holds fast to his wealth even at the time of his death, so the solstices or East and West points (Aśvinā) carried the sun in a secure boat across the celestial ocean. They carried him thus for three days and nights of the three previous cycles marked with the eclipse called Patanga, in three chariots in all, each marked with a hundred camping places, and in all with six horses (two asses for each of the 3 chariots of hundred Padas each). Śambara pursued the sun thus driven and fully drunk with Soma. A battle ensued with horses and in thick darkness. Prayers like cows went to rescue the Patanga as usual.[1] These eclipses between the two regions are periodical and the offsprings of those that preceded them. In the first cycle the Earth brings forth Day and the sky the Night; these latter become a couple: in the second cycle (or the first cycle, if the day and night are not eclipses) they bring forth an eclipse called Āditya of white colour and Agni of reddish colour'

In the third (2nd?) round, the couple, Aditya and Agni, bring fourth two eclipses called Vritra and Vaidyuta respectively. In the fourth (3rd?) cycle Vritra and Vaidyuta bring forth two other eclipses called Ūshma and Nīhāra respectively. Now in this recurrence of the sun's eclipses with black horses the Seven Bears kept so high, which usually appear at night and vanish at daytime in obedience to the law of Varuṇa, and also the moon going through the constellations, all these, ... when just now the sun withdrew his rays of light and when all the upper regions became dark, came in sight

1 (Vāk, speech, is addressed to Patanga, Rigveda X. 180, 1 B.Y. I. 5, 3).

as in night. Now about the name and gender of the offspring born. Neuter, masculine or feminine? Immovable or movable? "I (says the poet) sacrifice, I did sacrifice, and I shall sacrifice." The offspring is really female; yet they call it male; (for the name is Patanga, male). The ancients considered all these? they made the formula of eclipses and kept it in their memory. (We have to follow). Who knows this celestial tree which has its root in the sky and its branches spread below? Who will believe this?"

Then the poet proceeds to distinguish between lunations with eclipses and ordinary lunations without eclipses. The former he compares to the pleasure caused by music and musical instruments (हसितं रुदितं गीतं वीणापणवलासितम्), the latter to the quiet pleasure felt in secret pairing. Whenever there is laughter, or crying (as in the case of kidnaping or outrages) or lute-playing, it is pleasure with disturbance. Likewise in the pleasure caused by drumming or dancing the pleasure is accompanied with disturbance. Pleasure felt in secret pairing is commendable pleasure. Now ordinary lunations are the pairing of the sun and the moon with no sound. But in eclipses there is disturbance, as either laughter or music[1] indicative of the emotion of men and women violating marriage law. Crying is an indisputable evidence of violence. Vāsishṭha Rauhiṇeya is said to have affirmed that in such forced or faithless coupling as eclipses, there is either music or crying like that of a cow in rut, or of thunder, (" Vāsreva vidyut," Tait. Ar. 10, 12, 5.) Then the Āraṇyaka refers

1 Music of the Gandharvas to captivaṭe Gayatri (sun) Soma, the moon.

to a sage who devised the geometrical figure of the cycle of eclipses and commending secret pairing called the Sambhava rite, it praises the chariot of Agni (on the earth.)

In 1, 12 the Āraṇyaka addresses Indra and invites him to impregnate the Āditya with the seeds of Chandra. It then says that the seed has entered the sun resulting in the birth of " Varada Agni, a form of Rudra." Then it refers to the number of such Rudras and calls Indra to come with horses of **peacock colour making Mandra** (deep) sound and bearing **1,000**. Then mentioning such transgressions of Indra as his Soma-drinking by force, and his seduction of Gotama's wife called Ahalyā, it refers to the departure of the eleven Rudras of red colour and of the atmospheric region and the arrival of the eight red Vasus of equatorial (earth) region (already enumerated) and invites Bṛihaspati and Savitar to come through their path in the atmospheric or water region. Then it goes to speak of the 8 or 7 sons of Aditi, the seven intercalary months of the cycle of eclipses, their departure and return, making Yugas, and names them as:

1. Mitra, 2. Varuṇa, 3. Dhāta, 4. Aryama, 5. Amaśa, 6. Bhaga, and 7. Indra.

Evidently the 19 or 18 years of the cycle of eclipses is divided into 7 parts, and the parts are named as Mitra, Varuna and so on. Accordingly the cycle is called **Saptapurusha**, container of **seven purushas** (7000 days) to impregnate the year, and the seed is called Prājapatya, Prajāpati being the moon and the mother being the sun, (Sāvitrī)

Then in 1, 15, the Araṇyaka mentions the eclipses in the 8 (rather seven it ought to be) divisions as follows:

(i) "Let me shine in my own light in the region of the 8 Vasus."

(ii) "Let me shine in my own light in the region of the 11 Rudras."

(iii) "Let me shine in my own light in the region of the 12 Adityas."

(iv) "Let me shine in the region of the Satya Adityas."

(v) "Let me shine in my own light in the region of the three times seven Maruts called Abhidhūnvan, Abhighnan and Vāta."

(vi) "Let me shine in my own light in the region of the Ribhus."

(vii) "Let me shine in my own light in the region of the Viśve-Devas."

(viii) "Le me shine in the year (the cycle) that has returned."

Classified in this way the solar and lunar eclipses with their names can be tabulated as follows:—

8 Vasus,
11 Rudras,
12 Adityas,
2 (Satya and Rita Adityas of the Arctic Circle.),
27 Maruts,
4 Ribhus,
4 Visvedevas.
——
68

LIST OF ECLIPSES IN A CYCLE ACCORDING TO THE TAIT. Ar. 1.

No.	EARTH		AIR		SKY	
	Spring Vasanta	Summer Grishma	Rains Varsha	Autumn Śarad	Snowy Hemanta	Dewy Śisira
1	Agni	Prabhrājamāna	Svāna	Varāhu	Aroga	Bṛihaspati
2	Jātavedas	Vyavadāta	Bhrāṭ	Svatapas	Bhraja	Rudra
3	Sahojas	Vāsuki	Anghāri	Vidyurmahas	Paṭara	Sarama
4	Achirāprabha	Vaidyuta	Bambhāri	Dhūpi	Patanga	Ahah
5	Vaiśvānara	Rājata	Hasta	Svapi	Svarṇara	Ratri
6	Naryāpas	Parusha	Suhasta	Grihamedhas	Jyotishman	Agni
7	Panktirādhas	Śyāma	Kriśānu	Asimividvit	Vibhasa	Āditya
8	Visarpi	Kapila	Visvavasu	The four Ribhus	The seven Adityas bringing with them the years' produce	Vṛitra

No.	EARTH		AIR		SKY	
	Spring Vasanta	Summer Grishma	Rains Varsha	Autumn Śarad	Snowy Hemanta	Dewy Śisira
9	...	Atilohita	Mūrdhanvan	...	...	Vaidyuta
10	...	Ūrdhva	Sūryavarchas	...	...	Ūshman
11	...	Anavapatanta	Kṛiti	...	...	Nīhara
	These are Kshitivasus while Aroga, etc., are Antariksha-vasus	...	These are Nīlakaṇṭhas	Once on earth and once in air. These are Kapardins	...	These are Visvedevas
				(7 with 4 Ribhus); (7 on earth & 7 on air with 4 Ribhus)	7 or 8 with Kasyapa in the Pole	
Total	8	11	11	11		11

8+11+11+11+7+7+11=66

With Indra and Krishna at war with each other once in 20,000 days.

This number various, as it ought to, on account of the invisibility of some eclipses in the latitude of the Vedic poets,=68

It must be noted that Vasu is wealth. Golden or silver ornaments of women are also Vasus, because they are the wealth of women. They go by the name "Nidhi" when kept in safe custody for use in time of distress. According to the Amarakośa the Nidhis, nine in number, are (1) Mahāpadma, a big lotus flower of gold, (2) Padma, a small lotus flower, (3) Śankha, a triangular golden ornament like a Conch, (4) Makara, an ornament like an alligator in shape, (5) Kachhapa, an ornament like a tortoise in form, (6) Mukunda, a ruby, (7) Kunda, an yellow or green gem, (8) Nīla, a blue stone, (9) Vara, a diamond. Even now ornaments of the above type are worn by Hindu women. It is likely, therefore, that the Vasu-eclipses are so called on account of their resemblance to the golden ornaments with or without gems inlaid in them.

Groups of three eclipses, solar and lunar, are called Rudras or Tryambakas, gods with three eyes, if close to each other; if far from each other, they are called Dīrghalochanas, or Dhurilochanas. The latter words are names given to Viśvedevas, eclipses of the winter season. Rudras appear only in summer with rains or in the rainy season. Of the three eclipses forming a Rudra-eclipse the disappearance of one seems to have been spoken of "as a friend deserting his friend" in the beginning of the Araṇyaka. In the Brāhmi alphabet three dots placed in triangular corners represent the "I" sound. When there are only two dots owing to the disappearance of the third dot (eclipse), it does not represent "I" sound. In the altar eight bricks represent the Vasus and eleven bricks the Rudras.[1]

1 See the Author's "A theory of the origin of the Devanāgari Alphabet" in I. A. 1905-6.

In A. V. 10, 8, 9 we are told that there is a gourd of Soma-liquor called Yaśas with its bottom turned upwards and that the seven Sages (the Great Bear or the seven Suns of the seven intercalary months of the cycle of eclipses?) are said to be guarding it. In A. V. 12, 3, 41 we are told that a guardian diety (Nidhipa) is coveting the treasure for himself once in 60 years. Since cattle are also called wealth, I think that the Gopas of A. V. 10, 8, 9 are the same as the Nidhipas of A. V. 12, 3, 41. Anyhow the seven suns presiding over the eight or 7 intercalary months of the eclipse cycle which is called in the Tait. Ar. Aṭṭāra or Ashṭāchakra are the guardian dieties of Vasu, wealth, for Vasu-eclipses occur in the cycle. The word Aṭṭāra or Asāṭhchakra is a name given to the 20 years' cycle with eight intercalary months presided over by eight suns or Adityas meaning sons of Aditi. According to the Vedic poets a Parva or half a lunation consisted of $14\frac{3}{4}$ days and a lunation $29\frac{1}{2}$ days. The fractions such as $\frac{1}{4}$, $\frac{2}{4}$, $\frac{3}{4}$, and $\frac{4}{4}$ after fourteen days of a Parva were called Ekapāth, Dvipāth, Tripāth and Chatushpāth. As the Vedic poets used to count days in terms of syllables of a verse, assigning one syllable to one day, the first three fractions equivalent to one-fourth, two-fourths, three-fourths of a syllable standing for similar divisions of a day, they rightly regarded as inexpressible and only four-fourths which is equivalent to one day and one syllable as expressible. This is clearly stated in the following hymn (A. V. 7, 10, 27.)—

चत्वारि वाक्परिमिता पदानि तानि विदुर्ब्राह्मणा ये मनीषिणः ।
गुहा त्रीणि निहिता नेङ्गयन्ति चतुर्थं वाचो मनुष्या वदन्ति ॥

" Four are the divisions measurable by a syllable and its parts. Those Brāhmans who are wise know them. Of them three divisions are kept in the cave, and are not utterable; only the fourth division men express."

In other words one-fourth, two-fourths, and three-fourths measuring similar divisions of a day are inexpressible, and only four-fourths equivalent to a complete day is what can be uttered. As already pointed out, Kali, Dvāpara, Tretā, and Kṛita were originally the names of Parvas ending in one-fourth, two-fourths, three-fourths, and four-fourths of a day. Five hundred such Parvas or seven thousand days in round number constituted a cycle of eight intercalary months and seventy-one or seventy-two eclipses as a rule. In course of time the eight intercalary months were reduced first to seven and a half (saptārdhagarbhā Aditih), and then to seven months called seven Manus, the years of the cycle being at the same time reduced to 19, as already shown above. In the Amarakośa 1000 Yugas or Parvas are stated to constitute a day or night Kalpa of Brahma and are split into fourteen Manu-periods or Manvantaras of seventyone Yugas or Parvas each. Thus it is clear that Manu is a name given to seventyone or seventytwo Parvas with eclipses. It is not a word found only in the Amara and classical literature. It is also frequently met in the Vedas, undoubtedly in the above sense, when used in connection with eclipses.

Punning on the words " Dvipāth " and " Chatushpāth," the Vedic poets call lunations ending in half a day bi-peds or birds and men; and lunations ending with a complete day cattle or, cows and other quadrupeds. Whenever the poets speak of Indra letting out cows hidden in caves, we have to understand them to mean that the lunation with Krishṇa eclipse is a complete Parva and that the eclipse is got rid of with the aid of Indra. It is also called an Ukthya or expressible day. Ukthya is also used in the sense of 15. The same idea is expressed in the following famous verse of number-puzzles, which later commentators have variously interpreted in vain.

चत्वारि शृङ्गास्त्रयोऽस्य पादा द्वेशीर्षे सप्त हस्तासो अस्य ।
त्रिधा बद्धो वृषभो रोरवीति महो देवो मर्त्यान् आविवेश ॥

" Four are the horns ; three are the quarter divisions ; two are the heads ; seven are the hands ; bound in three ways the bull bellows ; the great god has entered into men."

Here the four horns are the four fractions in which Parvas end and of which one is complete being four-fourths or Chatushpāth ; hence there are only three Pādas or quarters, namely, one-fourth, two fourths, and three-fourths. The two heads are the two Parvas which end in a complete day and between which Parvas ending with fractions of a day come. The seven hands are the seven intercalary months due to these fractions in the course of 19 or 20 luni-solar years. The cycle is bound in three ways, that is, three cycles of 20 years each, making a great cycle of 60 years.

CHART OF ECLIPSES FROM 1840-1920

	Vaiśākha	Jyeshṭha	Ashāḍha	Śrāvaṇa	Bhādrapada	Aśvina	Kārtika	Mārgaśira		Pushya	Māgha	Pālguṇa	Chaitra
									1840	......	Feb 17 ○	Mr. 3 ●	......
1840	......	......	......	Ag. 12 ○	Ag. 27 ●	......	......	......	41	......	Feb. 6 ○	......	......
1841	......	......	......	Aug. 2 ○	......	......	......	......	42	Jan. 26 ○	......	......	......
1842	......	......	Jly. 7 & 22 ● ○	......	......	......	......	......	43	......	......	......	......
1843	......	......	......	......	......	......	......	D 6 ○	44	D. 21 ○	......	......	......
1844	......	My. 31 ○	......	......	......	......	N 25 ○	......	45	......	......	......	......
1845	My 6 & 21 ○ ●	......	......	......	......	......	N 15 ○	......	46	......	......	......	......
1846	......	......	......	......	......	......	Oct. 20 ●	......	47	......	......	......	Mr. 31 ○
1847	Ap. 15 ●	......	......	......	Sep. 24 ○	Oct. 9 ●	......	......	48	......	......	Mr. 19 ○	......

	Vaiśākha	Jyeshṭha	Ashāḍha	Śrāvana	Bhādrapada	Aśvina	Kārtika	Mārgaśira		Pushya	Māgha	Pālguṇa	Chaitra
1848	.....	.....	.....	.....	Se 13, 27 ○ ●	.....	.....	.....	49	.....	.....	Fe 22 M. 8 ● ○	.....
1849	.....	.....	.....	.....	Ag 18 Se 2 ● ○	.....	.....	.....	50	.....	.....	Feb. 11 ●	.....
1850	.....	.....	.....	.....	.....	.....	.....	.....	51	Jan. 17 ○	.....	.....	.....
1851	.....	.....	Jly. 13 ○	.....	.....	.....	.....	.....	52	Jan. 6 ○	.....	.....	.....
1852	.....	.....	Jly. 1 ○	.....	.....	.....	.....	D. 11 & 26 ● ○	53	.....	.....	.....	.....
1853	.....	Jun. 21 ○	.....	.....	.....	.....	.....	.....	54	.....	.....	.....	.....
1854	My. 12 ○	.....	.....	.....	.....	.....	Nov. 5 ○	.....	55	.....	.....	.....	.....
1855	My 1 & 16 ○ ●	.....	.....	.....	.....	Oct. 25 ○	.....	.....	56	.....	.....	Ap. 5 & 19 ● ○	.....
1856	.....	.....	.....	.....	Sep. 29 ●	Oct. 13 ○	.....	.....	57	.....	.....	.....	.....

1857	......	......	......	......	Sep. 18 ●	......	......	......	58	......	......	Feb 27 ○	Mr. 14 ●
1858	......	......	......	Aug 24 ○	......	......	......	......	59	......	Feb 17 ○	......	......
1859	......	......	......	Aug 4 ○	......	......	......	......	60	......	Feb 6 ○	......	......
1860	......	......	......	Aug 1 ○	......	......	......	......	61	Jan. 11 ●	......	......	......
1861	......	......	Jly. 7 ●	......	......	......	......	Dec 17 ○	62	......	......	......	......
1862	......	Jun. 12 ○	......	......	......	......	......	Dec 6 ○	63	Dec 21 ●	......	......	......
1863	......	......	......	......	......	......	......	......	64	......	......	......	......
1864	My. 6 ●	......	......	......	......	......	......	......	65	......	......	......	Ap. 10 ○
1865	......	......	......	......	......	Oct. 4 ○	......	......	66	......	......	......	Mr. 30 ○
1866	......	......	......	......	Sep. 24 ○	......	......	......	67	......	......	Mr 5 & 23 ● ○	......
1867	......	......	......	......	Sep 13 ○	......	......	......	68	......	......	......	......
1868	......	......	......	......	Aug 18 ●	......	......	......	69	......	Jan. 27 ○	......	......

	Vaiśākha	Jyeshṭha	Ashāḍha	Śrāvana	Bhādrapada	Aśvina	Kārtika	Mārgaśira		Pushya	Māgha	Phālguṇa	Chaitra
1869	......	......	Jly. 23 O	......	......	......	......	......	70	Jan 17 O	......	......	......
1870	......	......	Jly. 13 O	......	......	......	......	......	71	Jan 6 O	......	......	......
1871	......	......	Jn 17 ● Jy 2 O	......	......	......	......	Dec 12 ●	72	......	......	......	......
1872	My. 22 O	Jun 6 ●	......	......	......	......	Nov 15 O	......	73	......	......	......	......
1873	My. 11 O	......	......	......	......	......	Nov 6 O	......	74	......	......	......	......
1874	My. 1 O	......	......	......	......	Ot 10 ● & 25 O	......	......	75	......	......	......	Ap 5 ●
1875	......	......	......	......	......	Sep 30 ●	......	......	76	......	......	Mr. 11 O	......
1876	......	......	......	......	Sep 3 O	......	......	......	77	......	......	Feb 27 O	Mr 14 ●
1877	......	......	......	Aug 23 O	......	......	......	......	78	......	Feb 17 O	......	......

1878	.....	.....	.....	Aug 13 ○	.....	.....	.....	.....	79	.....	Jan. 22 ●	.....	.....
1879	.....	.....	.....	Jly 18 ●	.....	.....	.....	Dec 28 ○	80	.....	.....	.....	.....
1880	.....	Jun 22 ○	.....	.....	.....	.....	.....	Dec 16 ○	81	.....	.....	.....	.....
1881	.....	M. 28 J. 11 ● ○	.....	.....	.....	.....	.....	Dec 6 ○	82	.....	.....	.....	.....
1882	.....	My. 17 ●	.....	.....	.....	.....	.....	.....	83	.....	.....	.....	Ap. 22 ○
1883	.....	.....	.....	.....	.....	Oct 16 ○	.....	.....	84	.....	.....	.....	Ap 5 ○
1884	.....	.....	.....	.....	.....	Oct. 4 ○	.....	.....	85	.....	.....	.....	Mr. 30 ○
1885	.....	.....	.....	.....	Sep. 24 ○	.....	.....	.....	86	.....	.....	.....	.....
1886	.....	.....	.....	.....	.....	.....	.....	.....	87	.....	Feb 8 ○	.....	.....
1887	.....	.....	.....	Ag. 4 & 19 ○ ●	.....	.....	.....	.....	88	.....	Jun. 28 ○	.....	.....
1888	.....	.....	Jly. 23 ○	.....	.....	.....	.....	.....	1889	Jan. 17 ○	.....	.....	.....
1889	.....	.....	J. 28 Jy 12 ● ○	.....	.....	.....	.....	.....	1890	.....	.....	.....	.....

	Vaiśākha	Jyeshṭha	Ashāḍha	Śrāvana	Bhādrapada	Aśvina	Kārtika	Mārgaśira		Pushya	Māgha	Phālguṇa	Chaitra
1890	......	......	Jun 17 ●	......	......	......	Nov 26 ○	Dec 11 ●	91	......	......	......	......
1891	My. 23 ○	......	......	......	......	......	Nov 16 ○	......	92	......	......	......	......
1892	My. 11 ○	......	......	......	......	......	N 5 ○	......	93	......	......	......	......
1893	......	......	......	......	......	......	......	......	94	......	......	Mr. 21 ○	Ap. 5 ●
1894	......	......	......	......	Sc 14 & 29 ○ ●	......	......	......	95	......	......	Mr. 11 ○	......
1895	......	......	......	......	A. 29 S. 5 ● ○	......	......	......	96	......	......	Feb 28 ○	......
1896	......	......	......	Ag 8 & 23 ● ○	......	......	......	......	97	......	......	......	......
1897	......	......	......	......	......	......	......	......	98	......	Jan 22 ●	......	......
1898	......	......	Jly. 3 ○	......	......	......	......	Dec 12 ○	99	......	......	......	......

1899	......	Jun 23 ○	......	......	......	......	......	Dec 17 ○	1900	......	......	......	......
1900	......	M. 28 J.11 ● ○	......	......	......	......	......	N. 22 ●	01	......	......	......	Ap. 4 ○
1901	My. 3 ○	My 18 ●	......	......	Sep 28 ○	Oct. 27 ●	Nov 11 ●	......	02	......	......	......	Ap 8 & 22 ● ○
1902	My. 7 ●	......	......	......	......	Oct. 17 ○	Oct. 31 ●	......	03	......	......	......	M 27, A11 ● ○
1903	......	......	......	......	......	S. 21 O. 6 ● ○	......	......	04	......	......	......	M 19 & 17 ●
1904	......	......	......	......	Sep 9 ○	......	......	......	05	......	Feb 19 ○	......	......
1905	......	......	......	Ag 15 & 30 ○ ●	......	......	......	......	06	......	Fe. 8 & 23 ○ ●	......	......
1906	......	......	Jly. 21 ●	Ag. 4 & 20 ○ ●	......	......	......	......	07	......	Ja 14 & 29 ● ○	......	......
1907	......	......	Jly. 10 ●	......	......	......	......	......	08	......	Jan 3 ●	......	......
1908	......	......	Jun. 28 ●	......	......	......	......	Dec 7 ○	09	......	......	......	......
1909	......	Jun 4 & 17 ● ○	......	......	......	......	N. 27 ○	Dec 12 ●	1910	......	......	......	......
1910	My 9 & 24 ● ○	......	......	......	......	......	N. 2 & 17 ● ○	......	1911	......	......	......	Ap. 28 ●

	Vaiśākha	Jyeshṭha	Ashāḍha	Śrāvana	Bhādrapada	Aśvina	Kārtika	Mārgaśira		Pushya	Māgha	Pālguṇa	Chaitra
1911	......	......	......	......	......	......	Oct. 22 ●	......	1912	......	......	......	Ap. 2 ○
1912	Ap. 16 ●	......	......	......	Sep. 26 ○	Oct. 11 ●	......	......	13	......	......	Mr. 22 ○	......
1913	......	......	......	......	Sep. 15 ○	......	......	......	14	......	......	Mr. 11 ○	......
1914	......	......	......	......	A. 21 S. 4 ● ○	......	......	......	15	......	......	Feb. 14 ●	......
1915	......	......	......	Ag. 10 ●	......	......	......	......	16	Jan. 20 ○	Feb 14 ●	......	......
1916	......	......	Jy. 15, 30 ○ ●	......	......	......	......	......	17	Jan. 8, 24 ○ ●	......	......	......
1917	......	......	Ju. 19 Jy 4 ● ○	......	......	......	......	Dec 28 ○	18	......	......	......	......
1918	......	Jun 8, 24 ● ○	......	......	......	......	......	Dec 3 ●	19	......	......	......	......
1919	......	My. 29 ●	......	......	......	......	Nov. 8, 22 ○ ●	......	20	......	......	......	......
1920	......	......	......	......	......	O.27 N.11 ○ ●	......	......	21	......	......	......	Ap. 7, 22 ● ○

In the first twenty six Anuvākas of the Tait. Araṇyaka we are told of the cycles of 1,000 days and 7 intercalary months termed seven Adityas. From that we inferred that the author of the Araṇyaka is talking about a cycle of eclipses and classified the several gods associated with the seven Adityas and their 1,000 days as the names of eclipses occurring in each of those cycles. Now the Āraṇyaka is going to take us with another practical part of the subject. It is a diagram of the celestial circle to illustrate the cycle that has been so for taught. (See Diagram II) In 1, 27, 2 it says that there is a Chakra called the Ayodhyā pūr of gods. In that city there is a golden Kośa called Svarga pervaded with constant light. Brahma (moon) has also entered that city which is pervaded with **Yasas.** Around that city and up and down is constantly going one (the sun) who is learned and who has neither old age nor any disease and who is unknown to death and who sees both the gods and the Asuras. It also says that whenever the young maiden makes a Mandra or deep sound or when the faithful house-wife violates chastity, then Agni touches and pierces through (her). Then the Araṇyaka tells us in 1, 31 that there is Vaiśravaṇa with Tvashṭri on the other side of the water which prevents us from seeing him. His Chakra also contains 1,000 and he comes to us to receive the offerings made by our people to him in the form of money, cows, elephants, gold and horses. It says that this circle of one hundred Padas and eight Chakras or entrances is to be found in the hills called Sudarśana, Maināka, and Krauncha and that it is a Chakra drawn in Samhārya form.

What we have to particularly notice and understand in the above portion is the purpose of the Ashṭāchakra with Navadvāra of the gods and the transgression of the young maiden and her being touched and struck by Agni. We have also to understand who the Brahma and the learned one that are moving in that city are. Here as elsewhere, Brahma is the moon who is believed to create offsprings every new moon. The learned one is evidently the sun who goes round that city and also up and down that city. Accordingly the eight Chakras or circles of that city are evidently the eight monthly circles corresponding to the seven or seven and a half or eight months, that is, the intercalary months that occur in the course of 19 or 20 yers. In a year of twelve months, the first four months are called the part of the gods, the second part of four months is called the division of the Pitries and men. The third part of four months is assigned to the Asuras whose chief is Kubera Vaiśravaṇa addressed as Rājādhirāja. He is worshipped as the great Yaksha to avert calamities, spirits, and evils connected with eclipses. Now when the eight monthly circles of the cycle of 19 or 20 years is on the side of the gods and men in the celestial circle, the remaining part of 4 months must necessarily be on the other night part of the celestial sphere occupied by Kubera and his Asuras, Yakshas, Gandharvas and other beings wandering at night.

The seven monthly circles are the seven circles parallel to the equator. Inclusive of the equator, they amount to eight concentric circles. The extermities of each of these circles form East and West points of each monthly circle. If from the extermities of each of the seven circles lines are drawn to the zenith point, seven triangles are formed with the three lines, namely:

(1) the line from East point to the West,
(2) the line from East point to the zenith,
(3) the line from the zenith to the West point.

The sun is considered to pass from the East point to the zenith, from the zenith to the West point, and there set and rise again in the East point the next day and repeat the same process till the close of the month. In the same way in the next circle in the next month, and so on till he passes through all the seven circles. Then the Ayanas change. When the Ayana of five months with five monthly circles begins, then also the sun is supposed to perform his daily and monthly circuits in the same way for five months and thus close his yearly revolution, if it is an ordinary year. If it is an intercalary year of twelve months, he will make 32 yearly revolutions corresponding to the 33 yearly revolutions of the moon. Thus the sun is supposed to pass through these 36 angles in every ordinary or intercalary year.

The moon is also considered to pass through such circles of his own. Each of the three lines forming the seven or twelve triangles is called a Daṇḍa, each triangle being formed of three *Daṇḍas* or *Nabhyas*, as they are also called. It is an occasion of an eclipse, when the Daṇḍa of the sun and that of the moon coincide with each other.

So for it is clear. What remains now for us to to understand is the transgression to which the young maiden is said to be liable. As already pointed out, eclipses are looked upon as a kind of seduction of the eclipsed planet by a Yaksha or Gandharva noted for music with which the Gandharvas attract men and women and kidnap them. When an eclipsed planet sets before the eclipse clears off, it is called abduction and captivity of the planet. Hence Kubera, the lord of the Gandharvas, is worshipped and given rich offerings to keep his Gandharvas under his control. Agnivedha means fire-ordeal. It is believed that if a person is not hurt by fire through which he passes to prove his innocence, he is considered not guilty.

Now as a rule eclipses will occur only when the earth, the sun and the moon are on a line, that is, when they are revolving, as it were, round an axel passing through their centre, or along their edges like the stick of a lute placed over the sounding gourds. (Diagram I) When two or three round bodies are on a line or touch each other, they are said to strike (to make a Vedha of) each other. As Agni is believed to be in the centre of the earth, he comes in a line with, that is, makes a Vedha of, the two other bodies, the sun and the moon. If the sun comes between the earth and the moon, the Daṇḍas of the sun and the moon coinciding, it is a lunar eclipse and when the moon comes between the other two, it is the solar eclipse. The eclipsed body or person is then called Daṇḍya, liable to Daṇḍa. Always it is the eclipsed body that is considered guilty. If it gives up Daṇḍa or is far off from it, it is considered to be innocent. Thus Daṇḍa came to mean a fine payable by a guilty person. Daṇḍa also means a new moon, because the sun and the moon are in a Daṇḍa above or below the nodes.

It has been already pointed out how laughter, crying, flute, and dancing are associated with eclipses. The celestial sphere, that is, the visible semisphere is, as explained above, divided into eight circles called Chakras. They are called, 1. Bhūh, 2. Bhuvah, 3. Suvah, 4. Mahah, 5. Janah, 6. Tapah, 7. Satya. (See Diagram II)

The cord from the East point to the pole is divided into one hundred Padas. The Bhūloka is made to extend from the East point 11 Padas upwards. The Bhuvah extends likewise from 11 Padas to 22 Padas. Likewise Suvah extends from 22 to 33, the Mahah from 33 to 44, the Janah from 44 to 55, the Tapah from 55 to 66; the Satya from 66 to 77; and the eighth Chakra called Tripāth from 77 to 99 and then comes above it the Apāth (10th) with no Padas assigned to it. The Tripāth is the Arctic circle where the day and night are each of six months duration. In the lower circles they vary (vishurūpe), the day becoming longer in the period from winter solstice to summer solstice and the night shorter, while in the other Ayana the night becomes longer and the day shorter. In the circle where the Vedic poets lived long ago the day in the Northern Ayana was 14 hours (Ashtādaśamuhūrtah) and the night 10 hours (Dvādaśamuhūrtah). The Sūryapragnapti frequently refers to this topic. This corresponds to about 40° latitude, as already pointed out. The seven notes from Sā to Nī are also made to correspond to the seven Lokas from Bhūh to Satya. The exact pitch of the notes are compared to the natural notes produced by the seven animals, such as 1. the, Peacock, 2. Kine, 3. Goats and Sheep, 4. The Krauncha bird, 5. The Kokila bird, 6. The Horse, and 7. The Elephant.

That this classification or allotment is Vedic, is clearly hinted in the Tait. Ar. 1, 12. There Indra at the beginning of a new cycle of eclipses is addressed to come in his chariot of green horses with **peacock-feathers.** Corresponding to the note Ri, the eclipses in the second cycle are called kine and the victims immolated are also kine. Likewise the other animals are made to correspond to the names and victims of eclipses in the other circles. This is what the poet means when he says that if one hears 'īm' sound on the occasion of an eclipse, then one hears of guilt or crime (Tait. Ar. 1, 3) and that according to Vāsishṭha Rauhiṇa the note of the thunder is like that of a cow. (Tait. Ar. 1, 12).

Like the notes of the gamut, alphabetic letters are also connected with eclipses. The followers of the Tāntric cult write the 'īm' sound of their fifteen syllabled Mantra like three dots placed in a curved line, or rather like the Tryambaka or Rudra eclipse. The form of the 'I' letter in the Asoka alphabet is a Tryambaka. The city of eight circles and nine doors with 100 Padas which is just explained above is called Śrīchakra in the Tantra literature and is worshipped through out India. Further researches into the Sāma Vedic literature in connection with the distribution of Sāmachants in the cyclic sacrifices of the cycle of eclipses and into the Tantric literature in connection with the Śrichakra and the distribution of letters in the Charkra will not fail to throw some more light on the origin of the notes of the gamut and also of the alphabetic letters from the cult of the cycle of eclipses. The name of the fifteen syllables of the Panchadasi mantra, such as, I Kāma, Yoni, and the like seem to be the names of eclipses. Translated into English, the names mean, 1. Moon, 2. the sun, 3. Rudra or Tryambaka, 4. Vajrapāṇi or Indra, 5. Hasa, laughter, 6. Mātariśvan, wind, 7. Abhra or Indra, 8. Three Caves or Māyas meaning the setting of eclipsed planets? As the main point under considerstion is not the origin of the gamut or of the alphabet of the Śrīchakra of the Tāntric cult, but the divine city of 100 Padas, eight circles and nine entrances connected with the Vedic sacrifices and cycles of eclipses, I confine my attention to the consideration of that point alone.

CHAPTER V.

THE SRICHAKRA.

I have explained that the city of eight circles is no other than the Northern hemisphere of the celestial sphere divided into eight parts, as shown in the diagram No. II. That this is the city of the gods is corroborated by the Skambha hymn of the Atharvaveda 10, 8. While the general belief underlying the Vedic cult of sacrifices is that eclipses are caused by the Māya of Yakshas and Gandharvas, the poet of the Skambha hymn seems to laugh at the above belief and point out the true cause of eclipses by his illustrative example of water-pots. Hence the hymn is very interesting, and serves to prove the point under consideration. And it runs as follows:

"Salutation to that Jyesṭha Brahma which has seated itself on the past, future and all, and which has the heaven as its own. (1)

The sky and the earth stand held by it. In it there is the whole that animates, breathes and winks. (2)

Three are the circles of offspring running across; some more (four) are there round the sun; the Rajo-vimāna (three more circles or doors) stands prominent; the Hariṇis have entered the regions (the points of the compass)."

Note—The three circles of offspring together with four more are the seven circles. The Rajovimāna is the triangle, the same as the sun and moon, (Bindu and Trikoṇa) of the Śrīchakra. Jyeshṭha Brama is moon.

"This Vault is itself a circle of twelve spokes divided into three parts called Nabhyas; Who knows it? Around the rim of that circle are fastened 300 and 60 immovablé nails." (4)

Note—This is an yearly circle with 12 parts corresponding to 12 montns, the latter being divided into 3 parts of 4 months each. Are not the 360 nails 360°?

"O! Sun, know this—there are the sixt wins and one single-born—they want to have their place in this circle together with the seventh." (5)

Note—The difficulty is in making the day circle to extend so as to correspond to the seven intercalary months. With six months it is not difficult. But unless the zone is above 40° latitude, it cannot be extended to correspond to 7 months, that is, a day of 14 hours.

"The cave-like old Pada or Mahatpada (Jarannāma) is close to this circle above mentioned; in it is kept the movable and lie-like one." (6)

Note—This refers to the Arctic Circle and the movable one is the Pole.

"There is also the Circle of 1000 syllables with one circumference, front and behind. With one half of this the whole visible world is made; where is the other half?" (7)

Note—This refers to the circle of the intercalary year with 7 intercalary months in the visible half. The other half with 5 months must necessarily be on the other side.

"The circle containing the five months holds the front (of the year) and the two parts go jointed well; its arrival is seen; but its departure is not seen; the front is near, but the lower is far off." (8)

(Imagine) a liquor gourd with its bottom turned up and having a hole on its side. In it Yaśas (liquor or Soma) of all forms is kept. On it are the Seven Bears who have become the guides of the Mahatpada." (9)

Note—The Seven Bears are quite different from the seven suns or sages mentioned before.

" What is yoked first, what later, what on the convex half, what round the vault, in correspondence to what sacrifices are they spread? That I ask you; what is that Rigverse which shows this? (10)

What is that which rises and which falls down? What is that which is movable, immovable, and winking and becoming? What is that which holds the whole concave vault and what is that which with the other part becomes one." (11)

Note—The pole and the circles below it are referred to here.

"(This sphere) is endless, spread on all sides ending, though endless, finally the past and future of this the learned keeper of thy Nāka (heaven) is examining and is moving." (12)

Note—The **learned one is the sun himself.**

" Prajāpati moves in the interior and becomes many, unseen; With half the visible world is covered; what became of the other half, the Ketu (flag meaning the other half of the moon)? (13)

ऊर्ध्वं भरन्तमुदकं कुम्भेनेवोदहार्यम् ।
पश्यन्ति सर्वे चक्षुषा न सर्वे मनसा विदुः ॥

This is to be illustrated by one who is carrying water with a pot up (on his head). All see this with their eyes, but do not understand it with mind. (14)

दूरे पूर्णेन वसति दूरे ऊनेन हीयते ।

One covers itself with another on its level (Pūrṇa) in distance and one is made partially visible (hīyate) by another lower than itself." (15)

Note—Of two pots of the same size on a level, that in front covers that which is behind. If one is lower, the other becomes visible to that extent.

महद्यक्षं भुवनस्य मध्ये तस्मै बलिं राष्ट्रभृतो भरन्ति ।

"There is the big Yaksha in the centre of the vault (world); to him the Rāshṭra-bearers make offerings." (15)

Note—The poet means that Yaksha is not the cause of eclipses.

यतस्सूर्य उदेत्यस्तं यत्र च गच्छति ।
तदेव मन्येऽहं ज्येष्ठं तदु नात्येति किं चन ॥

"Where the sun rises and where he sets, that (circle) alone I consider as the Jyeshṭha (spoken of above); nothing goes beyond it. (16)

Those who speek of an old loarned being of Vedic body as moving up and down (arvāg) and round in that circle say that that being is the sun, the second is Agni (the earth with Agni in its centre,) and the third is the threefold moon (Trivrit Hamsa). (17)

While flying to Svarga, the wings of that Hamsa called Hari are expanded with thousand days each; he journeys having kept all the gods clinging to his breast and looking on all the worlds." (18)

Note—This refers to the Eclipse-cycle in which the solar and the lunar years come to the same point.

" The topmost circle warms itself with Satya; the lower one looks up with Brahma." (19)

Note—Satya and Brahma are the names of the Arctic and Antarctic circles.

" It (the hemisphere) breathes with Prāṇa horizontally with Jyeshṭha clinging to it." (20)

Note—This refers to the equatorial circle.

" He who knows the Fire-drills with which Vasu (the fire) is churned knows, being himself a wise man, that Jyeshṭha and also that Mahat Brahma." (20)

Note—In verse 9, the poet compared the sphere to a liquor gourd with its bottom turned up and open in the middle like two half cups. Here he compares the same to two semi-circular Araṇis. (Fire-drills).

"(In this hemisphere the topmost called 'Apāth' comes first) and it has the Svarga world located in it; becoming fourfold, it becomes enjoyable and contains all that is edible." (21-22)

Note—Of the 8 circles into which the hemisphere is divided as the city of the gods; the upper one is undivided, as its division serves no purpose. It is Svarga, always covered with light (Svargo Jyotisha vritah-Tait. Ar.) The lower eight circles up to the equator are Bhogya, enjoyable.

" Having become enjoyable, it gives plenty of food; those who worship it as the eternal call it the eternal, though now it becomes regenerated often. (In the Arctic circle) day and night exchange their form with respect to colour." (23)

Note—In the Arctic circle there is a day of 6 months and a night of 6 months. There is change only in their colour, but not in duration as in other circles.

" In that Arctic circle are kept a hundred, a thousand, a ten thousand, an Arbuda, or infinite masses of wealth (our days regarded as wealth); this wealth they (others) carry off before the eyes of the very god who likes it to possess. (24)

The crescent moon less than the Bāla (three days old after new moon) there is one; one less than the latter may or may not be visible. By him (the moon) thou, O! Sāvitri, art embraced as " my dear goddess." (25)

This auspicious Sāvitri, never growing old, and immortal is in the house of man (the earth), is given in marriage; He who took her hand and he who joined the hands (of the sun and the new moon) grew old. (26)

Thou, O! Sāvitri, art a female, or male, a youth or a maiden. Thou, O! Moon, growest old and deceivest with the stick (Daṇḍa) which thou holdest and passest along; and being born, becomest visible to the world (the earth)." (27)

Note—The moon is believed to be born again and again after new moon and is spoken of as the father of the next moon and grand father of the third moon or as the eldest of the succeeding moons and the youngest of the previous moons. The sun is spoken of sometimes as a maiden and at other times as a youth. New moon occurs when the sun and the moon are on a Daṇḍa, a line, the planes, however, being different. If the planes are the same, that is, when the sun and the moon are exactly on a Daṇḍa, as explained above, then eclipses are caused. The poet makes this clear in the next verses.

" One god (the moon) has entered the mind, becomes reborn, and is in the womb inside. (28)

(Of two spheres on a line) one comes up from the other or one is poured over by the other; Now let us know that from which this phenomena of pouring over is seen. (29)

This one as eternal and born of the same source and very old, facing all is the goddess, Earth, shining from the dawns and she responds with every winking. (30)

The goddess called Avi is pervaded with Rita (centre of the earth); from her colour these trees appear green. (31)

She never leaves that which is near and never looks at that which is near. See the creative skill of god, which never dies nor grows old. (32)

They give expression to all kinds of speeches and say that that place to which the sacrificial Vadanties (expressions) are addressed is that Mahat Brahma, moon in which gods and men are found as spokes sticking to the central nave or hub. I ask about that flower of the waters where that nave is kept. (33 & 34)

Who and how many are those who propell the waters, whom they look upon as oblation, and from whom the wind blows, and who give us these five similar points of the compass? (35)

Of these one occupies this earth; another the atmosphere; he among them called Vidhartri gives us the sky; and others keep the quarters all. (36)

He who knows the thread which is extended and in which these offsprings are sewn, he who knows the thread (which is at right angles) to that thread can know the Mahat Brahma. (37)

26

In the midst of those threads (the horizontal circles and those that are at right angles to them) the sky and the earth together with the all-burning fire are located; wherein are those who have the one for their husband, where was the Mātariśvan then? (39)

The Mātariśvan then was in the waters, the gods also entered the waters; the big Rajovimāna (of the moon and the sun) stood there; the Pavamāna (the moon) entered the quarters. (40)

(The sun) rose up by the Gāyatri measure (24) above the north or by Sāma measure, say those who know Sāma; then where was the Rajas? (41)

He who puts and joins the Vasus is the Sāvitri like a god of unfailing Dharma; Indra has no place in the war for Vasus. (42)

This lotus circle of nine doors pervaded with three qualities (black, red and white colours) is there; Those who know Brahma understand the Yaksha within it. (43)

There is the immortal self-born with no desire, bold, satiated with the juice (Soma) and inferior to none; He who knows Him is not affraid of death and looks upon himself as a bold young one with no old age. (44)

This long hymn is quoted here to show that the Vedic poets were not unaware of the real cause of eclipses. They knew that in Mahāmeru, the pole, the sun never sets and is called Kaśyapa. They knew that in the Arctic Circle called the Tripāth the sun never sets for six months and never rises for another six months. This they expressed by saying that day and night exchanged their places in that circle. They called the day in the Arctic Circle Mitra and the night Varuṇa. They knew that in the lower circles the length of day and night varied and that in the latitude in which they could see and saw not merely the concave arch of the celestial circle corresponding to six months, but also one-third of the vault more corresponding to two months more, that is, eight months in all; and that the remaining two-thirds of the spherical globe was on the other side where ignorant people placed Yakshas and Gandharvas with Kubera at their head. The imaginary Kubera grew enormously rich with the valuable offerings given to him by the people for keeping the eclipse-causing spirits under his control. The poet of the above hymn is quite familiar with the cycle of 1000 days called the cycle of eclipses and also the city of the gods with eight circles and nine intervening spaces and the cord of the quadrant divided in 100 parts. He knows that the sun's Northern declination is only 33 of the one hundred Padas. It is hoped that the above hymn will establish beyond doubt that they had a cycle of eclipses and with it predicted the arrival of Mitra, Varuṇa, Indra, Krishṇa, Śyāma and other astronomical phenomena personified as gods and with or without eclipses. The Rajo-vimāna seems to be a name given to the cause of eclipses. The Jainas say that Rāhu-vimāna is the cause of eclipses.[1]

The Purushasūkta is nothing but a description of the city of the Gods. Considered in this light it becomes quite intelligible. In. A.V. 10, 8 the city of eight circles and nine entrances is made to consist of three divisions: 1. Apāth; 2. Tripāth; 3. Chatushpāth.

Of these, Apāth, the Pole, is only one and the Tripāth below it is also one, but threefold, that is, made

1 See Sūryapragnapti.

up of thrice the area of each of the seven circles each of which is called a Chaṭushpāth. It is yet one and has two aspects, that of a day of 6 months and that of a night of 6 months. Below this Tripāth come the seven circles of bipeds and quadrupeds. These seven divisions are called variously. They are seven Dvīpas, called Jambu, Plaksha, Kuśa or Barhi, Krauncha, Śāka, Śālmali and Pushkara. They are also called the Salt sea, the sea of Sugarcane juice, that of liquor, butter or ghi, curds, milk and nectar. They are also the notes of the Gamut, Sa, Ri, Ga, Ma, Pa, Dha, Ni ; seven animals, such as man, horse, ass, cow, goats and sheep, dog or camel. The other side of the sphere has for its animals wild beasts. Now let us see how this idea is implied in the Purushasūkta.

(1) Purusha hath a thousand heads, a thousand eyes, a thousand feet. Pervading the earth on every side he has risen ten Angulas high.

(2) Purusha in truth is this all, the past and the future; and lord, too, of immortality, and also of what grows by food.

Note—His rising by ten Angulas seems to mean the ten Angulas of shadow cast by a gnoman which is also called Purusha. When the shadow is shortest, as on the day of summer solstice, then Purusha is called Vāmana, the dwarf incarnation of Vishṇu. A.V. 19·6 has 1,000 arms. It makes no difference, for all that is intended to be meant is a cycle of 3,000 days, which gives three intercalary months and another 3,000 on the other wing of the bird-like Purusha which gives another three months.

(3) So mighty is his granduer; yea, greater than this is Purusha. All creatures (that is, the earth with all creatures) are a Pāda of him ; his Tripāth is in the

sky; he rose up above Tripāth; his Pāda again was here (that is the pole).

"(4) Thence he moved forth to every side over what eats not and what eats."

Here the earth is one Pāda, foot, of Purusha: his Tripāth, *i.e.*, the Arctic Circle, is in the sky and unlike the earth which is mortal, the Tripāth and also the Pāda above it are immortal (Amṛita), because they remain unaged and unchanged always.

(5) From him Virāj was born; again Purusha from Virāj was born, he spread to West and to the earth in front.

Note—Here Virāṭ means ten as elsewhere. From Pāda he came out covering ten divisions, that is, the three divisions of the Tripāth, and seven divisions below it, ten divisions in all. This lower sevenfold division is made clear in the next verse.

(6) When the gods performed the sacrifice with Purusha as offering, spring was the Ghi, summer the fuel, the autumn the oblation. Then seven were the circles (Paridhis=Manḍalas=bounding circles); thrice seven were the kindling fuels.

Note—Paridhi taken in any other sense makes the whole hymn meaningless.

(7) When gods performing sacrifice bound Purusha as the Victim, that sacrifice they terminated with Barhi or Kusa (circle, that is, the 4th circle); and also the Purusha that came next;

(8) With that victim(Purusha) the gods performed the sacrifice, likewise the Sādhyas and the sages did so; from that general sacrifice, clarified butter was gathered up;

(9) They formed the creatures of the air, and animals, both tame and wild;

From that general sacrifice Rig-verses and Sama-chants were formed; the seven metres were also formed; and Yajus-formulas were formed.

(10) From it horses were born; also those that have two rows of teeth; from it there came kine and also goats and sheep.

Note—Animals such as Purusha, tame and wild animals, cows, goats and sheep and horse, Kuśa grass and Ghi all these are not merely sacrificial victims and offerings, but also symbols denoting corresponding circles of the Chakra—(Vide the Diagram II). These victims are sacrificed in atonement for the sin of mankind which was believed to be the cause of eclipses. The animals and the circles which they signify are believed to correspond to the forms of eclipses occurring in the different parts or seasons of the cyclic year.

(11) When they divided Purusha, (for making a Chakra corresponding to him), how did they arrange his limbs?

(12) What was his mouth, what were his arms, what are the parts of the circle corresponding to his thighs and feet.

(13) The Brāhman was his mouth, the Rājanya (Soma) was his two arms, his thighs became the Vaisya, from his feet the Śūdra was made.

(14) The moon came from his mind, from his eye the sun came out, from his mouth came Indra and Agni; and from his Prāṇa, Vāyu came out.

(15) From his navel came out the atmosphere, from his head came the sky, from his feet came out the earth, from his ear the quarters.

(16) Thus they formed the worlds also.

Note—The whole of the hymn is a symbolical expression of the diagram of the cycle of eclipses. Here Purusha means the cycle. The several limbs represent the several parts of the circles. The Pāda, Tripāda, and seven Paridhis are evidently the eight Chakras with nine Dvāras. The animals, the oceans, and the Dvīpas assigned to the seven circles denote the names of those circles. This is confirmed by the 16th verse of the Atharvavedic version of the hymn (19, 6). The verse is as follows—

मूर्ध्नो देवस्य बृहतो अंशवस्सप्तसप्ततिः ।
राज्ञः सोमस्याजायन्त जातस्य पुरुषादधि ॥

From the Brihat representing the head of the Purusha, namely, King Soma, 77 Amśus or parts came out."

These 77 with the 77-88-99 of the Tripāth, and one of the Pāda make up the 100 Amśus ("Satadvārattāra gamam"—hundred doors of the eight Chakras—Tait. Ar.). According to the Tait. Ar. this is a Chakra found in Sudarśana, Krauncha and Maināka mountains and is Samhārya, drawn to show the world at the time of Samhāra, destruction. The other form of the Chakra is a symbol of Srishṭi, or creation. It consists of 1,000 parts. It gives 250 to a quadrant. The Chakra of the Purusha hymn is of this kind. For the seven metres are said here to have come out of this Chakra. The seven metres amount to 24+28+32+36+40+44+48=252. According to A.V. 13, 1, 37 the cycle of Rohita eclipse is spoken of as having 1007 births. This difference of 8 or 7 over 1,000 is due to the difference between the actual and theoretical number of days of the cycle in consideration. As already pointed out, Indra's cycle is nearly 57 years. It means an eclipse recurring thrice in 57 years in the part of the Chakra denoted by the mouth of Purusha.

It follows therefore that the Purusha hymn is a general expression of the Ashṭāchakra of the cycle of eclipses.

The Saundaryalaharī speaks of the same in terms of Purusha's limbs as follows—

महीं मूलाधारे कमपि मणिपूरे
हुतवहं स्थितं स्वाधिष्ठाने हृदि मरुतमाकाशमुपरि ।
मनोऽपि भ्रूमध्ये सकलमपि भित्वा कुलपथं
सहस्रारे पद्मे सह रहसि पत्या विहरसे ॥

O! Sakti, thou art in combination with thy consort in secret in the thousand-petaled circle of lotus, having gone up there piercing through the Kula-path, which consits of—

(i) the earth in the Mūlādhāra circle (a circle at the hips),

(ii) the four water circles in the Maṇipūra circle (a circle passing through the navel),

(iii) the fiery zone located in its own place in the same (fourth circle of the) Maṇipūra,

(iv) the sevenfold Marut circle in the region of the heart,

(v) the circle of the sky (dyauh) above the Marut circle,

(vi) the circle of the Mind (the moon) in the midst of the brows."

Evidently the Śahasrāra contains the Tripāth and Apāt or Pāda divisions together with seven circles.

In A.V. 4, 30 the same Chakra is described as follows—

"I move with the Rudras, the Vasus, the Ādityas, and the Viśvedavas (eclipses named as such); I move with both Mitra and Varuṇa of Tripāth circle); I bear Indra and Agni; and both the Aśvins; I am the dominion, collector of Vasus (wealth); I am the first prophetess of the worshipable (seven). ....
I make the bow for Rudra to kill the haters of Brahma.

In A.V. 5, 28, we have an interesting account of the three cities, that of gold, that of silver and that of iron.

"He measures the nine[1] Prāṇās with nine for a long life of a hundred autumns, three in the yellow region, three in the silvery white region and three in the iron-black region with the circle of Tapas;

Agni, Sūrya, the Moon, the Earth, the Waters, the Sky, the Atmosphere, the Quarters, those angles between the quarters, the seasonal things with the seasons, all in concert, may lead me with the Trivrit. (9)

Let the three nourishments come within the nine; let Pūshan develop them with milk and Ghi (of the circles);

Let plenty of food, of Purusha, and of the cattle be present here in the (nine circles);

Let the Ādityas with the Vasus sprinkle over this; O! Agni, develop them, growing thyself; O! Indra, create them and let the nine circles be included therein:

1 The seven circles, the Tripāth and the Apāth.

May the earth keep thee with the yellow; may Agni nourish thee with iron; and make thee fit well with the white;

In birth it came as threefold; the golden one became dear to Agni; That which was of Soma pressed (eclipsed) fell down; one of the waters became the seeds of the creators (moons), thus the threefold golden may be for thy long life;

Threefold is the age of Jamadagni in the Pāda division; Pada region; threefold is that of Kaśyapa (in the polar region); threefold is the age of Amrita (of the Tripāth region);

May the **golden one keep the sky**; may the **white one keep the middle region**; and may the **iron one keep the earth**;

May these three cities of gods protect you. .... I make the ten Easts (East to West circles.)

In another Purusha hymn (10, 2, 25-33) the Atharvaveda describes the city of the gods as follows:—

With Brahma the earth is made, Brahma became the Northern sky; Brahma is the upper and the crosswise atmospheric region. Having well sewn its head, and also the heart, Atharvan rose up above the head; the Pavamāṇa (the moon) also rose up the head; That is verily the head of Atharvan, the cell of the gods; Prāṇa (the sun) keeps it up, also he keeps food and mind (moon);

Purusha became the whole,—what is created above and what is created crosswise; who knows the city of Brahma, from which he became Purusha, knower of the city,

To him who knows the city of Brahma surrounded with nectar,—to him Brahma, and those from Brahma give the eye, life, and offspring;

Him the eyes never leave, nor the life-breath before old age; He who knows the city of Brahma, on account of which knowledge, He is called Purusha.

Eight are the circles and nine the entrances which make the Ayodhyā city of the gods; in it is the golden cell called Svarga always lighted with light (the Polar region);

In that golden cell with three spokes placed thrice, is the Yaksha possessed of animation; that the knowers of Brahma understand;

That golden Prabhājamāna cell of yellow colour pervaded with Yaśas, that golden city which is impregnable, Brahma has entered."

Here the Golden cell (Hiraṇmaya-kośa) always pervaded with light is the same as the Apāth of A.V. 10, 8, 21. It is also identical with the Bindu of constant light of the Śrīchakra. Below it comes the eight Chakras, namely, the Tripāth and the seven circles with Chatuspāths (quadrupeds) assigned to them. The three spokes, the same as the Trikoṇas or triangles of the Śrīchara, are the three Daṇḍas explained above. The three Nabhyas of A.V. 10, 8 are the same. It is this Daṇḍa with which the moon is said in A.V. 10, 8 to deceive others causing them to consider him to be an old ascetic with no energy left in him to elope with maidens. The story of Arjuna's disguise as an ascetic with a Daṇḍa or three Daṇḍas, his marriage with Subhadrā, Krishṇa's sister, and his running away with her can be traced to this source.

Sāyana in his commentary on the Tait. Āraṇyaka and Lolla in his commentary on the Saundaryalahari (Verse 11) hold that the Ashṭā-chakra is identical with the Śrīchakra. This Diagram (See Diagram III) seems to be so constructed as to indicate a sandhi or parva by the meeting point of two lines and a Marma or an eclipse by the meeting point of three lines.

द्विरेखा सङ्गमस्थानं सन्धिरित्यभिधीयते ।
त्रिरेखा सङ्गमस्थानं मर्म मर्मविदो विदुः ॥

Besides being an object of worship and also an amulet, the Śrīchakra seems to have enabled them to predict by the numbers placed on the Vasukonas, the Rudrakoṇas and the Manukoṇas the solar or lunar eclipse to come after the first eclipse of the cycle is witnessed.

The followers of the Tāntric cult count the eclipses of a cycle not as Vasus, Rudras and Adityas, but in terms of angular points called Vasukoṇas, Rudrakoṇas, and Manukoṇas of the Śrīchakra. The verse in which the solar and lunar eclipses of a cycle are counted runs as follows—

बिन्दुत्रिकोणवसुकोणदशारयुग्म-
मन्वश्रनागदलसंयुतषोडशारम् ।
वृत्तत्रिभूपुरयुतं परितश्चतुर्द्वाश्श्री-
चक्रमेतदुदितं परदेवतायाः ॥

A Bindu, a triangle, an octagon, two decagons, one of fourteen angles, an octagon and one of sixteen angles, three circles, and three more circles ealled Bhūpura, four and four doors on each of the four sides,—this is the Śrichakra of the great godess. Here Bindu (dot) is the name of the moon; and Trikoṇa is that of the sun, signifying Savitri's productive power. The moon or Śiva, as he is called in the Tāntric literature, is given four Chakras and the sun, called Śakti, is assigned to five Chakras. One octagon, two decagons, and one of fourteen angles together with a triangle and a Bindu are the five Chakras of Sakti. The total number of eclipse-points are 8+20+14=42. The triangle and the Bindu are not taken into account. They are the moving bodies, the triangle being the sun and Bindu the moon. Those of Śiva are one octagon, one of sixteen angles, and two triple circles making four Chakras. The total number of eclipse points of the moon are 8+16+3+3=30. The author of the Saundaryalahari makes the solar points, however, 44 counting the Bindu and the triangle as two and adding them to the above 42 points. The commentator on the saundaryalahari makes it 43 leaving the Bindu. The angular points are not, however, disassociated with their Vasu, Manu, Rudra and other Vedic names. The Vasukoṇa is of the Vedic Vasu eclipses; the two decagons are two figures or diagrams showing the angular points of 20 Rudra eclipses, the remaining two Rudra points being added to a Manukoṇa figure of sixteen angles denoting sixteen lunar eclipses.

I now proceed to give a few examples of reference made to eclipses in the Atharva Veda:—

In A.V. 1, 1 Vāchaspati is said to be the strongest of the 21 eclipses. In the Tait. Ar. 1, 10 Bṛihaspati and Rudra are said to be the two sons of Dyāvāprithivī with two thousand days for their cycle and Saramā, the four-eyed dog, to be their daughter. Evidently the first solar eclipse in the cycle is Brihaspati, the second is Rudra, one solar eclipse with one lunar eclipse preceding and another lunar following ; Saramā is a group of four eclipses, probably two solar and two lunar. Vāchaspati is called upon to come again.

In A.V. 1, 4 " Ambayah and Jāmayah " signify the appearance of two lunar eclipses and one solar eclipse, one after another as the eyes of Rudra in the rainy season.

In A.V. 9, 3 the bull is called the bearer of 1000 and Indra is said to have drunk a pot full of Soma and Indras's arrival is said to have been previously marked by Bṛihaspati and Saramā.

In A.V. 9, 5 the offering of a goat is praised and the goat is said to dispell the darkness (of the eclipse) (lunar) and to ascend to the third heaven. The goat (Aja) is said to be bearing 1000 and to have kept the earth and heaven with 7 rays (7 intercalary months).

In A.V. 9, 10 the grey-haired one (the sun) is said to have swallowed the youthful moon running on the surface of water. The poet wonders at the skill of the god in virtue of which one that died the day before came to life again the next day. He thinks that the cause of the double seed is Nirriti. He refers to the first two Apāth and Tripāth divisions of the nine celestial Padas or divisions which Gauri is ṣaid to measure. He speaks of the arrival of Krishṇa with 1000 syllables.

A.V. 12, 2, 40-44 discribes a solar eclipse (Avih Krishṇa) as a bull attacked by a wild beast, and Agni is prayed to for repelling the attack, a black sheep or Sīsa, lead, called Chandra (silver) is the ransom for the

Kravyāda who has the sun under his clutches. He is said to have given back the sun in the West and gone on his way.

In A.V. 13, 1 a solar eclipse of red colour (Rohita) is said to have occupied the region from the earth to the sky and to be the initial point of another creation, that is, another cycle of eclipses (L. 25). On this occasion Indra is said to be at liberty to drink Soma. The cycle of 1,000 syllables and nine Pādas (entrances) is also mentioned. In line 56 the poet says that he is going to cut at the root of him who, while illuminating the earth with his foot, likes to pounce upon the sun in the West, so that the cause of the eclipse may not make another **shadow** again.

In A.V. 13, 2 the sun (Śukra) is said to have been freed from darkness and Atri is stated to have placed him again in the sky after freeing him from darkness. In line 7-8, the circle of 100 rims and seven suns or horses is mentioned. Regarding Atri's connection with solar and lunar eclipses, we read in Rig. 5, 40, 7 as follows—

" Let not the oppressor with this dread, through anger swallow me up, for I am thine, O! Atri. Mitra art thou, the sender of true blessings; thou and king Varuṇa, be both my helpers.

Mitra and Varuṇa are, as pointed out above, the gods of a day of six months, and a night of six months respectively in the Arctic (Tripāth) circle. Accordingly when there is a solar eclipse in the lower zones in the six months from autumnal to vernal equinox, there can be no solar eclipse in the Arctic circle, for there is no sun there at that period. This seems to have been the reasoning of Atri in rescuing the sun or the moon from the clutches of the demons of eclipses. This seems to be the meaning of the Rigvedic verse quoted above.

In A.V. 13, 3, the eclipse mentioned is Patanga. In line 9 a Krishṇa eclipse in the house of Rita is also mentioned. He is also said to have the seven suns placed in him; He is called Varuṇa in the evening, and Mitra in the morning; he is also stated to have his wings spread out with thousand days each. He is Indra when he burns in the midst of the earth and the sky. He is Atri in water. He is also Paṭara when high up in the sky. He is also Rohita. That these are all names of eclipses has already been pointed out. His Ekapāth is the Pole. His Dvipāth or Tripāth, as it is also called, is the Arctic circle. His Chatushpāths are the seven circles below the Arctic circle. We are told that Krishṇa eclipse is followed by an Arjuna eclipse, being the son of the night period from autumnal to vernal equinox.

Leaving out a few historical hymns like VII, 18 and 19, most of the hymns of the Rigveda are prayers made on the occasions of eclipses to appease the wrath of gods for the sin of man which is believed to be the cause of eclipses. Hymns addressed to Agni, Indra and Viśvedevas contain references to eclipses which are generally known as Vasus, Rudras and Ādityas. The Vasus are either vernal or autumnal; the Rudras pertain only to the rainy season with its bellowing bulls and milking cows set free by Indra; and the Adityas to the winter season. Hymns addressed to Indra are accordingly descriptive of partial or complete solar eclipses of the type of the Krishna eclipse which, as already noticed, is described in the Tait. Araṇyaka. Distinct are the references in some hymns and vague in others. The following are a few hymns in which clear references to eclipses are made :

R.V. I. 10, 2 " The **ram** (the sun) hastens with his **troops.**

R.V. I, 13, 4 **Manu** appointed Agni as priest.

R.V. I, 14, 11 Ordained by **Manu** as priest.

R.V. I, 23, 14 Pushan, the bright, has found the king (Soma), concealed and **hidden in cave.**

R.V. I, 24, 10 Whither by day depart the constellations that shine at night; (they are) set high in heaven above us (now by day).

R.V. I, 33, 10 The foe of Indra sank during **darkness.**

R.V. I 36, 10 The gods for **Manu's** sake ordained.

R.V. I. 42, 2 Drive, Pushan, from our road **the wolf.**

R.V. IV. 28, 3 The archer, Krisānu, aimed and loosed the string to **strike** him (the moon).

R. V. I. 43, 8 Let not those who **trouble** Soma hinder us.

R.V. I. 45, 1 Worship the Vasus, Agni, here, the Rudras, the Adityas, all, who **spring from Manu.**

R.V. I. 51, 1 Make glad with songs **that ram** whom many men invoke.

R.V. I. 52, 1 I glorify **the ram** who **finds the light of heaven.**

R.V. I. 62, 5 (Sarama) praised by Angirasas .... with the sun **dispelled darkness.**

R.V. I. 74, 3 Even he (Agni) who slayeth **Vritra.**

R.V. I. 78, 4 Thee (Agni), the best of **Vritra's layers.**

R.V. I. 130, 9 Waxed strong in might at dawn he (Indra) **tore the sun's wheel off dark red.**

R.V. II. 27, 14 O! Indra, let not during **darkness seize us.**

R.V. III. 32, 17 To thee (Indra) proceed the **dark.**

R.V. III. 39, 7 He (Indra) took the light discerning it **from darkness.**

R.V. IV. 19, 14 Him (Indra) the black undulating cloud bedeweth in this mid-air's depth, at the **base of darkness.**

R.V. IV. 28, 3 The archer, Krisānu, aimed and loosed the string to **strike** him (the moon).

R.V. IV. 29, 2 Indra swiftly with might **pressed down the wheel of Surya.**

R.V. IV. 30, 4 Thou (Indra) **stolest away the sun's car-wheel.**

R.V. V. 1, 8 That strength the Bull (Agni) with **thousand horns possesses.**

R.V. V. 29, 10 One car-wheel of the sun thou rolledst forward and one thou **settest free** to move for Kutsa.

R.V. V. 33, 5 Thou (Indra), laidest him **(Vritra)** in the pit in **darkness.**

R.V. V. 40, 5 O! Sūrya, when the Asura's descendant, Svarbhānu, pierced thee through and **through with darkness.**

R.V. V. 40, 9 The Atri found the sun again, him whom Svarbhānu of the brood of Asuras had pierced with gloom. This none beside had power to do.

R.V. V. 45, 5 This day come to us; may our thoughts be holy, far from us let us cast away **misfortune.**

R.V. V. 45, 6 Where with Manu conquered Visisipra and where with the wandering merchant gained heaven's water.

RV. V. 45, 7 Sarama went aright and found the cattle.

R.V. V. 45, 10 He (the sun) hath yoked his fair-backed **tawny** horses.

R. V. VII 20, 2 Waxing in greatness Indra slayeth **Vritra.**

R.V. VIII. 85, 14 I saw the drop (Drapsa) in the far distance.

R.V. VIII. 87, 2 Thou, Indra art the conqueror; thou gavest **splendour to the sun.**

R.V. IX. 60, 1-2 Sing forth and laud with sacred song most active Pavamāna, laud Indu who sees **with thousand eyes.**

Thee who hast **thousand eyes to see,** bearer of thousand burthens,[1] they have filtered[2] through the fleecy cloth.

R.V. X. 54, 2 Men called thy (Indra's) battles (eclipse battles) illusion; no foe has thou to-day.

R.V. X. 54, 3 Who are the Rishis, (The seven sages=seven months) then, who comprehended before our time the bounds of all thy greatness?

R.V. X. 75, 5 He (Indra) cast away the gloomy mists, the darkness.

I think that these examples of eclipses referred to in the Atharvaveda and the Rigveda are enough to prove that the Vedic poets had a cycle of eclipses and that

---

1 Cycle of 1,000 days.

2 Filtering means clearance of the eclipse.

those eclipses were the measures of their seven Soma sacrifices.

It may however, be questioned why no commentator either on the Vedas or the Brāhmaṇas or even on any Tāntric work has a word to say connecting the numbers 1000, 72, 33, 14, 7 and the like with any cycle of eclipses or pointing out that the Ashṭāchakra or Śrīchakra is a graphic representation of eclipses. So for as the Vedic texts themselves are concerned, there can be no doubt that such Vedic expressions as 'Three sheep', 'five sheep', 'cow, calf and bulls', and 'he-goat and she-goat', all mean eclipses, as already pointed out. Likewise the word graha itself is a general name to eclipses, Soma-graha denoting lunar eclipses and Aditya-graha solar eclipses, the earthly soma-juice being an imitation of the heavenly soma-juice drunk by gods and departed souls on the occasion of new moons or by demons of eclipses on the days of eclipses. The Vedic poets feared that on occasions of eclipses their fathers in heaven would be starved to death on account of the moon which is believed to be their food on the days of new moons being carried off and devoured by "the seven hunters, or seven wild beasts, or seven Chakravāka or moon-light-drinking birds, or seven swans living in the Mānasa lake. Hence they seem to have thought it necessary to perform a Śrāddha rite and offer rice-balls to the fathers on the days of eclipses. These verses are even now recited during the performance of Śrāddha rites together with the Vashaṭkāra formula of 17 syllables.

There were also other learned poets like the author of A.V. 10, 8 who strongly disputed the above view on the cause of eclipses. Thus there seems to have been two Vedic schools, one blindly following the ancient view and practising the symbolical sacrificial performances and a second school of learned men who discontinued the sacrificial acts as useless.

It is also probable that the close connection of the Śrīchakra with Phalic worship compelled the line of teachers from the time of the Vedas downwards to strictly maintain the secrecy of the cult and captivate the mind of their students with mysticism of figures and words. The ancient teachers seem to have feared that if the perceptible scientific or natural phenomena of eclipses on which the whole of religious, philosophic, musical, medical, ethical, metaphysical and mythological literature is based were laid bare, their culture would fall to the ground and people would become atheistic or would pursue the worst practices of the Vāmāchāra sect of the Tāntric cult. Whatever might be the reason for the loss of tradition underlying the cult of eclipses which forms the main theme of the Vedas, this much is certain that the Vedas are the attempted explanations of the eclipses and the spells and charms divised to avert the calamities that would happen to gods, fathers and men from Vritra, Śambara and other demons of eclipses. Now I proceed to point out briefly the origin of the Epic and Purāṇic stories from the phenomena of eclipses and also the rise of the Vedic philosophy from the cult of sun and moon worship.

The union of the sun and the moon with or without eclipses seems to have given rise to certain speculations on the nature of auspicious and productive unions. The union of the planets on the occasions of eclipses seems to have been regarded as an open union and ordinary lunations as secret unions with no fear of double seed and calamities. The author of the Tait. Ar. explains the secret of good union under five heads. Among them those for speech and learning are very interesting. In the case of speaking the lower lip is the mother and the upper one is the father. If the union of these two is well done in secret harmony, the tongue being the instrument, then good speech will be the offspring, In the case of learning the teacher must always be above the pupil and the pupil must not come in a line with the teacher. A good union of these two will result in good learning to the pupil, revision being the instrument.

The forms of solar and lunar eclipses seem to have given rise to the twenty-four Mudras of witchcraft. Figures like those of the sun (Gāyatrī, Sāvitrī and Sarasvatī) in the several eclipses of the cycle are formed with two hands and fingers, and the Gāyatrī is recited to exorcise evil spirits by all the twice born people in their morning, mid-day and evening prayers. The names of the figures are—

(1) A disc-like figure;

(2) The same with fingers not touching each other;

(3) The same with fingers draws apart from each other;

(4) A figure with two, three, four, five, six heads corresponding to the six solar eclipses in a line in the cycle;

Figures like an Anjali, like a cart, like a noose, like two things kissing each other, like a fist, like a fish, a tortoise, a boar, a lion, a cow's udder with two or four nipples, like a circle, a lotus and like a Linga. I have already pointed out how the gnoman's shadow in various months forms figures like those of fish 'and other animals. This explains the traditional story that the various incrnations of Vishṇu were predicted by the sages and that they would take place in each Kalpa in their own time to ease the earth from its burden. The name, Ayodhyā, given to the drawing of the cycle of eclipses, or Śrīchakra, as it is called, points to the source to which the story of the Rāmāyaṇa can be traced. The Krauncha bird is connected with 'Ma' (Madhyama sound) of the gamut and eclipses on the fourth circle among the seven circles. The first or the basic verse of the Rāmāyaṇa to which Vālmīki, its author, has given expression speaks of the death of the male of a Krauncha pair at the hand of a hunter (nishāda–Rudra) and also of Rāvaṇa of the Krauncha family. It is given two meanings: a curse and a blessing.

मा निषाद् प्रतिष्ठां त्वमगमश्शाश्वतीस्समाः ।
यत्क्रौञ्चमिथुनादेकमवधीः काममोहितम् ॥

(1) O! Hunter, may ye not live long years, for thou hast killed the love-blind male of the Krauncha pair in union.

(2) O! Consort of the goddess of wealth, may ye live long, for thou hast killed the love-blind male of the Krauncha pair in union.

The male of the Crauncha pair is Rāvaṇa because he is said to belong to the Krauncha family. The seven Sāla trees are no other than the seven intercalary months of the cycle. Ten solar eclipses are probably personified as a ten-headed being carrying away the sun maiden. His heads like eclipses reappear in each cycle.

Likewise the story of the Pāṇdavas can be traced to the same source. In a cycle of 18 years there are 108 seasons, of which 100 are real and the remaining eight or seven have become sages. Now just like an ordinary year the intercalary year also must contain 12 months. Accordingly if the seven or eight months are on this side of the circle the other five months must necessarily be on the other side of circle. The other side is night and wilderness. There is war between two sets, hundred on one side and five on the other. Unless the 100 are drowned, the five cannot occupy the territory on the eastern side. The hundred are the Kauravas, and the five are the Pāṇḍavas.

Likewise the story of Dussāsana's attempt to render Draupadi naked by pulling away her garment seems to owe its origin to the phenomena of clearance of a solar eclipse. Draupadī is said to be the daughter of Drupada. The word 'Drupada' means the foot of a tree. The Atharvaveda (IX, 9 and Rig. I. 164) speaks of a tree on which the sun and the moon have their abode as birds. One of them, the moon, is said to live on sweet Pippala fruits and the second, the sun, is spoken of as keeping on without tasting anything. The sun and the moon are also discribed in A.V. VII 8 as two young children at play, one in the East and the other in the West and sailing on the ocean. One of them is said to be a spectator of the world and the other to be passing through a series of births causing seasons. The moon is also described in a separate hymn as a flag coming in advance of the dawns and becoming new and new after every one of his many rebirths. He is said to be distributing his own body as shares of food due to the gods and departed souls. The same tree that is said in A.V. IX, 9 to have afforded shelter to the sun and the moon is described in the Tait. Aranyaka I, 11 as having its root in the sky and its branches on the world down below. The Vanaspati of the Āpri hymns seems to be the same tree. It is at the foot of this tree that the sun and the moon come under eclipses which are spoken of as the magical influence of Asuras. As already pointed out, the Vedic poets regarded the arrival of eclipses in three colours. An eclipse of a black colour is called Krishṇa, that of a red colour Rohita, and that of white colour Paṭara meaning clothed as it were with a white garment. Accordingly the pulling or falling away of white garment means the clearance of an eclipse of the paṭara kind. As Draupadi may be taken to mean Sāvitrī in Paṭara eclipse, can there be any doubt that the story of Dussāsana's pulling away Draupadi's garment is based on the doubtful clearance of the Paṭara eclipse of the sun? Likewise, the story of a sect of Brāhmans being under the fear of loss of Brāhmanhood due to the curse uttered by a learned woman called Nāchiyar (Lakshmi) for their causing her garment to slip down from her loins and of her withdrawing the curse on the condition of their going away from the locality of the woman within three days, can be traced to the same source. Here the Brāhmans are the moons. The learned Nāchiyar is the Sāvitri in Paṭara eclipse. The slipping down of her garment means the slow clearance of the Paṭara eclipse. Of course, no successive moons will remain near the locality of new moons for more than three days.

As regards the growth of the Vedāntic Philosophy of Jīvātma, the chain of births and deaths, Māya and

Moksha, I shall content myself with pointing out that the moon is, Jīva, as stated in Tait. Ar. I, the sun the Paramātma or Prāna, teacher and the goal, as stated in the same Āraṇyaka; the births and deaths of jivas, the off-spring of the moon on earth, are as many as those of the moon; eclipses are Māyas which are immemorial in origin and which, though ending, are ever recurring. Moksha, liberation, is within the reach of those who know it.

The most interesting point in the description of the Tait. Āraṇyaka of the Ashṭāchakra with nine doors is its resemblance to Pythagorean theory of transmigration of souls, and of heavenly musical circles and triangles.

It is to be noted that we have not the same cycle of eclipses that the Vedic poets had before them. The features of the Vedic cycle are (1) the occurrence of three to five solar eclipses (Tryavi or Panchāvi), in a year, (2) the occurrence of solar and lunar eclipses in rapid succession (Rudra and Saramā), (3) the frequency of complete or almost complete solar eclipses (Śambara and his Forts), and (4) a greater number of visible eclipses in their cycle than in our own. None of our cycles possesses such characteristics in entirety. The Grahaṇadarpaṇa prepared by H. H. the late Mummaḍi Krishṇarāja Voḍeyar Bahadūr, king of Mysore, in Saka 1765 contains a table of 82 eclipses, 22 solar and 60 lunar with 22 intercalary months for the cycle or the period of 60 years from A.D. 1842 to 1901, all visible in Mysore. Swāmikannu Pillay's Table II of new moons and eclipses gives a list of about 142 eclipses, both solar and lunar, for the same period visible in India (Vide Chart of Eclipses). The same author has also given another Table (IV-L) of eclipses in his Indian Ephemeris. According to this table the maximum number of possible eclipses visible in India and outside for the same period is about 302. Out of these about 124 are printed in Italics indicating their invisibility in India and the visibility of the rest (178) in India for the same period.

As the preparation of an accurate table of eclipses answering to the series of our own time or of the series of the Vedic period (B.C. 3101–B.C. 500 and 40° N. Lat.[1]) is beset with such difficulties, I have left it for expert astronomers and contented myself with pointing out the features of the Vedic cycle of eclipses.

To sum up:—The cycle of eclipses was very widely known to all ancient nations. The Chaldeans knew it. The Biblical people were very well acquainted with it. The Romans had a fair knowledge of the same. The Vedic Aryans were no exception to it. According to the Hand-book of Astronomy by Chambers (Vol. II, PP. 464–467) references made in the Daniel and other parts of the Old Testament to 70 years and 7 weeks, 1200 years and the Messianic cycle of 33 years are all indisputably connected with the cycle of eclipses. As pointed out above, more numerous and distinct are the references made in the Vedic texts not only to the cycle of 1,000 days and its multiples indicating the interval of recurrence of Krishṇa and Indra, but also to the cycles of 17, 18, 19, 20, 33 and 57 luni-solar years with intercalary months varying from 6 to 21 suns, cows or 'Prathamajas' than the references found in the Daniel and other books. Accordingly there is no reason why

1 The number of complete solar eclipses (Sambara's forts) counted by the Vedic poets is ninety nine. Are $99 \times 18 = 1782$ years?

the Vedic 1,000 and other numbers should not be taken to signify the cycle of eclipses, when such interpretation is, as already shown, not only consistent with and appropriate to, the context, but also satisfactorily clears the obscurities which defy explanation in any other way. Besides this there is also another positive evidence in support of this interpretation. The Vedas themselves couple these numbers with eclipses which in the texts are called sheep, goat, cow, calf or bull. The reason for giving such names to eclipses is the belief that the sacrifice of such animals on the occasion of eclipses will not only appease the wrath of gods causing such eclipses, but also be either sufficient atonement for the sin of man which is believed to be the main cause of eclipses or sufficient ransom or offering to Kravyādas or Yakshas who are believed to cause them.

The list of visible eclipses of a cycle given in the Tait. Āraṇyaka as twice the number of 8 Vasus, 11 Rudras and twelve Visvedevas or Adityas (62 in all) along with a diagram called Ashṭāchakra which is identical with the Tāntric Śrichakra is an additional evidence in support of my contention.

For these reasons I trust that Drapsa, the Vedic cycle of eclipses will prove acceptable to Vedic scholars as a genuine and reliable historical fact.

Diagrams.

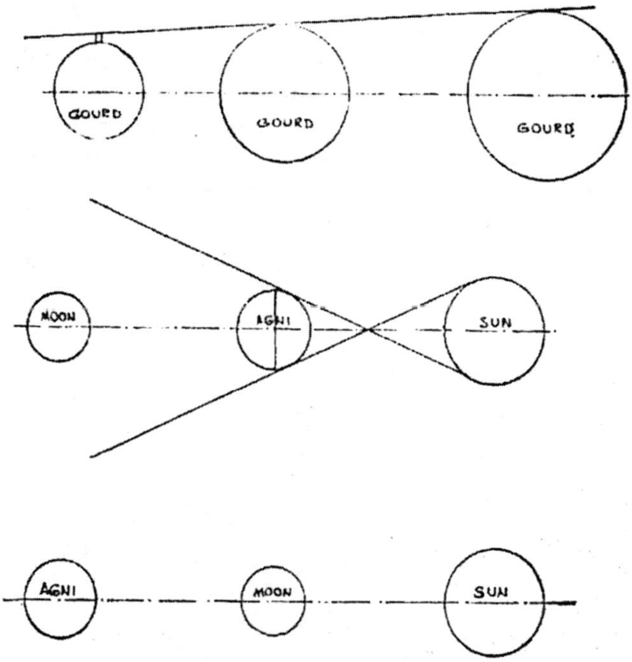

Diagram I

EXPLANATION

I. Of the three figures, those at the top and bottom are lute-sticks with gourds on a line.

That in middle shows eclipse of the moon,

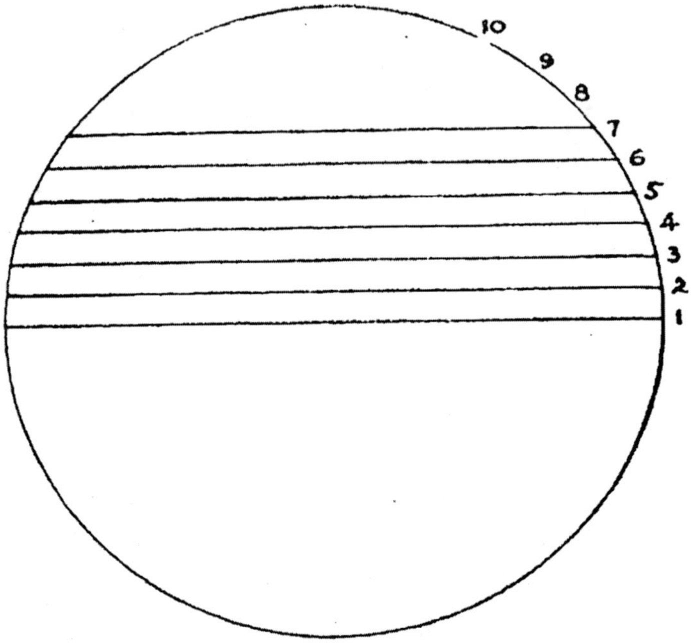

Diagram II

EXPLANATION

II. Celestial Sphere with seven monthly circles, one Tripāth circle, and one Apāth circle at the top.

1. Sa, Peacock, Bhūh, Jambūdvīpa, Lavaṇa-Samudra.
2. Ri, Kine, Bhuvah, Plakshadvīpa, Ikshu-Samudra.
3. Ga, Goats and Sheep, Suvah, Kuśa-dvipa, Surā-Samudra.
4. Ma, Krauncha bird, Mahah, Kraunchadvīpa, Sarpis-Samudra.
5. Pa, Kokila bird, Janah, Śākadvīpa, Dadhi-Samudra.
6. Dha, Horse, Tapah, Śālmalidvīpa, Kshīra Samudra.
7. Ni, Elephant, Satya, Pushkaradvīpa, Sudhā-Samundra.

8-9. Tripāth, a day of six months and night ofsix months.

10. Apāth, always shining with Kaśyapa's light.

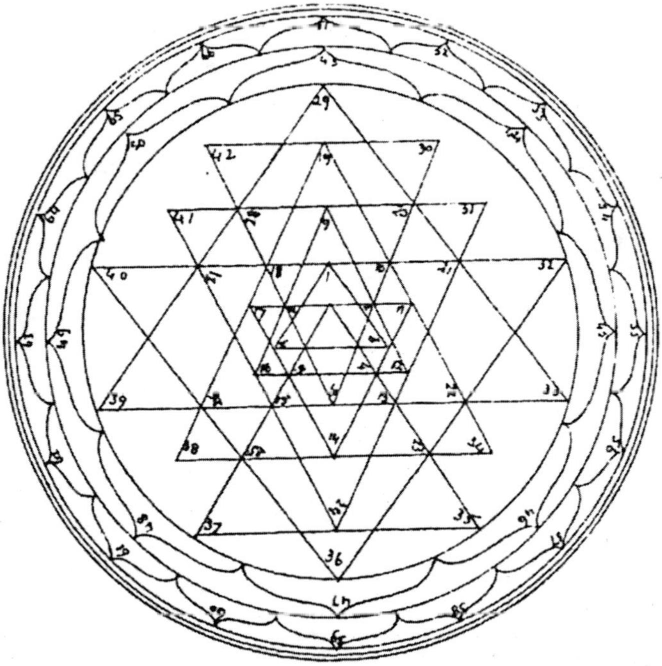

Diagram III

EXPLANATION

III. Śrīchakra of 72 Koṇas or angles—

Vasu-koṇas	......	1— 8
Two Daśakoṇas	......	9—24
Chaturdaśa Koṇas	......	29—42
Ashṭa Koṇas	......	43—50
Shoḍaśa Koṇas	......	51—66
Three Vrittas	......	67—69
Bhūpuratraya	......	70—72

Note—The angles in the original are marked with letters of the alphabet.

INDEX

The numbers refer to the pages.

A

D

E

F

G

H

I

P

R

V

W

Y

PRINTED AT SREE PANCHACHARYA ELECTRIC PRESS, MYSORE.

SOCIALISM

SOCIALISM
A Study

MAHAVIRPRASAD R. MORARKA

Socialism : A Study

Translated by: Neelabh

ISBN: 978-81-212-1184-0

Published in 2013 in India by
Gyan Publishing House
23, Main Ansari Road, Darya Ganj
New Delhi-110002
Phones : 23282060, 23261060
Fax : 011-23285914
E-mail: books@gyanbooks.com
Website: www.gyanbooks.com

Laser Type Setting by: Rudra Computer Graphics, Delhi
Printed at : Chawla Offset Printers, Delhi

CONTENTS

FOREWORD

It gives me immense pleasure to see that the wealthy people of the country are showing interest in and contemplating about the basic problems of society. Usually, they limit their interest to the matter of trade and industry. Under such circumstance, the present endeavour of Shri Mahavirprasad Morarka is worthy of appreciation. It would have been better if the structure of the book were a little more organised and the digressions were lesser than they are at present. On certain topics the ideas of the author are rather extraordinary, but he has expressed them with clarity and courage.

Socialism is a very important subject of the contemporary world and its literature is rich and vast. Wherever the principles of socialism are universally accepted, there are wide differences in the ways and means of their attainment. Some industrially backward countries have used undemocratic methods to establish socialism. Consequently, the success has been partial. India is the first underdeveloped country where socialism is being undertaken through democratic means. For its success, it is necessary to have comprehensive discussion on socialism. From this point of view also, this book should be welcomed.

The experiences of building socialism have demonstrated the fact that the establishment of true socialism is not possible just by changing laws and institutions. Until the moral and cultural tenets and values of socialism do not seep into the minds of the people, socialist culture cannot be built. This work cannot be accomplished through politics or state power. For this, a comprehensive public education would be required for. Unfortunately, the socialists across the world have paid little attention towards this obligation. Their endeavours have been limited only to the struggle for power. In India under the influence of Mahatma Gandhi's ideals, this is being attempted in the form of the *Sarvodaya* Movement. It is a new step towards socialism.

Jayaprakash Narayan
Bombay
July 26, 1962

PREFACE

This is my first foray into book writing. Politics is not my subject. The inspiration was sudden; a friend asked me one day, 'What is socialism? In India all parties wish to establish the socialist system, so what is going to happen in this socialism?' I had read many books on socialism in English, so, I suggested him one or two books, but he wanted a book in Hindi. I could not find any satisfactory, up-to-date book in Hindi on this subject. So, it occurred to me that I should write on it. I had liked a book by George Bernard Shaw titled *The Intelligent Women's Guide to Socialism*. It occurred to me that if this book were to be translated in Hindi, it might serve the purpose. I started with this but somehow it did not seem appropriate as India's problem is unique. Finally, on the suggestions of my friends, I decided to write a book independently. The impressions I received from George Bernard Shaw's book are present in my book to a large extent. Karl Marx's book *Das Kapital* also greatly influenced me and so its impression would also be visible. But, mostly I have expressed my own views. If the ideas have not been expressed clearly at some places, the reason is that I am not a scholar of this subject.

Capitalism is a very perverted 'ism'. Its place in the spiritual and metaphysical analyses of mankind is very low. Communism, too, having become the capitalism of the working class, is equally perverse. Fascism is completely despicable. If there is any system suited for the welfare and prosperity and happiness of mankind, it is socialism and I have presented the form and aspect of this system before you as I have read and understood it. If money or income is distributed equally among the people of the nation, that system can be called a socialist system. Only politicians can tell you whether it is difficult or easy to establish such a system, but if this can be done, it will be wonderful. This is a matter of feeling and imagination.

The language of the book has become a trifle uneven. The reason behind this is that I am no connoisseur of literature and the friend who was given the book to revise, did not deem it necessary to change a single word. Errors of punctuation are aplenty. At many places, because the expression was not clear enough, some vagueness or confusion has also crept in Hopefully, the sagacious readers will rectify these errors.

Actually, my objective is only this that all should think and mull over this question independently and thereafter form their opinion. They should not be swept away by the confusing ideas of any other person, like it is happening these days.

The people of India should exercise their voting rights after due thought and independent reasoning; they should not be misled by false promises and slogans. The aim of this book is to help people of India become intelligent voters.

The book had been written about a year and half back, but due to the expectation that the friends from literary circle would revise it and also due to the slow speed of the printers, the publication encountered many delays.

The foremost epochal figure of socialism is present in India. He is Jayaprakash Narayan. I had requested him to write an introduction to the book. There was some delay in writing the introduction because of the paucity of time, but I could not forego the temptation that the thoughts of the honourable Shri Jayaprakash Narayan along with mine in my very first book.

I am also indebted to my friend Shri Ramrichhpal Patodiya, who inspired and encouraged me to write this book.

Bombay
July 31, 1962

Mahavirprasad R. Morarka

INTRODUCTION

When this book was published for the first time in 1962, Shri Jayaprakash Narayan wrote in the introduction that this book, written on such an important subject as socialism by a business tycoon, should be welcomed. He expressed some reservations about the slightly unorganised structure and a few digressions in the book. In order to remove these shortcomings, the manuscript of the book was given to me by my friend Kamal Morarka (the son of the author) before the present edition was printed. I have tried to do this job to the best of my ability, yet I have taken care that the basic content of the book does not suffer even a scratch.

While editing the book, I felt how nice it would be if this book was prescribed in schools and colleges as a textbook of economics. If students have to take the help of a foreign language to learn the knowledge of trade and commerce in which our country and our languages were very rich, then it would be as ridiculous as 'carrying coals to Newcastle'. But, unfortunately, this is the present trend, wherein the efforts should be to teach not only economics, but all disciplines of knowledge and the sciences in our own languages. By doing this not only the needless waste of the time and energy of the students could be avoided, but at least one-third of the time would be saved in the curricula related to medicine and engineering. But this realisation has not yet dawned on the intelligentsia and educationists of this country. They are all caught up in the delusion of the Knowledge Commission of the

Government of India. Although we are unable to contribute any original thought, our educational centres have become the junkyards of the world and the scholars coming out of these institutions are but educated indentured labourers for the foreign countries.

At this juncture when the book is going to press, this feeling has become all the more acute. The reason for this is that the capitalist system is in such crises that it is difficult for us to even comprehend it. An economic depression which began in some large financial companies of America is grabbing all the economies of the world in its clutches. This epidemic has become for us like some unfathomable celestial calamity, and to ward it off we have no other recourse except to resort to spell-like conjurings. Our government or the institution (Reserve Bank of India) which regulates the financial system is trying to stem off this depression by blindly imitating the central banks of America, England, France, etc. We can see now that the analysis of capitalism, carried out by scholars like Adam Smith, Marx, Keynes, etc., is proving to be inadequate in reference to the present crisis and the essay 'Economics after Marx' by the eminent socialist thinker Dr. Ram Manohar Lohia has become more relevant.

The crisis of capitalism can be resolved only by following the path of socialism. This is the subject of this book, and it is interesting to note that it is an advocacy of socialism by a capitalist. Inspired by George Bernard Shaw's book *The Intelligent Women's Guide to Socialism*, this book can help us to understand the present economic crisis of the world. Some chapters of this book especially, 'The Magical Illusion of Extra Money', 'Various Forms of Cost Capital', 'The Capital Market', 'Speculation', 'Banking', 'Money', 'Currency Control', etc., are very relevant right now. In the entire book one will not come across the second- or third-hand bookish knowledge which is often found in the works of university teachers; rather, lively and energising knowledge infused with insights from direct experience. In the way Gulabdas Broker had enriched Gujarati language with the knowledge of the mysterious vales and meadows of the economic world through his literature, this book by Mahavirprasad Morarka would enrich the deep knowledge of the valleys of the economic sphere.

The third path of socialism advocated by the author after moving away from both capitalism and communism becomes all the more important because the author himself belongs to the capitalist class and the book, to quite an extent, is free of class prejudice.

January, 2011

Mastram Kapoor
79-B, Pocket 3,
Mayur Vihar Phase I
Delhi - 110091

1

THE BACKGROUND OF SOCIALISM

Change is the order of the world Why only socialism?

An age has passed since India attained freedom. The economic policy of the government is based on socialist principles. The Congress, Socialist, Praja Samajwadi Party, Communist and even the newly formed Swatantra Party and Jan Sangh support a socialist economy. The leaders of each party claim that only their version of socialism is true socialism. The government in office is of the Congress Party. Hence, the people have already had or are having some experience of the type of socialist economy which has been put into practice by the Congress. Even within the Congress there is no unanimity about the interpretation of the word socialism. However, one thing is clear, that the people in general are unanimous that only the socialist order is useful and beneficial.

Hundreds of books have been published in the world on socialism. Many are available even in Hindi. But in my opinion you need not read any one of them. The tendency to form one's opinion on the basis of the opinions of others was not prevalent in ancient India. This fault has crept in among us since the past few centuries and we are readily impressed by what others believe. If a writer or speaker puts forth his arguments forcefully and in a good language, we hold them to be true. One should independently form one's opinion as to what is socialism. What should be the blueprint of socialism? What type of socialism will benefit us and the entire nation?

What is socialism? There are varying points of view on this. It is not necessary that the opinion of others is better or more truthful than yours.

How much money or wealth should one possess? How much money or wealth should one's neighbour or friend possess? What is your opinion on this?

This larger question is still unanswered. Ever since childhood we have been taught that being rich or poor is a matter of fate. It depends on the will of God. It is God who makes one person rich and another person poor. This belief is totally wrong. The order of becoming rich or poor is ever-changing. The will, strength and manoeuvring skills of a man make him rich from poor or turn a rich into poor. The wealth of rich men would have remained unsafe and insecure, had there been no police or fear of prison. By seeing rich and poor all over the world, we have accepted that it is a matter of fate or destiny to be rich or poor. But, in reality, this is not true. Everyday new rules and norms are formed and put into practice according to which the rich and wealthy, kings and monarchs of yesteryear can be describe to be comparatively poor today; and the ordinary, common folk and the paupers of the past are accepted as the rich and socially respected people of the present-day society.

New laws are enacted and the old ones are amended by the parliament on daily basis. Why is this so? Because we are not satisfied with the present order. We endeavour to improve it. We are not satisfied even with the amended laws, and therefore we amend them afresh. This trend continues and the rules and regulations change every now and then.

With each amendment, many people lose some part of their wealth while many other people get enriched.

Dissatisfaction with these amendments and changes results in further changes which affect the reserves and pockets of other people in our society. Laws are again changed and altered. The rules and regulations, which could not have been imagined 100 or 50 years ago, have to be followed willy-nilly by every Indian. For example, who could envision in your grandfather's or great grandfather's time that ordinary peasants, workers or harijans would attain the same rights as those of rajas and maharajas and wealthy merchants? Who could imagine that workers in a factory would have to be paid fixed wages as a rule and that you could not terminate the services of a worker even if there was no work to be done? In this manner centuries-old laws are being annulled, and because of this we can do all those things which could not even be dreamt about in the times of our fathers and forefathers, for instance remarriage of widows, inter-caste marriages and having meals together regardless of caste or creed. Laws keep changing. They continuously amended to suit present social condition. The courts interpret these laws and on the basis of these interpretations, changes are brought about again in the social order. Because of these unending amendments a time comes when not even the shade of the original laws remains.

This change is inevitable. During elections each political party gives innumerable assurances that they would bring about this or that change. But it is you who have to think as to which change would be more suitable and beneficial for you, the society and the nation. You have to form your own opinion without being carried away by advertisements and long-winded arguments and try to ascertain what sort of change would benefit our social order, especially in the present economic system. Do not at any moment think that whatever God has willed will happen. You have the right to bring about changes through laws, through the parliament. The answer lies in what you decide to do. Anyway, in India, there have been many thinkers in the past who propagated an economy according to their own views and principles. Possibly, those economic orders were proper and beneficial for those times, but they could never become everlasting and permanent and should not have become so. The *Arthashastra* of Kautilya, the economic order of Yagyavalkya, Manu's social order, the economic system of Buddha's *Shoonyavad*, the *Din-e-Ilahi* of Emperor Akbar, the *Sarvodaya* economic principle of Mahatma Gandhi are the interpretations and analyses of the economy of their times. You must not allow your thoughts to be directed by any of these. Form your opinion after having considered and analysed each one of them. The question as to how much money or wealth one should have is still unsettled. Probably, it will never be decided and will keep changing forever.

2

WHAT IS SOCIALISM?

Saving and distributing The pillars of socialism

The basic principle of socialism is to distribute the national income in a new way. You might not have noticed but the fact is that the national income is being distributed daily, every minute, every second and this distribution will go on and on. This distribution will continue day in and day out until there are more than one person in the country. The question before us is not whether this distribution should take place, it will be done come what may; but what we have to think is who should get how much and on what terms.

The Western thinker St. Paul once said 'He who does not work, neither shall he eat.' He was probably short-sighted. He forgot that there are even babies in the society; he forgot that children cannot work and yet hunger overwhelms them the most. If they are not fed in time, then probably none will survive in the world. Hence, this rule that 'He who does not work, neither shall he eat' is not feasible.

Many people think that the wealth of the world can be saved and preserved like their savings. This is a great misconception. A lot of wealth which sustains our life cannot be preserved for more than a week; it is lost and is subject to decay. A gold bangle lasts a life-time, but one cannot eat the bangle to live and survive. No matter how hard you try to preserve food, whether vegetarian or non-vegetarian—bread, cereals, vegetable, milk or curd—it cannot last for more than a week at the most. A week because with the help of modern appliances like the refrigerator or the thermos one can keep it fresh, otherwise within a day or two all types of food will become stale. Thereafter, whatever stale food we eat, might poison our stomachs, cause us to fall ill and endanger our lives. The clothes we wear also cannot last for a long time. If we are working folk, the clothes wear away very soon. And of course they get dirty everyday by our sweat and the dust all around us and have to be washed and kept clean; so, it is the same with clothes. As for shoes, even if you have them rubber-soled, they may last a year instead of 6 months, but they will certainly wear away. Every year there are fresh crops. We cannot survive on the crop grown last year. The crops of the coming year are not ready as yet. So, we have to live on the produce of the current year. We have to sow, reap and carry away only this year's crop and make bread from its flour; we have to have clothes stitched and cleaned from what we have at present; and have shoes made for the present. Otherwise, we shall die of hunger, cold or heat; we shall not survive having nothing to protect us with.

What is termed as 'savings' is in its true meaning merely a transaction in future. For instance, suppose I have baked 101 loaves of bread today. I shall be able to eat only one myself. The remaining hundred I cannot keep safe and sound because they are bound to decay and be destroyed. They will not be fit for consumption after a week. Therefore, I can only enter into a transaction with the person who needs a hundred loaves of bread. If I give my 100 loaves of bread to someone and in return suppose he provides me five fresh loaves of bread every year in future then this is a transaction. This is not 'savings'. It is a deal or a contract between two persons. One party which wishes to secure its arrangement for the future and the other party which wishes to spend more in the present. A deal was struck between these two parties. So, the result is that I cannot save anything till I am able to find someone who wishes to spend then and there. Therefore, this concept that the whole country can save is impossible. The truth is that some fortunate persons, who have more than their needs, can secure their future, but that is only possible when they are able to find such persons whose expenditure is more than their present income. The conclusion is that Ram must spend what Hari has saved otherwise the savings of Ram will be futile, they will perish.

The entire nation has to make its own bread and survive. If the nation stops working or becomes inactive, it will perish within 15 days, even though its members have in their coffers hundreds of thousands of rupees and landed property and buildings. If you meet a rich man who laments that such and such poor person did not save and if he regrets or sermonises, then you should pray to God to enlighten such people. They are living in ignorance and do not know that 'savings' are not possible. The nation must not become inactive. Do not imbibe his meaningless words and preachings and then mortify a poor person by relating all this to him.

And so, the daily cycle continues. Every man must bake bread everyday. The question is how much bread each person should get. Suppose, for instance, a poor person toils the whole day long and has four or five children. He needs eight loaves of bread everyday and has only two. As opposed to this, another person has 20 to 25 loaves extra everyday. He is single and a priest and need not even work, yet he has extra bread everyday. The simple answer is that the labourer who needs eight loaves of bread for his family should be provided bread by the other person. This seems the right way to go about it, but how can it be done? If the labourer takes bread from the priest by force or by robbing him, he is liable to be punished by law and the police catch and send him to jail. Why does this happen?

According to the established social order only two loaves are his by right, whereas the right of the priest encompasses 20 to 25 loaves. Such is the law and the government has instituted the police to enforce it. Therefore, it is necessary to change the law. Laws are changed by parliaments and the people who change laws are elected and sent to parliament by the people. In order to fulfil their needs people choose only those people who they expect will change laws for their benefit.

You must have read in the papers that such and such son of a maharaja or a wealthy merchant filed an appeal in court saying that he cannot make do with a hundred thousand a year and should be given two hundred thousand yearly at least, and so on. There are millions of people in the country who do not get even Rs. 500 every year. Keeping such adverse circumstances in mind, people raise their voices and parliament enacts laws. As a result, some money is taken away from the prince and the merchant's son by means of income tax, wealth tax and other taxes. This money is distributed among others in the form of free education, subsidised charges for providing health and medical care, compensations, free drinking water in villages and cities and such amenities. In this manner the distribution of wealth goes on continuously, but its progress is so slow and sluggish that most people do not realise whether they are receiving these benefits. Those who save are few in number and those for whom the distribution takes place are innumerable. It is for this reason that this process of change is not considered to be worth taking into account, because even after paying all taxes, the prince or the merchant's son is left with Rs. 25,000 yearly for expenses whereas those ordinary people even after receiving all kind of facilities do not get even Rs. 1,000 yearly. Therefore, every person is dissatisfied with the present system of distribution, although all wish for such a distribution. All desire that the distribution should be more equitable. There should not be such an imbalance. But how much should be distributed and shared is something which people hesitate to specify or are unable to do so.

Then the question which remains is how money should be distributed. Everyone desires distribution and distribution is imperative. Yet which is the most proper way by which wealth may be satisfactory distributed?

3

LEISURE AND LABOUR

Synchronisation of leisure and labour
Labour is money

Indeed, work or labour is necessary before you can have money in the country to distribute. If money will not be earned by work and labour, what would we distribute and how. Grain will not grow without tilling the land. For this the peasant will necessarily have to work. Thereafter, the baker will have to work and only then we will obtain bread. Thousands of years ago our forefathers lived in jungles or in the mountains, but we are no longer living in that era. It is not possible for us to live like that. Without working tirelessly, we shall die of hunger. If somebody is not working and is sitting idle, then it is certain that some other person is working as much as the two of them would have worked, otherwise none will have anything to eat. Keeping in mind this natural axiom the Western thinker St. Paul had said that 'He who does not work, neither shall he eat.' Nature has assigned to us the responsibility to work. Just as the distribution of earned money is certain, similarly the distribution of work is also necessary.

It is not essential that only those who work should eat. A person can and does produce many times more than he needs, otherwise the children will go hungry and the old and indigent who are no longer capable of working will perish of hunger. We come across many such hard-working persons who had no other means except their hands, nurtured their families and children, looked after their aged parents and also paid hefty sum to the landlords as revenue or rent.

In the present age when electricity and machinery have been invented, man can decidedly produce a thousand times more than he produced a hundred years ago. In this way, with the development in mankind's thinking power and the help of inventions like machinery, electricity and steam, etc., man's labour has been reduced and he gets more time for leisure. This leisure, or time of rest, is a thing of comfort and convenience. It needs to be distributed in the same manner as money and work. Suppose by working 10 hours a day a man supports 10 persons for 1 day. Those 10 persons can share the work in a way by which each of them works for 10 hours daily and rest for the remaining 9 days or each of them can work for 1 hour daily and rest for 9 hours. Mankind can evolve such formula according to their will and desire. It is also possible that out of those 10 persons 3 persons work for 10 hours everyday. Whatsoever they shall produce will definitely suffice for 30 persons; and so, 7 persons will not have to work but also eat as much as 14 persons while keeping those 3 persons in regular employment. By another arrangement everyone works daily for the maximum number of hours. They work till they are fully educated or till they attain adulthood and then stop working and rest for their remaining life.

In this way work and leisure can be distributed in many ways. These systems of distribution of work and leisure are called by various names, viz., (1) slavery, (2) serfdom, (3) feudalism, (4) capitalism, (5) socialist order and (6) communist system.

The history of revolution is indicative of these facts. The struggle which man undertakes to change the conditions of his time or the struggle which he wages to attain the system which he feels beneficial and proper is called revolution.

Anyway, we are discussing right now how to distribute the income which accrues from labour? It is not difficult to distribute work and leisure. One cannot differentiate much between the labour and leisure of each person. But a marked difference can be brought about with the use of modern inventions such as electricity and machinery. In a rich man's day, you cannot increase the time beyond 24 hours, but you can easily add an extra Rs. 24 crores to his corpus.

4

THE INDIVIDUAL AND SOCIETY

The desire for wealth among humans
The duties of man towards society

Suppose the entire wealth of the nation is entrusted to you with an instruction to distribute it in an appropriate manner. How will you distribute it? Forget your family and close friends for a while because you are the distributor. You will have to distribute that wealth without fear or favour, treating everyone as equal. You will have to don the mantle of a judge who is not swayed by extraneous considerations. We all understand that it is very difficult to do so. None of us will be able to do this judiciously. The same problem will arise, we will be swayed by the opinions of others and the result would be faulty distribution. This problem is not a new one, it has been tried world over in the past, but none has succeeded in distributing the wealth appropriately.

Take a small example. In India there is a tradition of joint Hindu family, in which every member of the family contributes his income in the common family fund and each member of the family spends from this common fund according to his needs. In India centuries ago, the tradition of joint family was an ideal form of communism. But is this tradition still exist? Can each member of a joint Hindu family say that he is receiving money equitably to spend every year through a proper distribution from the common fund? It so happens that man's ambitions and desires inspire him to set aside some money for himself from the common fund. Other members of the family also resort to such tactics. As a result it can be said that when proper distribution of wealth within one family is not feasible, what can be said about the distribution of the entire wealth of such a large country.

Let us leave the family alone where distribution may be possible, but the bread baked in your house is not available to your neighbour, while the water you get is equally available to your neighbour. The road you walk on can be used by your neighbour or your enemy. The street lights put up by the municipal corporation benefit the entire populace in equal measure. Every person can avail the medical facilities provided in the big hospitals of the cities. Many such facilities are provided to you and all of us by the government or the municipal corporation. To provide these facilities, the corporation levies taxes such as house tax, water tax, etc. Your house is larger, you pay a little more tax. Our house is smaller, we pay less tax. There are many people who enjoy these facilities without paying a single penny. You cannot say that since you do not go out at night, therefore there is no need of light on the streets, and that you will not share the expenses. You cannot claim that since you are poor, you do not require police to petrol the street. You cannot even say that you have never been to the public hospital for medical treatment and neither will you ever go; and so, why should you share the expenses of the hospital? You never read books from the library and never send your children to municipal schools, so why should you share the expenses incurred on all this by the government or the municipal corporation? You may not be availing of all these facilities but all these amenities are for the welfare of society; they are equally useful to everyone, equally available to everyone. So, these and all such amenities can be called community services. You may or may not be a consumer of these facilities, yet you will have to invariably share the expense in a fixed ratio.

This ratio will be fixed in accordance with your income or the size of your house or its value. But even those persons who pay the smallest ratio of the expenses or do not pay anything at all can equally benefit from all these facilities. Whosoever you may be—young or old, rich or poor, sinner or saint, white or black, miser or generous, carpenter, tailor, cobbler, policeman, soldier, beggar, thief, robber—all have the right to enjoy all these facilities.

5

CURRENCY: THE MEDIUM OF DISTRIBUTION

The background of communism
Equal right on may amenities
Varying needs of all people

Having read about communism in the newspapers, you might have probably thought that it is a monstrous economic system imposed by the Russians by which the personal liberty of mankind is totally impinged upon. But, in truth, this is not so. Communism is in reality a venerable and easy path of fiscal distribution, which has to some extent existed in India in some form or another since millennia. The streets, the water supply, electricity, schools, bridges across rivers, etc., are symbols of communism in the present era. Suppose they are not considered as common for everyone by rights and instead, for instance, gates are put up on roads and bridges and every person who passes through them is charged a fixed amount, or torchbearers are deployed on the roads who collect a fixed charge from travellers using these torches, or wells are dug up for water in a similar fashion and money is charged per pitcher from people who take water from these wells, and so on, even then arrangements can be made and some sort of system can possibly run. Possibly, it is running even now in some villages. But civilised society has decided that all such amenities should be equally available to common folk and that every person must share the expenses in a fixed ratio. Undoubtedly, this distribution is the most respectable and easy method of communism. Every person who contributes in running these amenities probably does not even feel that he is doing so. The person who buys tobacco worth 50 paisa does not know that out of this amount almost 12 paisa has gone towards these common amenities and it has been taken from him. People contribute in some measure towards amenities like roads, bridges, police, schools, hospitals, water, electricity, etc. Even when they become aware of this, they are not perturbed.

But when they will come to know that many things like observatories or museums or zoos are being run by the taxes paid by them, they will definitely be dissatisfied, because 40 percent of people have no interest in astronomy or museums or zoology. Yet, knowledge is useful for the nation. The knowledge that is accumulated by fewer than 10 percent of people who are interested in these subjects and who study them becomes part of the heritage of the nation. Although all have equal right to use these amenities, only a few people use them. Still, their expenses are met by the pockets of all. The people also do not have any special objection to this. These things have been absorbed in daily life to such an extent that often we begin to think like children that such and such a bridge has been made as if by Almighty; it was always there and will always remain there. We only realise the real worth and facility of the bridge if it is not there and we have to swim to cross the river or arrange for a boat. The crust of the matter is that there are many facilities of life which we consider as our birthright. This is a burning example of communistic distribution or the civilised way of wealth distribution.

If this is so then why not share all other things in the same manner? Bread, clothing, milk, *sherbat*, fruit—all things be distributed according to the communistic system? If it were possible then this would be appropriate, but there are many difficulties in doing so. The interests and likings of man do not follow a set pattern, and so it is not surprising if there is a difference in opinion and taste. You wish to eat bread baked from wheat, whereas I am fond of rice, another person wants fruit and a fourth would have only milk. So how can things be evenly distributed among people? It is a difficult task. Suppose you wish to award a prize to every person in a village and that too, a prize of their choice. First you ask Govind. He says he wants a radio. Gopal says he wants a cycle. Old Kundan on being asked says that he has difficulty in hearing, he wants a hearing aid. Young Vinod desires a toy to play with and Laxmi wants a Benarasi sari. Now, tell us how is equal distribution possible?

If you give each of them a radio, a cycle, a hearing aid, a toy and a sari, the result would be that they will possess things that they do not need. They cannot give away these things to others, and so all these things would go waste. Suppose the cost of each of the above things is Rs. 200, then the simplest way is to give each of the above persons Rs. 200 and they can buy whichever thing they desire or need.

So we come back to the same question as to how much money should each person get? In communistic distribution, it is impossible to include any other thing except public services. So, the question what should bc the basis of distribution and how can we distribute money satisfactorily remains unanswered.

Money is very useful. With it you can get things you need. But not those things that other people think are necessary for you. During wedding friends and relatives give presents. It is found that three or four tea sets, five wrist watches are given at one go. However, a coffee-percolator that was needed was not gifted by anyone. A fountain pen was required but no one gave it. A shark skin suit was wished for, but was not received and the things required in single number were given in multiple numbers. If money had been given instead of gifts and in ratio of their value, the person receiving the gifts could buy things which he required. There is an old adage that money is the source of all trouble, but in reality the money is a very easy and simple medium through which all problems are solved. Of course the issue of distribution of money is definitely serious and difficult issue and requires deep thought.

6

VARIOUS SYSTEMS OF WEALTH DISTRIBUTION

Distribution according to worth
Satisfaction on what is received through fate
Equal distribution

One scheme which is put forth and which is acceptable and liked by working people and labourers is that a person should be allowed to keep whatever money he earns through his labour. Many others say that a person should get in accordance with his worth so that all idle, useless, weak and slow persons do not receive anything and they may perish. Able, strong and hard-working people should get everything and only they should survive. Some are of the ancient fundamental opinion that those who have the power should take what they can and those who can carefully possess and keep what they have should do so. However, the persons who are of this opinion do not express these thoughts openly in modern times. Some people opine that ordinary humans should be allowed to keep as much as they need to survive so that the circumstances in which God has given them existence are not tampered with, while the rest should be taken away by the opulent society. Those who believe in this notion also cannot speak out openly today, although they were quite vocal 50 years ago. Some people say we should be divided in classes and categories, and there should be equal distribution among all persons of each class in accordance with the ratio assigned to that particular class or category. The conclusion is that labourers should get Rs. 50 per month while the skilled artisans, professors, religious preachers, judges and thinkers should get Rs. 200, 1,000, 5,000 and 10,000 per month, respectively. Their wives and children may obtain from them as much as they can. Some even say that things should be allowed to go on as before. But the socialist says none of these schemes is satisfactory. The only satisfactory scheme is to distribute equally among all whether it is a woman or a man, an old person or a young man, whatever work they may be engaged in and whatsoever family they may belong to.

I crave your pardon if this last scheme or any other scheme mentioned above does not appeal you. I would then request you to suggest what sort of system may be put into practice. I am only placing before you the schemes which have been put into practice in the world and have been tried out. You are not bound to accept any of them. You are totally free to suggest an even better and more appropriate alternative, but please do not shy away by saying why get into this unnecessary bother. Because this is not an unnecessary bother. It is a matter of great significance. If you will not suggest a way, others will do so and it is quite possible that they may suggest what is of interest to them and not to you. They may secure their own portion, while you may be deprived of yours through the scheme of distribution that they suggest. So, it is very necessary that you decide as soon as possible as to which form is the best form of social distribution. The world is rapidly changing. The economic state of India, too, is undergoing a change at a very fast pace. If we do not solve these problems with discretion and proper thought we may have to repent in future. Hence, the schemes which have been tried out in the past are before you. In the following chapters we are going to present blueprints of all the seven schemes enumerated above and you should, after reading them, decide for yourself as to what can be said in favour of and against each of them.

7

VALUE ACCORDING TO DEMAND

The first scheme is that every person should get as much as he earns or produces by his own labour. This scheme seems very proper and attractive when one looks at it. But when it is put into practice many difficulties arise. First, it is almost impossible to determine how much has been produced by whom. Second, work or labour is entailed not only in producing but also in transforming or taking care of each other.

When a peasant tills and sows his land, he and other workers have to clear the field, till it, sow seeds and look after the crop. When the crop is ready, they reap it and bring the produce home. And by the grace of nature the grain has been produced. Now how can we specify that who has produced how much of the final product? Similarly, suppose a machine produces 10,000 needles a day, then how will you calculate that how many needles were produced by the man who was operating the machine; what was the share of the person who invented it; and how many needles were made by the engineer who designed and set up the machine? Besides, how will you determine the share of the combined labour of all the factory workers who contributed to make the needles? In this manner it is almost impossible to tell how much each person earned or produced.

Take, for example, the production of a table or a chair. Before a table or a chair is made, there is the share of the person who planted the jungle and the person who cut the tree, sawed the log and made sleepers of it. Later, there is the share of the carpenter who shapes and cleans and makes the structure of the table or chair; and there is also the share of those people who make the nails and screws which were used in making the table or chair. Then there is the share of the people who polished and painted and the shares of all the people who make the various parts made of iron and brass which were used in the making of these articles. There is the share of the people who produced the cloth and upholstery which were used in it. Now, just try to ascertain how many peoples' labour has gone into the making of the chair or table and how much has been produced by it. It is not possible to accurately ascertain the share of each one of them. Many people opined that to divide it among the workers concerned, whosoever has spent time in making the chair or table should be made a shareholder in ratio of the time spent on each function or process.

This, too, is very problematic issue. It is not at all possible to assign a value to the time spent on it. Ordinary workers who are coolies are ready to work for Rs. 1 per day. The skilled workers get Rs. 5, 7 or 10 a day. There are many workers who get Rs. 50 or 100 per day. Many of them get even more. Out of the persons needed to make the table or chair the woodcutter in the jungle or the upholsterer may be available probably for Rs. 2 per day. The engineer who first designed it may be getting Rs. 50 an hour. You will not hesitate to pay Rs. 1,000 per hour to a skilled surgeon who operates on your appendix. You will willingly pay such hefty fees to a famous barrister who pleads your case. The gist of the matter is that it is not possible to pay equal wages to each and every person. Their wages are based on the famous theory of supply and demand. Why is a woodcutter available at Rs. 2 per day, while a renowned doctor or a barrister is not easily available at Rs. 1,000 per day? The answer is the same. Woodcutters and labourers are plentiful while there is a dearth of doctors and barristers and therefore they are more expensive.

Besides, it is very difficult to ascertain how much time each person has spent in making a product. A labourer or a coolie may be paid in the ratio of time spent, but how is it possible to compare a doctor with him when the doctor has spent 10 years in obtaining the knowledge to become a physician. During all those years he did not earn or produce anything but spent a large amount in obtaining medical knowledge. All this, too, must be taken into account. It is quite appropriate for him to say that it takes half an hour to visit a patient and examine him but behind this half an hour there is the 10 years of study. Similarly, an artisan can say that in learning a certain skill and becoming adept at it he has to work for 7 years. This, too, goes into the calculation of his workmanship. He strikes a blow with his hammer. The skill behind this blow can never be seen in the blow of an apprentice or a novice carpenter. The singer has practised for years to train his voice. If he entertains you by singing for 2 hours, then the years of training and practice behind that singing must be taken into account. Everyone will accept that, there is a difference in each one of them. But there is a disagreement among various people as to the degree of difference there is. The same difference is there between the works done by a clever man and that done by a fool. Decidedly, the work, done by a clever person should be valued at a higher rate than that of the fool, but as to how much higher, this can be decided again on the law of demand and supply.

Besides these examples, there are certain other instances where the person working does not seem to be earning or producing anything. A carpenter makes a gate for a field so that animals do not enter the field and spoil the crop. The work of the carpenter is obvious and can be seen. Now suppose, a boy has been employed on that very field to keep the birds and locusts away by means of a sling and he protects the crop by shooing away the birds all day by creating a noise or by the beating of drums. His work is equally important, but it is not openly visible. A postman does not produce anything; he just moves around distributing letters. A policeman merely patrols the streets and produces nothing. A soldier produces or makes nothing, although he destroys things at times. A doctor only advises that take this medicine, take that pill. Barristers only give advice; they do not generate or manufacture anything. The work done by any of these persons cannot be measured or weighed. Still it is necessary to properly evaluate the services of all these people.

Besides, there are manufacturers of physical things, advisers who explain how these things be made, suggest the quantity in which they should be made, etc. Everywhere the help of many is needed. In simple terms, production requires thinking and working, and these two are complementary to each other. But in Indian villages thinkers and workers are the same. The goldsmith, blacksmith, tailor, carpenter and mason think for themselves, and then decide what is to be made.

This, too, should be considered that the person who thinks a lot may not become a good workman, and it is also not necessary that the person who is an excellent craftsman may also be a good thinker. This may be a reason why things crafted by him are slightly inferior.

Similarly, consider a family. Women look after the children, cook and do other household jobs. But no permanent wealth seems to be created out of their work. Still, it does have a value. A man gets more wages than a woman because it is supposed that he shoulders a larger burden of rearing the family. But despite this if he spends his salary on wine and gambling rather than on his family, then also the law does not take him to task. His wife is also helpless. But if she were not the wife but a servant, then she could claim her share legally and through the courts. Likewise, if a woman takes out some money from the household expense account and buys a sari on the sly or spends it on some other addiction, the husband can do nothing, whereas if it were the servant, he could have sent her to jail for this.

Now you yourself decide why a man's time is more valuable than a woman's, or why anyone else's time is more valuable than that of some other person? You must have reached the conclusion that the scheme that 'one should have only as much as one has produced' is not possible or viable, because it is difficult to ascertain as to who has produced how much, and it is not possible to calculate their share.

8

WHY SHOULD A PERSON NOT GET AS MUCH AS HE DESERVES?

The second scheme is one should get as much as one deserves. Those who are comfortably placed profess and believe this to be justified. One gets or has got as much as one is worthy of. Those who are hardworking and thrifty do not lack anything. Those who are lazy, who do not work, who are dishonest, drunkards, gamblers, addicted to vices are the source of poverty. These people also tell us that the labourer whose conduct is good always gets work. The worker whose conduct is bad, who squanders money, indulges in drinking and gambles, etc., does not get work, and therefore, he becomes lazy and poor. But this does not prove anything. What we only come to know from this is that our vices and weaknesses make us poor. But they forget that there are many addictions and weaknesses that make us rich. There are many hard-hearted, selfish, cruel and miserly people who are always on the lookout to derive undue benefit from the circumstances of their neighbours become wealthy quickly. On the other hand, people who are generous, philanthropic and friendly remain poor. And in today's social conditions many are born in poverty, whereas many are born in plenty; and before they can form their nature and conduct, they are already divided into the categories of rich and poor. Therefore, the scheme that one should get what one deserves is not even fit to be considered. Every person can see that it has only an adverse effect. Because of this a few people who are lazy and do not work will become rich while a large number of hardworking and laborious persons will remand poor.

It is natural for you to say that if the situation today is not that a person should get as much as he deserves, then it needs to be implemented. You will immediately put forth the argument that man should be rich or poor in ratio of his virtues and vices. This poses many hurdles. What is the measure of virtue and vice? How can a man's virtue be valued in terms of money? For example, take two persons—a blacksmith and a school teacher—from a village and determine their worth. Now decide whether the blacksmith should get less than the school teacher or the school teacher should get less than the blacksmith. Also decide what should be their margin? What should be the ratio—half, three-quarters, one-quarter, one-half, two times? Possibly, you may be able to decide who should get more, but what should be the difference is a difficult question indeed. Think patiently—the teacher has studied in college, but that is no measure of his worth. His mother and father sent him to study; that is definitely his qualification. Of course, the teacher can read good books and teach. That too is his qualification. But the smith can forge a horse-shoe, make a spade and plough. That is his qualification and worth. Deciding the qualification and disqualification on these lines will again lead us to the same dilemma. Now, since we do not have the measure of their worthiness, let us determine their unworthiness. Suppose you have managed to ascertain their faults and shortcomings. First, it is rather difficult to accurately determine this; but if you come to know of them, then will it be proper to reduce their wages merely on the basis of this criterion? Undoubtedly, when you will decide their worthiness on the basis of their virtues and faults, then their evaluation in terms of money would appear to you to be meaningless.

To what extent is it proper that two wrestlers be pitted against each other in the ring on the basis that the winner of the match gets Rs. 500, while on the other hand the fervent advocate of peace and learning the monk does not even get Rs. 100? You might hold that the reward of the wrestling match is right and proper while the alms given to the monk are not. So, the determination of worthiness and unworthiness is beyond man's ken. And, because of this the second scheme that one should get according to one's worth will have to be abandoned by us.

9

MIGHT IS RIGHT

The third scheme can be summed up as 'might is right.' If the scheme that those who have power can grab and those who can keep secure should retain is put into practice, then there would be no peace and security anywhere in the world. If everyone is equal in strength and shrewdness, then the struggles and skirmishes would be all the more violent and as a result nobody would be able to grab anything. Children and older people are less strong, while the young ones are more powerful. There are many kinds of mental traits in people, good as well as bad. Even the plunderers wish to lead a peaceful life after the distribution of loot among themselves. But if the process of grabbing things by force is in vogue, then the person more powerful than you will subdue you whenever he gets an opportunity and finds you weak. Consequently, you will have to be constantly watchful and so the possibility of a peaceful life would never be there. Loot and robbery are by law illegal so leave them aside. But in daily life the grocer and merchant and draper can charge whatever price they may wish. When things are in short supply they can charge you twice or four times for them. This, too, is a form of loot. The landlord of the house you are living in can demand and collect exorbitant rent from you, although the value of that house may be negligible and you may not have the wherewithal to pay the rent. To stop such civilised loot ever new laws were enacted. During wartime when there was a serious shortage of things, profiteering and black market operations were legally declared punishable crimes under the Defence of India Act. Even today the prices of many things are fixed and it is a crime to charge more. There is the Rent Control Act to prevent arbitrary increase of house rent. This scheme based on might and shrewdness can never succeed. The reason is clear. It would lead to anarchy and this whole affair of 'might is right' will create ever new problems. So, this path too is not worth treading, we have to find a simpler way of income distribution.

10

CAPITALISM

The fourth scheme envisages 'accumulation of wealth in a few hands while members of the remaining general populace get only enough for their upkeep'. This can be called the 'capitalist tradition'. It can be said that to a large extent this scheme or tradition is prevalent today. Suppose one out of ten or one out of fifty persons from the general public was chosen to be the chief and given the right to keep the entire earnings of the group. The remaining 9 or 49 people work round the clock to produce and earn and their leader, the chosen person, leads or rules. Now, the question is who should be that leader or person, and how can he be chosen? There can be an election—probably it takes place in present times. But 40 to 50 years ago, due to the obscurantist customs and tradition a person could become chief, headman, leader or king on his mare strength or wisdom. Later on, his sons and grandsons would perpetuate this leadership as a matter of heritage. By and large this custom is still in practice. This sort of distribution may have been very useful when human civilisation was at a very primary stage. It was the duty of the leader to encourage science, art, literature, research and trade, etc., and judiciously spend the money which he had accumulated from his dependents or subjects. Even today a strong argument to keep this custom alive lays down that if the wealth of the rich were to be distributed among the poor it would not make much difference to their poverty due to their large population, while the rich would definitely become poor. As a result capital or accumulated wealth would decline and no useful institutions, industries or professions would be generated in the country. In India, till 12 years ago, when there was a rule of a foreign power, this capitalist system received great impetus. The reason is clear. When a few people rule over the large populace, it becomes necessary to appoint a few people of one's choice out of the larger public to lead them. The British adopted this method to rule over India. Kings, nawabs, zamindars, jagirdars, merchants and traders, industrialists, bankers and tycoons-they have all become what they are because of the exploitation of the general public or subjects. If this is so, then why is this system good in certain respects? It is an appropriate question. Leave India and take developed nations like America, England, France, etc., where there is complete civic freedom. The above custom is also prevalent there, but with some changes. The fact remains that if every person has Rs. 100 each, nothing can be achieved. They can merely run their monthly expenses, that is all. But suppose 49 persons have Rs. 80 each instead of Rs. 100 and the 50th person has Rs. 1,000 and they accept him as their leader or patron and out of Rs. 1,000 a permanent fund is set up and trading is done. Then, all those persons stand to gain while they do not suffer much by getting Rs. 80 per month instead of Rs. 100.

Who should or can be this 50th person? The question still remains. In Western countries this too is solved in a way by legally electing the person. This custom was very attractive. If the leader adheres to ethics and a given policy, then probably this system would be considered the best up to now. But human nature is such that it invariably wavers from the given path. When one person becomes the authority-you may call him leader, chief, king, or whatsoever you wish-then he becomes prone to carelessness and deviation. And under the influence of this carelessness and deviation, the wealth which was meant for public benefit, he spends on fulfilling his own desires and selfish interests. The army, he should deploy to defend the country or protect the people from plunderers, robbers and goons, is used to perpetuate his autocratic and despotic rule. In view of these circumstances people raise their voice against this system itself. That is the reason why this custom is becoming unpopular in those civilised countries. The main reason is that the government of capitalist parties is not able to rule properly despite good intentions and views. As a result, they resort to all sorts of malpractices to keep their power stable and secure. Corruption spreads within schools, colleges and universities. One cannot get one's own rightful salary without paying a bribe.

According to their point of view if everyone will have an equal amount of money there would be no wealth left over for the country to have a permanent fund of savings. For this they try by all means to give the least amount of wages to the working class and keep on increasing the coffers of the rich and opulent. They say that with the accumulation of capital the country will become strong and its income will increase.

But people are beginning to feel that this argument is baseless and hollow. The only logic by which this capitalist scheme of things is encouraged and which has some elements of truth is that capital formation without this system is not possible. And, without capital formation nation building is not possible. No matter how many economic plans are made, but without capital all schemes are bound to fail. Their argument is that you may change capitalists but you cannot change capitalism. If the nation has to progress, then this progress can be achieved only through the capitalist system. Instead of the kings and emperors, nawabs and barons, etc., who were the capitalists of the previous era, you will find government officers and their kin or many other people in more-or-less the same position. Has this question not been settled that capitalism is inevitable. Many people say that the socialist system of India, the socialist system of the present government is another form of capitalism. In essence the distribution of wealth even through this system is not satisfactory.

11

CLASS SYSTEM

The fifth plan is class division. The nation or society is supposed to be divided into many classes and the income of each class is almost fixed. Suppose there is a class of labourers, coolies and porters who do physical labour; they get Rs. 70 per month on an average. The second class is that of professors, doctors, etc., who get almost Rs. 1,000 per month. You will say, this is the system that is prevalent in society today. Probably you are right. But it is not so. There is no law or regulation that a doctor should get more wages than a coolie. But because the number of doctors is less and there are numerous coolies, so this situation has developed.

This too is not necessary that the worker deserves less than the intellectual. It is often seen that some intellectuals are extremely talented and yet they are always steeped in poverty and suffer hunger and deprivation. The public does not recognise their worth and greatness in their lifetime. But there are workers to be seen who are able to save a lot after meeting all the expenses of their daily lives. So, this cannot be ascertained by monthly income alone. Because of social circumstances and the demands of civilised society the people of the upper class have to necessarily bear certain expenses. Therefore, compared to the workers who do not have these problems, the intellectuals would be termed as poor. A professor who earns Rs. 1,000 every month spends Rs. 1,050 on meeting the expenses of his family and social responsibilities, while a worker getting Rs. 70, only spends Rs. 50 because his responsibilities are limited and other members of his family also work. In such conditions whom will you call poor? The kings of yesteryear decidedly feel themselves to be poorer today because they find themselves unable to meet their expenses.

It is also not necessary that people of the class which has the most resources of income or are wealthy are powerful. Undoubtedly, money has great power but still a millionaire has to be fearful of a village police constable who is a person of the ordinary class. The people who have power may not necessarily be rich. Even in honour and respect it is not necessary that the rich take a front seat. In the present times even an ordinary person who has the populace behind him, that is a person who is popular among most people, is the first to be honoured. The Prime Minister of India, Pandit Nehru, is considered to be the foremost among people and is loved by all not because he is a rich man, but because he has served the people and shown them a new path and worked for their upliftment and is still doing so. Today he is very popular. When you talk about the power of the rich, it is true that their power is the power in true sense. If they wish, they can dismiss whomever they want from work; if they dislike a merchant or shopkeeper, they can stop their trade. A highway robber may show you a gun and force you to submit all that you have. But this is not power in the real sense; it is a threat of force. There is no relation between true power and money. All those people, from the prime minister to the ordinary police constable are higher than the rich who have to obey their orders.

So, even the income distribution through this class division is not satisfactory.

12

STATUS QUOISM

The sixth scheme is status quoism, that is, everything should go on as before. There are many who wish so. They are apprehensive about change, although they are not at all happy and satisfied about how things are presently going on, but they fear that change may be a change for the worse. So, they say 'let things go on as they are.' This is pure escapism. They know that things cannot go as they are; change is the law of nature. It must take place. You cannot keep things as they are. Take a room for instance. Close it and leave it as it is. Do not clean or dust it. Do you think it will remain as it was? Never! Probably, there will be more decay by keeping it as it is compared to cleaning and caring for it and keeping it in use. So, the conclusion is that everything in the world is dynamic and active. One cannot stop or hinder their activity and movement.

The conservative supporters of status quoism gradually become indolent. If a man is inactive or less active, then as a rule some other person will have to do his share of work. The result is that some lazy rich become richer and the majority of poor become poorer. And so a dangerous situation develops. In the 11th century in Saurashtra, there were millions of defenders of the Somnath temple yet the Sultan of Ghazni, who had come with only 25 thousand cavalry, ransacked the temple and took away the loot. This was the fault of the inertia prevailing in India at that time. It was because of this tendency that India was subject to foreign rule for 200 years. Had the active leaders of India and before them the brave soldiers in 1857 not shed this inactivity and raised the slogan of 'change', India would have been in fetters for a few centuries more. Status quoism is truly a terrible tendency. It is against the law of nature. Things cannot go on as they are. Change is inevitable.

Now, what sort of change should be there? This is what we all have to deliberate upon? Suppose you are driving a horse-cart and the horse is suddenly startled and nervously takes flight. There is a large pit just ahead of you. The horse will certainly fall in that pit and your life will be in danger. You may say, what can we do, what is willed must happen. In these circumstances what can one do? This is a grave mistake. This tendency towards inaction is surely bound to land you in that pit and put an end to your life. You will have to make an effort and try to calm the horse or guide the horse-cart along the sides of the crater and try to stop it from crashing till he calms down. It is the same situation in trade and commerce. To save the nation from falling in a pit, it is necessary that the trade and commerce of the nation be run with the public good in mind. 'The government should not enter into trade'—this maxim is put forth by those very conservative elements. If the government will interfere into the field of trade how will trade go on? These are the dissenting voices of those people. In other words they wish to earn money by any means, fair, foul, good or bad. Let the rest of the people go half-fed or hungry, but the government should not put any hurdle in our path. This is status quoism. Do you think this system is right? Never!

13

EQUAL DISTRIBUTION OF INCOME IS SOCIALISM

Now, only the last scheme, that is the socialist form of income distribution, remains to be examined. You must recall that we had considered a question in this regard that what is the best way, the most satisfactory and useful way in which wealth may be distributed. In this respect when we had deliberated that every person must get as much as he produces or earns and should not get anything without earning it, we had found this way to be foolish. When we considered that one should get according to his worth, this too was found to be unsuitable. When we examined the proposition that he who has strength and power may take as much he wishes, we found that this also will not do. And when we considered that things should go on as they are, we found that this is not possible either, things cannot go on as they are. So, the thing now to be thought about is what would be the best way of distributing wealth.

Before we think and decide about this, we must ascertain what poverty is? The lack of something for a person is poverty. A poor man can not necessarily be unhappy and a rich happy and content. As far as food and clothing are concerned, once you get them even then wealth or poverty is not the criteria of happiness or sorrow. If a person is so poor that he is not able to appease his hunger or obtain clothing to protect his body, then that level of poverty is truly a source of unhappiness and suffering. But after getting all these things that are necessary and useful for life, if a person has more or less money in hand it does not denote that he is happy or unhappy in that ratio. You must have seen people who had ten times more money at the age of 60 than what they had at the age of 25. Yet, they were not 10 times more happy. Many rich persons consider themselves unhappy than most poor people and in fact they are. Happiness and sorrow are matters of self-realisation. They are not specially linked to physical and economic well-being or wealth. Money can help in appeasing hunger and obtaining clothes to cover the body, but it cannot generate inner peace and spiritual well-being.

Poverty is the biggest universally accepted curse of mankind. No person wishes to be poor or destitute. But the interpretation of poverty is different with each person. The kings and emperors of yesteryear, who enjoyed estates with incomes of millions of rupees yearly, now have incomes of only five or ten hundred thousand a year. They now consider themselves to be poor and destitute. And the peasants of yesterday whose houses and lands were auctioned off in order to realise arrears of rent, get enough to eat in peace today. Or they have an income of Rs. 100 to 200 a month and so they feel that they have become rich. So, it is a proven fact that there is no certain measure by which rich and poor people can be measured. Man has a tendency not to see one's own self but look at others and follow in their footsteps or copy them. So this feeling or misconception that he lacks something arises out of seeing others. A person was quite content and happy earning Rs. 200 a month, but ever since he saw a radio, gramophone and refrigerator in his neighbour's house, he began aspiring to obtain those things. The birth of desire and the failure to obtain those things is to 'experience the want', and he is poor in that sense of the term.

Poverty is not as bad in itself as we are afraid of. The biggest fault in poverty is its tendency to lead people astray and defile their character. Certainly, when the lack of something is felt, man endeavours to obtain it and if he is not able to get it by fair and just means, he adopts other illegal and unfair ways. Thievery, gambling and such other vices emanate from these tendencies. When a woman sees that she can barely earn Rs. 2

or 4 by working hard every day and this money is not sufficient to satisfy her needs and that if she adopts a path of debauchery and unvirtuousness she can earn Rs. 50 to 100 and get all those things that she desired, then it is quite possible for her to go astray. In this way mankind's tendencies are polluted and defiled and become depraved. It is not necessary that these faults are only found in poor people and not in the rich. Even the rich are prone to these vices. But as I said the interpretation of poverty is the feeling of want. And so the rich who have a feeling of want can be placed in the category of the poor. Poverty is the biggest hurdle in path of character building. When the economic condition worsens to the extent that man is not able to obtain food and clothing, then a whole community sometimes takes to crime. In India even today there are hundreds of thousands such people who are criminal by profession. The government is aware of this situation and now gradually those tribes and communities are being rehabilitated.

It is the responsibility of the government to provide food and clothing to every individual. India being a welfare state, this responsibility has been assigned to the government through the constitution and there is some progress in this direction. Well, if large part of society becomes morally corrupt due to poverty and destitute, then the remaining opulent part cannot avoid being affected. If there is an outbreak of an infectious disease in your neighbourhood, it is bound to spread to your area too no matter what preventive measures you may take. It is because your servants and other persons doing menial jobs reside in that locality, and so you cannot keep yourself entirely isolated. If the education being imparted to their children for their mental development is not proper, then you may prohibit your children from playing with them or coming into their contact (although such restriction may have an adverse effect on their immature minds) but they cannot be prevented. They come into each other's contact on the sly or by other means and their virtues and vices are mutually transferred none the less. If women of that area are free as far as sexual contact and behaviour is concerned then sooner or later they will creep in to your area through some means, through you or your near and dear ones; you cannot keep yourself in quarantine forever.

So, it is necessary that no person in society should be so poor that the process of character building is hindered or people are led astray. This is as much your responsibility as any other person's. So, now the question arises as to who should get how much in the distribution of wealth; how much is enough for a man so that his instinct will not get distorted. Poverty is bad, this of course is true. And it is easy to measure 'poverty', especially extreme poverty. If a person does not have enough food to eat or clothing to cover his body, he can be termed as 'extremely poor'. But is 'opulence' a good thing? Again, its measure is not difficult. He who, after satisfying his needs, has enough to spend unnecessarily is 'rich'. This situation, too, is as bad as poverty. The patient of this ailment becomes a prisoner of worry day and night. He is first of all fearful of imagined diseases and concerned about his health all the time. Then he is apprehensive that no one should deprive him of his wealth by forcibly taking it away from him. Then he constantly worries that his servants and managers do not cause loss by their negligence and incompetence. As a consequence they are all the time engulfed in mental worries and they are not able to organise the lives of their wives and children. If a wealthy man has Rs. 5 million and five sons he is constantly worried that each of his sons will get only one million. Furthermore, they have been habituated to such a style of spending that in time income from the million in each son's share will not be enough for him. Soon his sons become addicted to vices and lose whatever wealth has been left to them.

You must have seen such wealthy people, too, who spend very wisely. They keep accounts of their income and expenditure in such a manner that they are not afraid of anyone or anything. And mostly they are free of worries. The living style of such rich men is like that of middle class persons.

The rich have to resort to all sorts of masquerades to maintain their position of imagined status in society. They have to take a part in art, literature and other cultural programmes just because all rich and prominent people of society partake of such activities, although they may have no interest in them. They may get bored of listening to good music, classical music, but they sit for hours in gatherings nodding their heads off in praise and of course pay thousands of rupees as donations. Similarly they have to be in the forefront of all such social events covering science, art, music, sports, debates, religious story-telling, etc., and they have to bear the expanses of all these events, though they may not understand one bit about these activities and programmes. So eventually the accumulation of money in one place becomes almost like public capital or a matter of public right.

How much is enough for a man? This question is embedded in this very thing. Every man's ambitions and tendencies are different and varied, so a fixed amount of money may not suffice for each one of them. The life style or pattern to which a man has become habituated determines the amount of money enough for him. But there is marked different in life styles to which people from different walks of life are habituated. So, if we define poverty as the state of not having the object of one's desires, then there would be no one in the world who would not be categorised as poor. Because man by himself is incomplete, he feels the lack and want of something or another. Only the Almighty is complete and perfect. Keeping this is mind our ancient sages had preached that the thirst for physical and worldly goods is never ending. There can be no feeling of happiness and contentment in it. Man may try of forget his pain in the intoxication caused by wine, but that is not its cure. One may keep on drinking, but the thirst for wine will not be quenched. The more you drink, the more you will want. That is the reason why the sages had said that real happiness lies in renunciation and giving up. On his very base the Jains and Buddhists built their doctrine of total negation and non-essence and spread it all around and it still exists in India in some form or another. The question is whether one attains real happiness through renunciation and *vairagya*? We are not ready to believe that in essence the person who renounces or a monk is happy. Even if we come across a person who is happy having achieved this extreme state of grace of total renunciation that path is definitely not meant to be followed by the general public. Then what is the solution to the problem? Who should get how much?

The principle of socialist economy is to distribute everything equally. You will say this is not possible and never will be, no matter how principled you may become. Equal distribution cannot be done, will not be done then why take on this unnecessary botheration? But now we have seen that there is no other better way. So why not fully examine this? What are the advantages of equal distribution? Whether or not it is possible to equally distribute wealth? What are the pitfalls of not distributing equally? We have to understand all this.

So, equal distribution is necessary.

14

CURSE OF SOCIETY PASTIME FOR THE RICH

Consider a joint family which has many children, and you are the head of the family. Suppose you have is Rs. 500 to spend. First of all you have to prioritise your spending—which item should be bought first and which item should be purchased later? In this way a list of things that are needed will be prepared, and you will acquire these things. Suppose there is nothing to eat in the house. Naturally, you will buy food grain first and thereafter clothes for everyone. Then suppose you have to pay rent for the house you live in. If the house is your own, then you will have to pay its taxes, and bear the expenses of its repairing and cleaning. And thereafter, if you got some money left in your pocket, then only you will buy toys for the children and jewellery and perfume for your wife. Suppose, the head of a family has no idea about his needs, that is he is not clear about his priorities. First, he buys perfumes or other toiletries and attractive gold necklaces. You will say what a fool he is! There is nothing to eat in the house and he is throwing away his money needlessly on perfumes and jewellery. You know that money spent on perfumes and jewellery is not wasted, even then his actions will be termed as foolish, because he has no proper knowledge of his priorities. The children are hungry and need milk and bread. What will happen if you get them a beautifully published Ramayana or a Gita instead? It is not wrong to spend money on Ramayana or Gita, but his first duty was to arrange milk and bread for the children. The books could have been bought later if he had surplus money, otherwise not. Suppose several members of the family have the right to spend money from that very joint fund. One brother is fond of perfume and he spends money on lavender, another one is fond of cockfights and so buys cocks of high breed one better than the other. Consequently, the remaining two brothers will not get enough money to buy food and clothes for their children and wives. Obviously, there will be a conflict. The joint family will break up and people will admonish those two brothers who spent money on perfumes and cocks.

Suppose, the entire nation is a huge joint family and people living there are all members of this large family, consisting of adults or minors. Now you see that there are hundreds and thousands of people in the country who do not have enough to eat two times a day and they wear worn out clothes. Yet, there are some people who suffer from constant indigestion because of overeating and doctors and physicians have to be summoned daily. They take part in horse- racing and spend millions of rupees in buying horses of the highest breed. They drink high-quality foreign wine. And yet, they have enough surplus money to go on foreign trips twice a year and spend hundreds of thousands there on enjoyment and pleasure. What an unequal state of affairs this is!

So many people in the country yearn for food, while some people spend money beyond one's ken. Now, what will you do? The family get divided after a conflict. But when the same disease ails the entire nation, what is the remedy? This is the question that will pop up in our minds.

Agreed that diamonds and pearls are necessary for the world, to breed good horses is also fine, to encourage fine silk cloth is also not improper and the doctors and physician, too, should be kept in employment. All these are necessary, but what is more necessary is that everyone gets enough to eat, clothing to cover the body and houses to live in. Then only gaiety and entertainment can be excused to some extent. These people claim that they are providing food for so many people through their pursuits and inclinations and have we ever taken this into account? They claim that because of their life style hundreds and thousands of people are employed in manufacturing excellent apparel, looking after the horses, and mining, polishing and cutting diamonds, while and many other are engage in collecting pearls by diving into the oceans. Countless thousands are busy doing all this. They provide them with bread and upkeep.

What a strange claim to make? It may even be true. But cannot the murderers and pirates make a similar claim? Those accused of murder and loot can also say that they provide employment to countless persons—the entire police force and then the doctors and surgeons, the provider of wood for the pyres at the cremation grounds, the woodcutter who has cut and brought the wood, the weaver of the shroud, etc. Hundreds of persons are employed in these pursuits and gain employment. Those who claim these are murderers and dacoits but they say that if they were not murderers and dacoits, all these people would become unemployed. What an argument? I see no difference in the above two claims and opinions.

Therefore, if the very money, through which the rich provide employment by spending it in pursuing their hobbies, is distributed equally among all, then it will provide employment in the right sense. The nation will have more food to eat, more clothes to wear, more houses to live in and invariably this will lead to greater encouragement to the arts and sciences.

15

SOCIAL INEQUALITY : THE CURSE OF THE HUMAN RACE

Yes, you may ask that will everything be put right by equally distributing things among those people who do not have enough food to eat or clothes to wear. Is one certain that they will not misuse the money thus received? Will they be aware or careful about their priorities? Is it certain that they will not spend the money thus received on gambling and such vices and again become poor and in want of food and clothing as before? This state would be very deplorable. What is the guarantee that they will follow the right path? How can one say that after equal distribution the new employment would not be of an objectionable nature? These questions are proper and apt and there is no easy answer. When the very base of man has become awry, the very background has become out of joint, then nobody can take responsibility or guarantee anything. None can say what will be the result, whether it will be the desired one or not. But the only way for this is to breed proper men, improve the race of mankind besides education and training.

The atmosphere and general ambience has to be improved for the ideology of mankind to adhere to the right and legal path. Improving the race! What a fanciful proposition! Are they horses, bulls, cocks or parakeets that by mating them one can breed good animals and birds? That too in the sense that they may run faster, be strong enough to carry heavier burdens or fight as soldiers. There is no development of the intellect in them. So, there is no need to see to any other kind of racial characteristics in them. Because you cannot have a couple mate at your will nor regulate the birth of a son or other children. Marriage is an institution by which one cannot force the couple to produce children of a given kind. And again it is not necessary that the children turn out to be like their parents. In ancient India racial breeding was prevalent to such an extent that the sons and grandsons and great grandsons of a man would traditionally have the same characteristics of faults and virtues, generations after generations. Take any of the sages like Vasishtha, Jabali or Bharadwaj. These names are not of a particular person but of a family. The son would be known by the very name of his father and in all respects the father was active and present in his son's body.

This birth control or self-regulation has almost disappeared in India; still, compared to other nations, it is present here to some extent. Even today the son of a smith would be a better smith than any other educated and strong lad. The son of a tailor would be a better tailor and maker of clothes than any other person, etc. etc. We can discern many things by which the purity of race can be seen in India. The reason is clear—marriage here is still regarded as a *sanskara*, a sacred ritual. The marriages of boys and girls are arranged by their parents, they cannot marry at will. So, the fundamental race is maintained and kept going.

There is very little individual freedom in the custom of marriage in India. Only those boys and girls are paired off who have similar economic and social status. So the difference among various social classes will remain as it is. Therefore, today a labourer cannot even dream of marrying the daughter of a rich merchant neither the son of a rich merchant dream of marrying the daughter of a labourer passing down the road. So the question is how to improve the race. The main reason is a pollution of mental faculties which I mentioned earlier. That is, poverty or the deficiency in fulfilling the needs of mankind. Once this tendency is put right then things are not so bad in India that a cure is not possible. This fatal tendency that we rich or highly placed persons always speak the truth, while our servants or drivers tell lies, their statement is not as trustworthy as ours is the basis of all difference and without eradicating it, any talk of improving the race is useless. First, the race of the rich needs improvement which has become intoxicated by money and power. It is imperative to rid their minds of the vapours of arrogance and then only, with time, the racial characteristics of the second and third classes would improve.

Actually, this whole idea of improving the race is useless. The only reason why the racial characteristics of the people are not being able to be formed is because of unequal distribution of money. If the economic status of every person becomes the same or almost the same in society and class distinctions do not exist any longer for ideological relationships, then it is possible that with time the racial characteristic would also become similar or almost similar.

The caste system in India is more responsible in keeping alive this difference and inequality. It is possible that at some time in the past the class distinctions engendered by the caste system may have been useful. If the caste system is considered in its essence and its original ideals are adhered to and put into practice then it may prove useful. But that time is past. Today's era is purely materialistic. Who has the inclination to repeat the worn out and obscurantist caste system after seeking out its essence, which is present before us in its most corrupt form.

The cause of non-improvement in the racial characteristics of mankind also lies in the system of unequal distribution of wealth.

16

THE LEGAL CODE

The rich manage the legal system
Peoples' representation in the legislatures

Justice is done in the courts of law. It is said justice is blind; for it, rich and poor are all alike. There will be no discrimination before justice. But is it actually true? The laws present in the books are indeed applicable for the whole country but just see how they are conducted and regulated.

You have a small piece of land. Your neighbour has occupied a portion of it wilfully. Now you will have to lay a complaint before the court of law. You filed a case after paying the court fee and engaged a lawyer or barrister after paying him good fees to plead your case. You and many of your neighbours went to the courts regularly and endeavoured for months to prove you right. Eventually, after two years or ten years suppose justice was done and the land usurped by your neighbour was given back to you. In this whole affair, leave alone the harassment which you had to face, you may have had to spend twice what the land was worth in court fees, legal expenses, remuneration to the barrister and running to and from the court. Now, just consider, if you do not have the resources to spend this amount, you will have to sit back helplessly or even after filing the case, the barristers engaged by the opposite party on a hefty fee, are capable enough to prove your just case wrong. Perhaps, the judgment declares that the land was your neighbour's right from the beginning. It was never yours. You are talking nonsense. Now say, whether your neighbour who is richer than you was one step above you or not in practical terms. In this instance was justice meted out in equal measure? Or, will it be meted out?

If you have offended a rich man with or without reason he will immediately have his lawyer serve you with a notice and file a case against you in court. The result will be that you would be prevented from your work and business and you will have to bear needless expense in the bargain. Under duress you will bow down before that person and compromise with him. Such is the social condition where, although everyone has the equal right of justice, yet he is not able to receive it.

In my experience there are debtors who owe money in truth, but by falsely complicating matters and creating obstructions, force the opposite party to file a case; and then, they with great pride say, 'so what! It'll be 10 years before the case is settled and even if a decree is issued he'll not get half the amount in hand.' Just see, a person can keep the case going for 10 years by his cunning; so, even after one gets vindicated after that time, what is the use of it? Every where only the good and capable barristers, who charge hefty fees and who can be engaged by only the rich, can make the wheels of justice go slower or faster.

All this is about civil cases. In a country where poverty is so widely prevalent, it should not come as a surprise if the magistrates and authorities also succumb to a little greed. Although that sort of greed may not be considered illegal by law, still it is improper all the same. It is seen that the rich employ the sons of magistrates dealing their cases at a large salary or assure the magistrates of a life-long post-retirement employment? Then is it any wonder that the judgments passed in those courts favour them? This happens often. Unfortunately, all these recourses are not open to a poor person. He has to suffer because of his fault and many times even when he is innocent. This is to say that justice is not being meted out equitably, and it will not be able to be regulated till the whole situation is rectified by equal distribution of wealth.

Besides, it is also true to some extent that the institutions which enact laws are being manipulated by the rich. Take parliament or the Lok Sabha for instance. Elections are held. A member has to deposit Rs 500 as deposit before the election. Later, almost Rs. 15 to 20 thousand are spent on one seat alone. A member represents 7,50,000 citizens, so there is no question of spending less. In the same way in the elections held for the membership of state legislatures, at least Rs. 5,000 are spent on each seat.

If the elections are lost then all the expenses and allowances paid to members is barely enough for their normal expenses. In these circumstances who can dare to stand in elections against the rich and opulent? The parliament formed out of these successful candidates enacts laws? It is natural that their individual economic condition can not represent the economic state of the country. Their ideas and ideals, concepts and motives will be of a different kind. In India, it is proper that elections are mostly party based and so if a person is popular in his area and yet without means, still he can get a chance to be elected, because the expense is borne by the party he represents. Take the Congress party or the Praja Samajwadi Party, the same situation prevails within the parties themselves. It is most unlikely that the expenses in having oneself nominated by the party are within the means of poor candidates. So, the conclusion is the same that if the Lok Sabha, Vidhan Sabha, etc., that is the law-making bodies have to be truly made representative then giving the people adult franchise will not be enough. Only after the equal distribution of wealth among them will they be able to take part in elections in true terms.

At the moment adult franchise in India does not hold much meaning when the majority of its population is illiterate and ignorant. It is like a leap into the dark. Any person nominated by a party does more and more publicity. Whosoever is able to con and manipulate these illiterate and ignorant people gets the most of numbers of votes. Even among parties there is mutual distrust and bickering. The party nominates one person. Some other members of the party wish to have another candidate nominated. Now decency has vanished to such an extent that the party members openly oppose the candidate nominated by the party. Violation of discipline has increased in such an extent in the country that there is no limit to it, and its fundamental reason is the unequal distribution of wealth.

17

EVALUATION OF ACTIONS

Action, inaction and misconduct Division of labour along with distribution of wealth is imperative

Does the rich or opulent person remain idle because of his wealth or does not work due to laziness? If he works and labours then why do you deride him for being rich and wealthy?

No one has the right to malign a person who works. Provided he works for the country, for society or for his family. We will term as work all work being done to earn something.

Suppose there is indigestion because of overeating and the doctor has advised walking 4 miles to cure it. Now, one is walking 4 miles morning and evening. Quite a labour you would say. Similarly, one went to play tennis, watch horse racing, and see the cinema or a play. No end to it. These rich people do not get a second's leisure in the 24 hours they have at their disposal. They labour by day and night, are all the time immersed in work, not a minutes' free time. But still, since they are not producing or earning anything, therefore none of this will be called work. Work has been aptly described in the Gita by Lord Krishna. The action which results in no recompense and by which one gets nothing in return is called non-action. After action, one's conduct or bad action is accepted in some senses. By bad action a few benefit while a large number of people suffer harm. But non-action is the symbol of inertia, of indolence.

The desire in the rich and wealthy to such work or undertake such action by which they keep themselves fit for inaction and indolence is certainly very strange. But why blame the rich and opulent for this? Do other people not spend their days thus in laziness? They do, sometimes the persons coming from the middle or lower classes become so lazy that they are not able to do anything day and night except eating and sleeping.

But this fault also lies in that very system. Equal distribution of labour is invariably tied to equal distribution of wealth.

18

FAULTY EDUCATION

The wealthy are worthy of worship
The poison of dissoluteness and non-education

Consider schools. You send your children to school to be educated. Is the education that they receive there of the right kind? Have you ever paused to think about how these teachers, in whose minds seeds of education have not been able to develop because of hunger and poverty, can impart education to the students related to proper fulfilment of duties? The majority of schools are run by the municipal corporations or by the education departments. Many schools have been established by donations given by the rich or are being run by them.

The teachers in these schools cannot even mention equal rights or equal distribution of wealth, leave alone teaching the students about them. Often the rich and wealthy become objects of worship and are described as great and ideal men of India. The teachers barely get enough to live by and are afraid constantly of losing their jobs. The measure of their ability is examination results. So the only desire in the students and teachers is how to pass the examinations in ever greater numbers.

Leave alone the standard of examinations whatsoever they are. Even they are not successfully achieved by the children. So they try to find out the teachers who have set the papers in order to obtain the papers from them before the examination or after the examination they try to find out to whom have the answer books been sent to, so that by applying pressure get better result from them and pass with better grades. As soon as the results are declared, teachers think that they have done their duty and the students think that they have now become educated citizens. What will be the result of the sweet poison of non-education and slavery which your children have to drink in the name of education? You may decide for yourself.

Even today the glory of sacrifices made in ancient times is lauded and sung. Servants were devoted to their masters to such an extent that they were ready to shed their blood for their masters and lose all that they possessed in doing so. Their glorious tales are still taught in our schools with great fervour and zeal and are studied and mulled over. Even now we are told in schools that it is God's will to be rich or poor, man has no say in it. Sayings and adages made on the basis of legends of 400 years ago are accepted as gospel truth or God's words. When the greatness of the rich and wealthy is taught in this manner right from childhood, these children develop an inferiority complex by the time they grow up and begin to believe that the rich and wealthy are better than them.

Nowadays it is also being taught in our schools that every child in India has equal rights and these economic distinctions have not been made by God but by individuals themselves and that this order can be altered at will. But the influential rich class will immediately dismiss from the school the teacher who propagates this. You must definitely have seen that in school the scions of the rich are treated with favour. Leave aside a few, but most schools cannot dare to take disciplinary action against a rich man's child who has committed an offence, while a poor man's child is expelled from the school on that very ground. The teachers who set the examination paper often reveal the contents prior to the examination because of some inducement or another and the students who give these inducements or are in a position to give them are the children of those very rich and wealthy. Similarly, the children who try to have their marks increased in order to pass are also of the same class. Whereever I am using the word 'wealthy' or 'rich', it should be understood that I mean the well-to-do, opulent persons.

19

PROPAGATING POLICY

The aberration of newspapers
Ancient examples of religious discourse

In the present age besides the schools the human mind is swayed to a large extent by newspapers. You must have noticed that owners of 80 to 90 percent of all the newspapers that are being read in India—in the villages and towns and cities—are millionaires. The entire business of journalism is in the hands of millionaires. How can they ever raise their voice against the rich?

When some illegal or unlawful work done by the government is exposed, they raise a huge hue and cry, call to the heavens, do not let the government take a breath. But when some section of the rich class—whether they are industrialists of the cloth mills, owners of sugar factories, tycoons of the jute industry or barons of the chemical industry—does some illegal activity and the people are fed up and terrified at their misdeeds, then not a hair stirs in the heads of the editors of these newspapers.

Whenever there is a rail accident and some people die in it, there will be a strong public outcry. The newspapers will write that rail accidents are deplorable. They will express their sympathy in long-winded articles and messages. But if after ascertaining the causes of these accidents, the government should declare that the wages of railway men have to be increased, their hours of work need to be decreased and the fares and tariff raised, then these very newspapers will shout and holler that a grave injustice is being done, etc.

The influence which grips the human mind because of education or regular reading of newspapers is adverse. You cannot set it on the right path. The reason being that the schools, the education department and newspapers are all controlled by the rich; they are run by them or exist because of the advertisements given by the wealthy. So their partisanship is self-evident.

The third influence on the human mind in villages and towns comes from going to the temples or taking part in religious discourses and meetings. People in villages gather in the evening at the chaupal, that is, the community centre or at the temple and talk and discuss. Most of this is an exposition of the rich and their goodness or a detailed account of how the kings of old looked after their subjects. The idea of *Ramrajya* or Ram's order is considered to be worthy of praise, although it was purely bourgeois feudalism. India has always had a tendency of hero worship. Keeping the individual virtues of a person in mind, the citizen of this country starts worshiping that person. Ram, Krishna, Buddha, Rana Pratap, Shivaji, Lokmanya Tilak, Gandhiji, Pandit Nehru, Sardar Patel, etc. There have been many great men and, enamoured by the personal virtues of these people, the people of India consider them worthy of worship for all time.

This is not to mean that those great men are not worthy of worship or praise, I only wish to say that what has been done by which person among these with regard to alleviating the economic condition of the common man. Class division, whether it is based on the caste system or on knowledge or on the tendency to gather and amass, has been encouraged by almost every one of them. Even today the story of the Ramayana is recounted at the village community centre. The entire audience praise and laud Ram and *Ramrajya* in a state of bliss. In no one's mind the question arises as to what was the thing which should be considered as such a great ideal. In that very *Ramrajya* after the untimely demise of his son a Brahmin came and complained in Ram's court, then King Ram picked up his bow and arrow and killed the shudra Shambook who was meditating in the jungle. And then all started praising Ram with one voice. Even if we accept that it was unlawful for Shambook to meditate, then does this story not reveal that there was a huge distinction between men at that time? So what sort of an ideal was this *Ramrajya*? If you went and asked Shambook's parents then you would really know the real state of affairs. What is meant is that schools, newspapers and religious discourses—these three main agents—form the opinion of the masses and all three are directly or indirectly controlled by the rich. Because of these reasons the question of equal distribution does not even arise.

The biggest problem which has been prevalent in India since centuries is of the caste system. Man has begun to place himself at a higher or lower pedestal because of these castes and his feeling has entered the very blood in such a manner that people are not able to remove it from their minds even in this progressive age and era. Even today the most influential slogan is that of caste. It is openly used in elections although India has been declared a secular country by the constitution. Still the banias would vote for banias, jats will flock together and because Brahmins are in a minority in the south they cannot be elected, not getting votes from non-Brahmins. This disease is rampant in the whole country. Where the leaders preach that untouchability will have to be eradicated, caste discrimination would have to be erased, there those very leaders take care and see that in every committee, in every service such and such a caste is represented or not; why is there only one Rajput in the ministry and only three jats and one Brahmin and two harijans. They are all the time concerned that the representation of castes is based on ratio. Then, how can this mindset based on feelings of caste be changed.

Suppose, everyone had the same amount of money, then could the teachers be such cowards? Could the newspapers ignore honesty and integrity? Would there be such praise of kings and emperors and the wealthy merchants in temples, mosques, religious meetings and community centres? Never! They are all a reflection of the inferiority complex that has gripped us all.

20

THE INTOXICANT OF RELIEF

The illusion of being the well-wisher of the poor Crushing the seeds of revolt

A question arises as to why do people accept it when the glory of the rich is so forcefully propagated and it is so undesirable? Why do they not rebel against it? Quite rightly put, but this capitalist poison is being given to us to drink in such a manner that we are not able to feel that there is some fault of the rich in this and so we go along with it. Man is normally influenced by the visible and obvious, very few people endeavour to think about the result and understand it. Visibly these wealthy people seem like gods. Wherever relief is to be provided to the victims of flood, they are the first to offer donations. No sooner than an epidemic strikes the people, relief work will begin by these persons. As it is, they spend some portion of their money to run clinics, free boarding houses, schools, hospitals and build wells, reservoirs etc., Impressed by their generosity evident in this form, people readily accept them as their benefactors and what to say of revolt against them, they are, on the other hand, praised everywhere.

These people do not know why after all there is plague and other epidemics, or why are clinics and hospitals or rest houses needed. They do not even realise that the basic cause of all this is the exploitation of poor common folk by the wealthy. This exploitation takes many forms. If the poor do not get proper, nourishing food and their children do not get milk, then their bodily powers will gradually wane and they will suffer from serious diseases like tuberculosis in large numbers. When these poor people do not get enough wages to live properly and they fall prey to diseases even then the rich try to only see that their wages are not increased. A rich man will buy a silk sari worth Rs. 500 or a diamond necklace worth Rs. 10,000 for his wife or lover. But he will never pause to consider that this very amount could save 10 or 20 poor children from death. You might believe that death is in God's hands. *Jas apjas jeevan maran vidhi hath* (fame, infamy, life and death are in fate's hands) so argued the saint-poet Tulsidas and these lines must have been heard from the lips of many people. It is the common people's belief that the days and breaths of man are decided by God and come written by fate from there. No one can increase or subtract a day or a minute from that total. It is a strange way to reassure one's mind. When a thing has passed, it is immediately said 'it was ordained to be so—who can erase what fate has inscribed.' What pessimistic language! Words of the defeated and vanquished! In the present age it has been proved that if man can plan and organise the course of his life properly, then life and death can be controlled to a large extent.

In the cities as well as the villages, wherever there is extreme poverty, the incidence of child death is very high; whereas in good localities where proper care is taken about eating and drinking and cleanliness, child demise is negligible. In those countries or those parts of our country where poverty prevails, the average lifespan of people is low. As the economic condition of the country improves, the lifestyle of people is raised and the average lifespan goes up. Thirty years ago the average life of humans in India was 27 years. Census records for the whole of India prove that the degree of child death at that time resulted in the determination of this statistic. This average lifespan was probably the lowest in the whole world. The basic cause was economic dependence, a very low living standard, improper and inadequate arrangements for food, etc. In 30 years there is so much change that the average age now is being determined at 36 years. Although the price of food grains has more than quadrupled and the prices of other commodities have also increased, still, because the standard of life is improving somewhat, the average lifespan of man has gone up. Did fate will that the average age of the Indian people would be 36 years after 1960 and was to remain 27 years 30 years back?

All these diseases, epidemics and starvation owe their existence to the grace of these rich and wealthy. If any kind of equality by wealth distribution was acceptable to them, such a situation would never have arisen. Similarly, why does man need free boarding houses?

These wealthy persons provide some sort of place for people to stay or live free of charge which is usually of a dirty, low-quality nature. Dharmashalas or sarais built by them are present in cities and towns where people can stay free of charge for 2 to 4 days. At many places the rich have had fooding joints opened where people are provided free food. Such small conveniences have been provided by the rich at various places for common folk. The obvious effect of all this on peoples' minds is that these rich persons are very good, generous and philanthropic. And it is only then that they begin to be praised. Nobody thinks on the real fact that if there are no dharamshalas visitors will stay at some hotel or lodging house by paying the proper rent. If there are no free kitchens, a person will, of his own accord, go to some food vendor or a food joint and eat by paying the requisite money. He will become self-dependant and will rely on his own endeavour, whereas it is quite possible that because of the charity of the rich the chances of his becoming lazy and indolent increase. All this is because of the so-called generosity of the rich.

Behind all these acts of generosity which the rich undertake ostensibly, lurks their tremendous self-interest on the basis of which they gain popular adulation and acceptance. What happens in reality is that the people fall into the illusion that these persons are very good and benevolent. Consequently, no one thinks of revolting against them. There are some good people among these rich, but they are not aware about the real effect of the so-called charitable and generous acts done by them on the general public. If they come to know then, as a consequence, they may do good deeds, but not such deplorable acts. In fact, these people are not bad or evil in themselves. It is only their capitalist system which is bad, and about which I have made all these observations.

Observing the rich and wealthy, other ordinary people also wish to do the same. If the rich buy saris worth Rs. 500 for their wives, the common folk, too, wish to do likewise. So, they try to find simple and easy ways to obtain money and the easiest way they can figure is gambling. In every city, in every town some sort of gambling or another goes on. In all these betting games ordinary persons always lose. The daily wage earning labourer or salaried worker loses Rs. 10 to 20 every month at the hands of this vice. There is rarely a person who may at some time get a thousand or five hundred rupees as winnings. Hoping to win like this, people keep on betting and the well-to-do persons of the middle-class or the rich persons who take the 'bet' always profit from this. This, too, is a vice by which the poor become poorer and the rich become richer.

God has given the boon of a keen intellect to many poor children in the same way as he has given it to the rich and wealthy. Still, because of the lack of opportunity, the intellect of the poor is not able to develop fully and because they are not able to complete their education, society is deprived of this great wealth. The rich child gets an opportunity to develop his mind and intellect through high education and it taken little time for him to shine. But there are as many stupid or intellectually inferior children who are not able to either complete their education or develop their minds because of natural faults despite getting opportunities. Children with such faults are also found among the poor, but of course no one cares for them. But the intellectually sharp children of the poor are left behind because they do not get opportunities.

If equal distribution of wealth were possible, this hidden wealth of the nation would flower and bear fruit and would not be wasted thus.

21

THE POWER OF MONEY

The present blueprint of equal wealth distribution The root cause of the greed to become rich The area of talent development

There are many pitfalls arising out unequal distribution of wealth. Not only in trade and business, see anywhere else, in the army or police, too, officers' emoluments are several times more than that of an ordinary soldier. As a result there is much dissatisfaction. Similarly, in the courts of law, the schools, government offices, everywhere, great discontent prevails because of unequal distribution of wealth and man is not able to work properly according to his intellect or ability. Equal wealth distribution is a tried out thing, nothing new. It is fully possible. At present it exists to some extent in various classes, but not among persons. If one mason or a carpenter gets Rs. 8 or 10 a day as remuneration for his labour, the other masons or carpenters must be getting almost a similar amount. So all the masons and carpenters of India would be in the same income bracket. All postmen must be in the same classification of income. But as compared to a postman or teacher, the income of a doctor, a lawyer or a businessman is many times more. So this unequal distribution of money becomes painful. Now, if this method of equal income could be extended to all classes, the dilemma could be solved.

It is possible that there may be some person of exceptional intellect and ability who can earn several times more than a worker of his own type, and class and perhaps there may be someone who may not able to earn half as much as a worker of the same type; but these would be few in numbers. Most persons would be of average ability or brains. If equal distribution of money is applied to the workers of an industry or to the government servants in public service, then this scheme may become a success.

You may ask that even when everyone gets the same amount, then too it is possible that many people, by saving money, may gradually become rich while others would become poorer. What is the advantage? Even if you have distributed equally once, to what avail is it? You cannot have equal distribution of man's intellect or his ability. Then this whole discussion becomes futile. Man will remain as he is. How will it be possible to alter his state?

Maybe you are right, but I think that when equal distribution of wealth has taken place and there are no rich or poor, then the ambition to become rich which prevails today will also become non-existent. The argument about saving is baseless as it is, but if man could become rich by some means and even then not get any special benefit or right then he will have no inducement to become rich. On the other hand man would definitely try to use his ability, intellect and talent to attain happiness and that would be available to all through equal distribution of wealth.

Equal distribution of wealth is possible and there is no obstacle in its permanent function. The long-standing dispute would also end as so to how much each person should get. A large area will become available for all to display their brilliance and talent.

22

THE FEAR OF EQUAL DISTRIBUTION OF INCOME

We are not used to it
Labour, too, should not be centred in anybody Nobody has a right to poverty

The question which arises is that when equal distribution is to take place, then what is the meaning of getting a vast area to achieve the most out of one's ability or talent. Why would people put their ability or talent to use? When a person of a lesser intellect will get the same as the person with a higher intellect why would anybody endeavour unnecessarily? Why would he apply his mind? This apprehension is quite correct.

The difference between man and man, as far as income is concerned, can be eradicated by making an effort but ability or talent can never lie hidden. Many capable or talented persons almost live in penury in India despite labouring to their utmost, whereas people of lesser talent and those who are downright stupid, ruthless or crooked are millionaires. Their coffers swell up by millions without any endeavour on their part. Suppose a great scientist of the country invents a new machine which is useful to mankind and your parliament or government gives him Rs. 2,00,000 as an award (though it is fairly impossible in India) still that scientist is absolutely poor and destitute as compared to those traders who have bought and hoarded the entire stock of wheat in their godowns and sold at a very high price or those who give vegitable oil (vanaspati ghee) instead of real ghee or those who mix flax along with cotton while weaving cloth or who sell adulterated foodstuffs and spurious medicines. Or those, who find out in advance that the cotton crop will be lesser that year or who have made their newspapers financially profitable by false advertisements or who by many hateful actions have amassed millions and billions of rupees while a highly educated, talented and ideal person who endeavours to increase the power and reach of man's knowledge, gets very little money. Hence, wealth has more importance over virtues and everywhere in society the rich are venerated more than the virtuous. If money is equally distributed then the virtuous and capable man will shine because of his talents. He will be venerated in future. People will try to follow him considering him to be an ideal. What is money worth without virtue and talent?

The ideas and feelings prevalent in society will have to be changed. After the equal distribution of wealth, the class division of mankind will be based on their talent, virtue or qualities. That would be an ideal society; still there would be classes or categories in human society of high, medium and low people. But their measure would not be money: it would rather be intellect, talent or qualities. Those who will establish ideals in the country in the fields of art, science or craft will be hailed as great and honourable. Those who will do their work in the normal fashion with diligence for their own and the nation's good like the common populace will be placed in the middle order and those cunning, stupid or vicious persons who would create hindrances in the work of the higher and middle classes for their own selfish interest, will be termed as low people. The day the touchstone is logical, the importance of money will be reduced. Good qualities and natural talent would have a higher value, only they will be venerated and that society would be more useful, venerable and worthy of honour.

23

THE VALUE OF EACH JOB

Who will do menial jobs? No work is low and demeaning Leisure is more attractive than money

What is the inadequacy in the scheme for equal distribution of money?

Whenever this scheme is given due consideration, the objections which are mentioned or come to mind are that this scheme is new, we are not used to it, so how will it run? What would happen if it was put into practice? This fear terrifies everyone. Now, the main question or objection which remains is that if people were not given some incentive or award for extra labour or hard work then why would they do it. What is the incentive for a person to do anything?

One answer to this is that in the work done for the nation no one has to do more labour or hard work than the other person. Who says that you do more labour than the others? On the other hand it is desirable that along with the wealth of the nation the work of the nation should also be distributed in that manner. There are many people who do not get peace or happiness till they are fully immersed in work and work as hard as possible. The answer to such persons is that you work hard for the satisfaction of yourself, why put its onus on society. Why take this plea that you are working hard for the nation while you work for your own desire and satisfaction?

There are many others who do not like to work at all; they wish to while away their time in laziness. They should be declared criminals in accordance with the above social ideal. Just as a man is liable to be punished if he steals something which is of public use and belongs to the nation, similarly the man who deprives the nation of his labour and shirks from work is a criminal and should be penalised.

Same people will say that they will work less, even though they remain poor; they will not work more. They are happy in their poverty, they say, and why should anyone bother them needlessly. No person in the nation can or should remain poor because of his own free will. Poverty, whether it is forced by the capitalists or chosen of one's own accord, is unwanted and undesirable.

No man in the nation should have the right to live in a filthy environment and neighbourhood even if they so desire. Those who say, 'we don't want to be well to do' should be legally forced to do their share of work and contribute to the rebuilding of the nation.

In the present age man cannot work more or less according to his own inclination and desire. There are many difficulties in this path. In the industries, factories, etc., everywhere work is done collectively. If the hours of work in a factory have been fixed at 8 hours, no worker can work for 10 hours even if he wishes to. The reason for this is that other fellow-workers are needed to run the machines. And so, many kinds of work are dependant in such a way on each other that one or two or even five workers cannot work more than the fixed time even if they want to.

If the whole nation were to tighten its belt to work more, then the result would be entirely different. There are many instances in the world where a whole nation endeavoured to work for 12 to 16 hours a day and as a result many complexities and struggles resulted from it.

Well this is to say that every person of the nation has to only fulfil his share of labour. Now you will say that when everyone is equal then who will decide what work is to be done by whom? No one will be ready to do ordinary, menial or dirty jobs. Everybody would want to do clean, respectable or decent jobs.

As far as work is concerned, no job is dirty or demeaning. We term them so because of our own feelings and viewpoint. For instance, doctors or surgeons undertake operations and dissections and even dissect dead bodies and examine each and every part of the body. Many doctors examine sputum, blood, urine and faeces, etc., and look at these from very near through the microscope everyday, still no one says that the surgeon or doctor does menial work. Similarly, take the nurses. While tending to the patient they clean him, bathe him and help him in all his bodily needs like passing the urine or the stool. They change the patients' clothing and do all sorts of jobs for him, yet their profession is no less respectable than the work of a secretary or a typist in an office.

Since cleaning of latrines has been done for centuries by a certain caste of people—the bhangis or the mehtars—it is considered to be a menial job. Since the traditional scavengers live in dirty surroundings and are very poor, so the work has been attributed to them and is thought to be lowly. When the urine or faeces of children are cleaned in homes by their mother and other elders, then no one considers this work to be menial or lowly, and there is no objection in doing it. This feeling in people is a hurdle only because it is considered that this work is bad and demeaning. You will have no compunction in cleaning your own house, dusting it or cleaning the bathroom and latrine, but the way mehtars are seen carrying a broom and basket on the roads while doing it, is considered lowly and you would decidedly feel bad while doing it. Besides, all these jobs which are considered to be low can be eradicated in this modern machine age. In the cities wherever there are automatic flush latrines no bhangi or mehtar or cleaner is needed to clean them. Similarly, the eradication of many jobs which are considered low today by people is being thought out by man and his inventive faculties. Even the son of a millionaire will drive a car with pleasure and without any objection but just ask him to wear a driver's uniform and cap and then drive a car and see how displeased he will be.

The conclusion is that no job is bad or menial, but the people who do certain job traditionally are considered lowly by society which is why other persons hesitate or object to do those jobs.

If everyone gets the same amount of money, there are no ifs and buts; then it is but natural that every person will like to do the job of a clerk in an office or some such work nobody would like to become a cleaner and take up a broom to clean the gutter. And similarly women would like and desire to become typists and secretaries and actresses and would not wish to become cooks or house cleaners. What incentive do they have to take it up?

The question is a serious one. Besides money, the next thing cherished by man is leisure or free time. If the clerk in a library or an office who works with pen and paper is offered to do 8 hours of work everyday with 1 day per week for rest and suppose the cleaner is asked to do only 2 hours of work per day and his duty be fixed at only 4 days a week then you may rest assured that people would like to become cleaners rather than clerks.

When there is equal distribution of money or there is no difference in income, then the next incentive for man is leisure. Suppose you work in an office and the hours of your duty are 8. In the remaining 16 hours you will sleep or rest for 8 and use 4 hours for washing and bathing and eating etc. so, you will have only 4 hours per day for your hobbies or entertainment. You must also be wishing to spend that time with your wife and children while your wife who has more leisure than you desires that you spend that time with her in going to the cinema or for a walk or in having a good time. You have only Sunday or a public holiday to do all this. But if you had to work only 2 or 4 hours every day, then you would have the remaining time to fulfil your hobbies, etc.

A blacksmith wields a heavy hammer in front of a burning furnace; a labourer walks miles with a heavy load on his shoulders, while a clerk or an officer sits under a fan doing paper work. Undoubtedly, there will always be a difference in jobs. Everyone would like to do comfortable, likeable jobs and none would probably wish to do heavy labour when the distribution of money is equal. If the income in every profession is equal then should a person not opt for a popular job? As I said above, this deficiency will have to be rectified by increasing or decreasing the hours of work.

Eight hours a day 6 days a week should be the time allotted for popular comfortable jobs and comparatively lesser hours of duty for jobs requiring heavy labour with more leaves and holidays should be the norm. If this is applied then automatically all jobs will become popular. Money is only relatively attractive. If your neighbour or your enemy has millions of rupees and you have none then obviously you will desire to possess millions. But suppose this equal distribution of money is put into practice then there will be no charm in amassing money. In that situation the greater attraction would be in lesser hours of work or freedom for more hours of leisure.

Man is by nature a lover of freedom. He can rarely bear to be bound to some responsibility. Therefore, there will be more freedom in those jobs which are heavy or are considered menial.

24

NO MASTER

Work done to fulfil one's duty
Following one's hobbies
Freedom without chains

It is the law of nature that man has to do some work or another. Not to work also requires the inspiration of a special kind of action. If a man does not work his work will have to be done by another person. If the whole nation stops working then mankind will perish at the hands of famine, so the responsibility to work has been given by nature to each one of us. The only freedom is to think which work we will do and which we will not do.

If, according to its popularity, the relation of work with leisure or free time is established, then the problem is solved. Labour or the duties of one's employment are the type of work by which the benefit that a nation receives is shared by all.

Even during leisure people work in accordance with what they like to do; they pursue their hobbies; for instance many people like to do gardening, produce vegetables etc., many go into poultry farming and produce eggs, others play football or golf; many women take up stitching and knitting as a hobby. In short, many things done for one's pleasure or hobby also go to increase the nation's wealth.

Man spends his leisure resting. Although no wealth is generated by it, yet rest and repose is very necessary to rid the body of its weariness. But if man does not rest he will lose his ability to attend to his duties the next day and even if he wishes to he will not be able to work during leisure.

Man does three kinds of work during the day: (1) the work assigned to him as a duty to fulfil his responsibilities; (2) work of one's desire during his leisure; and (3) rest to rid his body of weariness.

The work which nature has assigned to man includes all the above three. Why do we object to people working like a slave or a horse? Is it not because some other person is making you work as if he were the master. The soul rebels from this. What right does a person have to enchain another person?

But with the equal distribution of wealth, all work will be 'for the pleasure of one's one self.' No one criticises the rule of nature. Everybody is ready to accept the natural laws. Only one is not ready to become someone else's slave. Often, even work becomes like a vice or addiction and then that person cannot sit idle even for 5 minutes. Automatically, there is a feeling of pleasure in work and man takes up work for his own happiness. But this is possible only if there is no one to rule over us and we are allowed to work in accordance with our own free will.

With equal distribution of wealth no one would be able to rule over us, nor would there be any kind of slavery. All will perform their tasks under the regulations of nature or destiny.

25

THE UNMANAGEABLE POPULATION

Unlimited expansion of population
Forms of natural equilibrium
Birth control: a necessity

In the equal distribution of wealth the difficult problem now is of population. A couple has 10 children. This family produces very little and takes equal share, whereas another couple has just one child or no child at all. The mental faculties of man have a streak of love for progeny. All parents wish to see their children happy and prosperous. Those who oppose equal wealth distribution say that without doing any or much work the 10 children of the first couple will flourish, while the only child of the second couple will be brought up out of the same national wealth. If this feeling spreads among all then the population will increase to such a level that the entire earth would not be able to feed them.

These critics do not know that nature always maintains its equilibrium. If a part of the human race is bent upon making itself extinct, nature steps in and prevents it. The effort to increase the population disproportionately and engulf the earth can never be successful. The rules and laws of nature are strange. No one has been able to fully understand them. The population of India is 400 million where as the combined population of India and Pakistan 20 years ago was 330 million. With the unchecked explosion in population there would be a dearth of houses to live in, in any city or town. The supply of food is also a problem. If the rule of nature is to destroy the Indian people in this manner by putting it through ordeals, then we should steer away from such rules and laws. There must be some way to rectify the situation.

It is also true that the normal rule can never be applied all the time while the increase in population is a curse on the one hand, it is a boon on the other. The more people we have as workers, the more the national wealth will increase. In real terms wealth is the work force of man. The nation which has this in the largest quantity is the wealthiest.

The equilibrium of nature is such that if the progeny of existing couples are successful in surviving then at the present rate their numbers would increase to 2 billion in 400 years. This is not possible. A cod fish lays millions of eggs at a time or a bee gives birth to thousands of bees at a time, but they are destroyed at the same rate. The birth rate of a life form conforms to the rate of its destruction. This is the natural law and this is what is called the equilibrium maintained by nature. In slums and filthy surroundings and in the nomadic and so-called criminal tribes where the resources of good health are minimum or non-existent, a maximum number of children are born and in the same ratio deaths of infants and children occur among those people. In the country where there is extreme poverty and the economic situation is the weakest, there the most number of children are born and also die. To prevent the race from becoming extinct nature produces in a large quantity so that after balancing births and deaths enough people survive.

This question cannot be ignored just by relying on the laws of nature. Therefore, birth control is a must. Couples should not have the right to produce more than two or three children. After that it should be legally made imperative that the either the man or the woman undergoes an operation. Or else, men should regulate and control relations with women by modern methods and use external devices, etc. It is necessary to check the population thus by passing laws. You will say what a contradictory statement. On the one hand it is said that the nation which has the most human wealth is the wealthiest and on the other hand it is being proposed that a law should be enacted to prevent the increase in population.

True. Increased population is indeed the wealth of a nation. But when? And in what circumstances? Just consider. A husband and wife are well-to-do. They have employed two servants to look after them. Both the servants easily do all the work. Another couple has employed nine servants to all their various jobs. Their work, too, is done with ease. Now tell, whether those nine servants do some special work? No. work increases or decreases by itself. Actually, those nine people not only do the work done by the two servants of the first couple but of 9 + 2, in all of 11 people. Even then in doing the work of 11 people they feel somewhat inconvenienced and so they require one or two servants more. Not only in the homes but also in offices or government establishments, too, the situation is the same. The more workers you have the lesser is the work accomplished, rather there is obstruction in the job. To do every work you need a fixed amount of wages. Less work is done if it exceeds this limit. This is also true of the workers in factories and workshops. In India surplus and unproductive workers are found in great numbers in every field of work; so, instead of producing their own share of wealth, they tend to hinder or obstruct the others is producing their share of wealth. Citing the example of surplus labour, most workers grow lazy and as a result the expansion in population becomes only a curse.

Another country which has a greater population than that of India is China. Twenty years back China's situation was many degrees worse than India's. Later, after the revolution the communist system of Soviet Russia transformed that country. And in this way by putting the ability of man to use constantly the population was made almost a boon. I do not agree with the economic policy or the political system prevalent there nor should you have anything to do with it. It would be another kind of feudal tyranny in which slaves or serfs (every citizen is slave or serf) have no right to weep if they are unable to speak or express themselves.

If every citizen does his share of the work assigned to him, and is productive then decidedly an increase in population is not a curse. Of course, there is a limit to everything. If the population increases to such an extent that the nation is not able to arrange enough food to keep it alive and people start starving, then it is better that they should not have been born in the first place. The population of India has reached such levels today. Nowhere, in any town or city, can one find houses to stay in or enough food to eat. How long can one survive by importing food from abroad? This, too, is a dilemma. So, today it is necessary that birth control be made compulsory. A law should be made that no couple shall produce more than two or three children. You might say that this is a blow on our religions, an effort to lower our ethical plane which has come to us through many centuries. It is written in the Vedas and the Upnishads that one should never sever the thread of the people; this is an order of the Almighty. It is said in Taittireeya Upnishad: '*prajatantu ma vyavachhetasi*' etc. At many places in our religious scriptures it is forbidden in clear terms to prevent or obstruct the reproductive force. It is correct that in the Vedic age the population was so meagre that all such instructions and sermons were needed. In those days it was considered a great sin to abort a child. Abortion is still a heinous crime morally and legally. Social responsibilities will be analysed only after the birth of human beings, but we also have a definite responsibility to prevent the terrifying explosion of population for the future welfare of mankind. So, the issue to be considered is this that is it better to produce children and then leave them to die of hunger, cold, heat or not to have them take birth in the first place? Out of the two which is the more or less heinous offence? One cannot turn away from reality by saying 'what God has willed will happen.' It is certain that one cannot care for and look after the number of children born.

In Rajasthan in some tribal children, especially female infants, were killed by parents knowingly after some time. Even if you do not commit this horrible crime of murder, yet, if you are not able to organise things in a manner by which children may be fed and kept clean and free of diseases, then this too is an indirect crime. How many parents can say today that have fully secured all their children? Children are born, grow up and keep dying. Those who survive, they grow up to full age. They do so by themselves. Parents and patrons have little share in this. There is nothing to gain by deluding oneself. It is certain that you have no right to produce children whom you cannot feed or properly look after.

The question still remains that when the wealth of the nation is to be distributed equally, then those who have more children will get more in that ratio. They will also be blessed by the love of children. Those who have no children, their hunger for this emotion or sentiment will not be fulfilled. So, equal distribution of money is not right. The answer to this is that national wealth and property is needed to do different jobs. For example, there is a carder who gets Rs. 100 or 200 a day for carding cotton and there is a pilot who flies an aeroplane worth almost Rs. 1 crore. None of those things are their own property—neither does the cotton belong to the carder nor the aeroplane to the pilot. Both are national property. The workers, each in his own time, act as trustees of the properties. Similarly, the care and welfare of children should be a national duty. Children may not get a share in the equal distribution of money, but if the responsibility of their upbringing, education and health is shouldered by the nation, no one will have any complaint about equal distribution of money.

In India married people or parents get a relief to only some extent in income tax, but that is not enough. All the money which is justifiably spent on children should be likewise tax-free. There is a proposal to make the care for children through such schemes the nation's responsibility. It is the nation's responsibility to care for children who are its future citizens, its heritage and wealth.

26

THE WORLD TERRIFIED OF THE PRESENT ECONOMY

Live and let live
the vision of heaven
The real socialist

Now you must have come to know that there is no difficulty in the equal distribution of wealth. This plan is applicable for all time and is possible. You must have seen that no one is satisfied with the present economy. Without analysing or fully understanding the word socialism you must have realised that if money is distributed equally among all, it is the best alternative. In case you do not feel this, then suggest another way which is satisfactory and beneficial. It is possible that the way shown by you is better.

If you have no better option of offer then you, too, should become a socialist. The Socialist political party has no affinity or affection with this ideology. There are many kinds of people in that political party just as they usually are in all political parties. There are those in it who give long speeches, yet if you invite them to a party at your house, they may quietly slip off with an item or two of cutlery. But there are many good people also. When you tell them that brother, what we have understood about socialism is that according to it the entire wealth of the country is to be distributed equally, whether it is an old man or a young man, male or female, thief, robber, washerman, potter, drunkard and gambler—equal wealth distribution will have to be for all, then after listening to you they will only laugh as if you have said something quite stupid or meaningless.

Then you ask them that brother, you are a socialist, you tell us what sort of socialism you want? Then their answer would be cut and dried, that every man should get equal opportunity. That is socialism. They say that even if the capitalists remain there is no problem, but everyone must get an equal chance to become a capitalist. You will ask, brother, how is this possible, without equal money how will equal opportunity be possible? Most socialists are not able to answer this question. You must have read many books. If not, then see the *Arthshastra* of Kautilya or the economy of the Mahabharata period. There you can find a picture of socialism. *Das Kapital* by Kart Marx is also an important book worth reading. When some leaders raised their voices after being terrified by the present economy, many isms were included in them such as communism, terrorism, anarchism, nationalism, factionalism. In fact out of all these, not one ism has been successful in running socialism as socialism in the world anywhere.

In any way, right from the beginning of creation, from the Vedic times, socialism is infused in the principle of 'live and let live'. The Vedic hymn to God is '*sarve santu niramaya*', that is, all the people should be happy and not feel any kind of want or adversity. This feeling was echoed and publicised by many a later saint and philosopher. Sages like Aristotle and Socrates and Plato declared that God has made everyone equal and all have the right to live life equally. In the Gita even Lord Krishna says, not only humans, but all life forms are equal in this sense.

Vidya vinaya sampanne brahmane gavi hastini
shuni chaiva shwapa ke cha, panditah samdarshinah

(whether it is an educated brahmin or a cow, elephant, dog or a man of the lowly castes, learned people see them equally.)

This is the very source of socialism. This was the socialism of those times.

The dream of bringing heaven down to earth is evident from this desire that all physical matter and things should be enjoyed and consumed equally. The mission of heaven, too, depicts such a picture. Whoever has learned that all physical stuff should be distributed equally is a true socialist, whatever path he may have had to take to reach this conclusion. And the person who does not know this is not a socialist no matter how loudly he professes to be a socialist or declares this from atop the Himalayan peaks and may well become a martyr in this exercise. Still, he is not a socialist. So, by now you must have realised the difference between the so-called socialists and the real ones.

Different parties publicise differently to bolster their political stand, but it will have to be seen as to who is making a true effort for the establishment of logical and factual socialism.

27

PRACTISE WHAT YOU PREACH

Alms are no remedy
New aspects of alms giving
Change the laws with discipline

Now even if you are convinced that equal distribution of money is a very good thing and it is needed, still, if you are rich, opulent or a person of the upper-middle class you are not going to distribute your wealth among the poor. This much is certain. Because, had you done so under the influence of such ideas, how could you have become wealthy and rich in the first place? There are tens of situations which arise everyday where one's sentiments are shaken and ruffled. If you are poor you will definitely wish that this distribution should take place at once so that you may get something. Some of you will say among yourself that this writer has supported and written so much about equal distribution of wealth, but it is matter of enquiry that when he himself is the owner of so much wealth, why does he not begin by distributing his own wealth. They will try to fit the lines of the saint-poet Goswami Tulsidas on us: 'many are those that preach yet practice is beyond their reach'. True, the scheme of equal distribution of money is necessary and you must have also understood that to solve the present economic problem satisfactorily this is the only remedy. But the dilemma lies in the implementation of the scheme.

This is not some work to be done by a particular individual or a group of people or even the entire wealthy community. The most they can do is give up their wealth. If they do so it will be alms-giving which would be a virtuous act and its recompense will be the reservation of a seat in Heaven for the alms-giver. As a rule the status of the giver is higher than the status of the recipient. So, naturally, there will be a feeling that the giver has obliged the person receiving the money. Again, it is a moral crime to take something without doing any work. Such an act can only be called beggary and money received thus will be alms. In a country if the number of recipients exceeds by far the number of givers, then it is difficult to think of any scheme related to it, what to say of its rise and progress.

To live on alms is the greatest curse of mankind. Not only today, but thousands of years ago sociologists of this country had laid down strict rules in this regard. They had declared that if anyone except the celibate student or the person who has renounced the world begs for his livelihood then he is a thief and in whichever land such thievery takes place and the king does not stop it then that king is a worshipper of thieves and has no right to rule.

'yen bhaikshyachara dwija.
chaur-bhakta pradau hi sah'

(He who allows beggary is a disciple of thievery.)

By this quote from our scriptures it is proven that to beg or give alms is a moral crime. Even if legally it is not punishable, still it is a moral degradation nevertheless. Then how can equal distribution of wealth take place? If we threaten and loot a wealthy person or pick the pocket of a person who has money, it is a crime and we cannot do it. The police will arrest us and put us in jail and the whole scheme of equal distribution of wealth would be bogged down. There is only one way to go about it. You will have to enact many new laws through the parliament and the central and state governments. And at the same time abrogate many old laws in force since centuries. This process shall continue for many months or years. With the promulgation of each act there will be some decrease in the wealth of the rich and a similar increase in the wealth of the poor. There will be ideal by-laws, keeping in mind every aspect of every regulation, so that the balance is maintained and in this way every one shall get equal opportunity and wealth will be distributed equally.

The first rule of every civilisation is that every citizen of the nation should be law abiding. Wherever there is an infringement of law and every person does what he wishes, there anarchy and brute force prevails and mankind heads towards extinction, no matter how many ideal men exist. If a person murders someone who is near and dear to you, then by law that murderer will be given a death sentence by the authorities, but you cannot say that he has killed someone near and dear to us so we shall kill him in return. If you do so you will land into the same predicament which that murderer is in. There is a person who is very lazy and does not work but idles and eats. He is a blot on the society. Undoubtedly you will feel angry but if in this fit of anger you berate him or even give him a slap, you will have to bear the consequences. This means that if in a society someone commits a mistake or a crime or even harms you, then you can only take legal action against him, you cannot take law in your own hands. By doing so you yourself will become an offender. The cornerstone of civilisation is to abide by the law and rules and regulations.

By becoming a socialist or by the establishment of a socialist economy the world is not going to change. There is going to be no change in normal laws or in their compliance. The crimes which are considered social, moral or individual will remain the same. Courts shall dispense justice as before.

People say that after socialism one would not have to pay a fee to the doctor or lawyer; there will be no tickets in the railway; things will be free in the shops; every man will do his job automatically without a charge; socialist ministers will not ride in cars; they will walk; socialist landlords will not charge rent to the tenants of houses; and so on and so forth. People have formed hundreds of false notion about socialism. Unfortunately, they have not understood socialism one bit.

Rather it is my belief that such people have not understood any ism, and not even civilisation or mankind.

28

INDIVIDUAL AND PRIVATE PROPERTY

Difference between the two
True aspect
End private property

Before establishing a socialist economy we will have to fully comprehend it. Until we are not able to understand its very opposite point—capitalism—it is difficult to understand the elements of socialism.

Before putting socialism into practice in the nation, it is necessary to understand the capitalist economy which we are thinking of replacing with socialist economy. The meaning of socialism is not to destroy or eradicate capital in any way. Those who describe this to be the aim of socialism are on the wrong path. Actually it is the capitalist's propaganda that the socialists are bent upon destroying the nation's capital. Individual capital has as much place in a socialist economy as in a capitalist economy, rather many times more.

There is a vast difference between individual capital or wealth and private capital or wealth. For instance, you have an umbrella. This is your individual property. The food served before you, the clothes that you are wearing, these are all your individual property. But the field in which grain is produced has come down to your family from your forefathers and you have been getting its revenue every year in the form of either rupees or grain from the person who has been tilling it. This field is your private property. The factory which you set up and you spent money to install machinery in and hired people to run those machines—that factory or mill is your private property. You earned a million rupees through such enterprises and lent them to many people; you have taken promotes or *hundis* from them or kept their land and property as mortgage—that money is your private property.

In normal spoken language very little difference can be made out between individual property and private property, but in reality this is the cause of all trouble. It is here that socialism and capitalism take two separate paths. That is the reason why through a few examples it was explained to you what is called individual capital or property and what is called private capital or property. So in short you should understand capitalism as a system in which the entire land of the nation comes into the possession of a few people's hands, and they use it as they wish. They give you the permission to remain on it on their own terms. Such people are called zamindars, jagirdars, landlords, owners, esquires, etc.

Legal experts say that the land does not belong to any special person; all the land belongs to the state. Whoever rules the state has the right to acquire it at any time and take its possession, probably this is the legal situation, and it is right. The state or rulers do not acquire this land in reality and therefore it remains as the private property of the people and this is the present situation. There is produce from that land. This produce accumulates with those people who possess that land. This accumulated property is called capital. This capital also becomes private property. As a result, all the factories and industries which need land and money to be established they, too, become personal property.

Now, because the factories and mills cannot run without labour, so the owners of these factories employ people who are called workers and they are not the owners. The owners give them only that much as wages which allows them to live and work. Children are born in large numbers and they, too, become workers so that these factories and mills keep running. Goaded by this selfish interest, the owners of those factories feel that it is their duty to give as little wage as possible to the workers and take as much work as possible from them that they finally wither away and die. After knowing all this, people will accept that the inequality in wealth keeps increasing as one goes along. The worker or servant becomes poorer day-by-day while the factory owner becomes richer in that ratio. The oppression of extreme selfishness reaches such limits that dissatisfaction, misery, poverty, crime and serious ailments spread widely, and finally a terrible rebellion takes place. Government has, therefore, to employ police, armed constables and soldiers to quell and suppress this revolt and maintain peace. This is called the capitalist system. In this system it is the duty of the government to protect private property, which is there in the forms of land or factories or capital and punish the people who pose a threat to it through the police or the soldiery. The government properly protects the agreements and contracts signed with others to increase private property and wealth.

As opposed to this, socialism declares that it should be the first duty of the government to see that everybody has the right to equal money, no one should have the right to possess private property, every contract and agreement should be treated as a national contract, nobody should accept such agreements by which one person dies before his time by working day and night and feeling deprived, while the other sleeps peacefully and spends extravagantly—all because others labour for him. This is what is termed as socialism, which will eradicate personal property and wilful contracts and if you have noticed, during these 10 to 20 years some progress has been achieved in this direction. The struggle between capitalism and socialism has been going on in India for the past century or so and its impact is being felt after independence. Because of the peoples' wrath capitalism is relenting to some extent and has become tolerant of many socialistic institutions. As I said earlier, there is a great difference between individual property and private property. One should not commit the mistake of considering them synonymous. The example cited is this—your umbrella is yours, you can use it as you wish, but you cannot use it to strike somebody on the head with it (although, this may be a necessary use for umbrellas at some places). You may evict a tenant from your house along with his wife and children without any regard to the inclemency of the weather. But the platform outside your house obstructs the traffic on the street. By demolishing it the way can be made clearer and it shall benefit the people using the street. Yet your dogs sit on that platform in the afternoon or you like its view and so you do not permit its demolition. Even though it may cause no end of trouble to the passers-by and motorists going down the street or cause any number of accidents, you still do not want your platform to be demolished. It is to be noted that this action is much more objectionable than hitting someone on the head with an umbrella. But in a capitalist system the protection of private property is a must and therefore you have a right to do all this or even more objectionable acts. But you have no right to use individual capital in this fashion wilfully and so if the socialists say that private property should be abolished as soon as possible, then they are right in saying so.

The capitalists and socialists both claim that they are working towards the interests of mankind as much as possible. Both present arguments in their favour. Prior to independence in India, capitalism reigned supreme. Under the British, the working class or the poor people could not dare to object. Ever since the Congress raised its voice against the foreigners and at the same time made the people aware of their predicament, people came to realise that 'capitalism is poison.' As a result, even the premier capitalists of India or the supporters of capitalism cannot openly profess that they are capitalists. This word is now taken as an abuse. Its other form which remains of it is leftwing or rightwing. The rightists think their system is more satisfactory. The leftists profess that their system is better. Both, while supporting their individual stands, say that in reality their respective systems are socialist.

Anyway, the aim of our examination is to ascertain which socialist system is better and to what degree. So, let us first examine capitalism.

29

THE INDIRECT FORM OF INCREASE IN CAPITAL

A certain share of the capitalists in every purchase and sale
Nationalisation of private property
The guarantee of enterprise run by the nation

The first question you should ask yourself is where this unequal distribution of money in daily life pinches you.

The answer is clear. You go to the market to buy things which you need. There you purchase items of food like flour, cereals, cauliflower, peas, ghee, oil, etc. Thereafter, you buy things to wear, such as shirts, dhotis, coats, caps, etc. Similarly, you buy tickets to the cinema or plays for your entertainment. Or if you fall sick you call the doctor and pay him his fee and have yourself treated. Or if there is a legal problem, you pay a fee to the lawyer, and so on and so forth. If you think and analyse, you will realise that out of the money you have spent to pay for all your purchases and for the services you have enjoyed, a part is being paid to some person who has done nothing for you. Now, you will agree that no person happily pays more for something which in reality costs less than the amount paid for it. And when after analysis you realise that those things and services are necessary for you and you have had to pay an amount or a part of it for the capitalist's profit out of each item purchased and service enjoyed, then you also realise that this is the basis of the capitalist system. Without doing anything, those capitalists have received their share out of everything that you have bought and paid for.

Socialism declares that this sort of system should be put to an end. There should be no private property. There should be nationalisation of all land, factories or capital. That is to say that all factories, mills, etc., should not remain the private property of any individual but should become the property of the state. Then there would be no profit payable to some other person out of the things you buy. Those who object to this, say, how is this possible? How can the nation run all the factories and mills? And what is the guarantee that it will be able to run it properly and efficiently.

Just try and think why this cannot be done. There are many industries which the nation is running. See, if they are functioning properly or not. The Post and Telegraph Department is run by the nation. Railways are run by the nation. Many departments like the army, navy, air force, police, telephones, roads, bridges, electricity, gas, etc. are under the supervision and ownership of the nation, and we are not uncomfortable or unhappy in any way at this state of affairs. Then why should the nation not take into its possession other industries, factories and services as well, and run them properly?

In the work done by the nation the principle of profit does not exist. Take the Post and Telegraph Department. You pay 15 paise and your letter will be delivered from any part of India to any other part. The postcard will be delivered at 5 paise. The expense which the Post and Telegraph Department incurs in reaching a letter from Bombay or Calcutta or Delhi to Trivandrum is much more than the expense incurred in reaching a letter from one part of Bombay to another part of the same city. Yet an average has been ascertained and 15 paise has been fixed as the general charge. Now, if this very work were to be done by a private company then perhaps it may only undertake to deliver letters from one part to Bombay to another on which it would not be spending 1 paise per letter and refuse to deliver letters to distant cities or charge more for them. However, a private company will not work without charging an appropriate profit. The result would be that on an average you may have to pay 16 or 20 paise for having a letter delivered for which you now pay 15 paise.

It is necessary that in order to successfully run essential industries and services which are in the hands of the nation, such laws and regulations be passed that no one else can do that work. Because if, for instance, a private company also undertakes the job of delivering letters, then it will manage to deliver letters to all the nearby places at 10 or 8 paise per letter instead of 15 paise. Then the nation will have to pay more for those letters which have to be delivered at a distant destination, because it will incur a loss by losing revenue on those letters which it would have delivered to a nearer destination and so offset the extra expense on letters delivered to a distant destination. So it is necessary that no one else be allowed to run a service or industry which is run by the nation. If this happens then other things will likewise be available at one price everywhere. The clothes you wear, the food you eat, the house you live in, etc., all these things will be available to you everywhere at a proper price due to the law of averages in case they were produced or rendered by a nationalised factory or farm or department.

In a capitalist system this is not possible.

30

STATE REVENUE RECEIVED FROM TAXES

State administration based on taxes
Most of it goes back to the capitalists
Payment of taxes
Firm steps towards socialism

Besides the items of food, etc., bought from the market you spend a lot of money on daily expenses.

Telephone rent, electricity charge, water charge, etc., are paid by you. In the house rent, municipal taxes are included and you have to pay them. Many citizens treat these taxes as useless and unnecessary and hesitate to pay them or pay them unwillingly. But on careful consideration it will be evident that municipal taxes are very useful and how much benefit you receive in lieu of them. Numerous conveniences like roads and the footpaths on sides, street lights, water pipes and faucets, gutters, etc., are managed through these very taxes by the municipal corporation. On seeing these conveniences the taxpayer definitely feels some satisfaction. Municipal Corporations carry out these works at cost and do not keep any part of the tax as profit. Similarly other taxes like the income tax, etc., also go to provide the people with many conveniences whose cost is equal to the tax paid. After seeing and knowing all this, you will decidedly feel that when next year the income tax officer orders you to pay the tax you will have no objection in paying it immediately.

But it is sad that while apprising you of the whole state of affairs, we have to also say that the capitalists grab their share even from the municipal corporations and the state out of the various taxes.

You must wonder how this happens.

In the same way as other things happen. While land and capital remain private property, the municipal corporations or the government have to purchase something or another from these capitalists which includes their share of profit. Municipal corporations will buy bulbs for the street lights. They will have to be bought from the people who have private property and manufacture bulbs. Similarly, hundreds and thousands of things are purchased daily by the municipal corporations and the government, and in each of these the share of the capitalists is assured. In this way a large part of the house tax and other taxes finds its way to the pockets of these capitalists.

To rectify this situation the government levies taxes at a higher rate on the rich and at a lesser rate or none at all on the poor. Because of these taxes the rich have to hand over half or two-thirds of their income to the government. In lieu of this if the government or the nation renders no special service to the rich, then too only a half or two-thirds of the income of the rich is a sort of nationalisation. In the beginning they had felt bad when their income was thus appropriated; but now this arrangement has become so natural and accepted that no one even talks about it.

Every year on the last day of February, the Finance Minister enacts a new law and presents a new budget by which several taxes like the income tax, the super tax, etc., are levied on the people. This yearly regulation is known as the Finance Act. But in reality it is Poverty Act, because through this Act peoples' money is taken away from them every year.

There is no restriction on this in the Constitution of India or in any law or custom. This totals almost half to two-thirds. Even if it were raised to 90 or 99 percent, there is no one who can stop it. It is entirely on the government's will that which person or persons will have to pay taxes and at what rate. It is not necessary that all people are taxed equally. This method adopted by the government is the first step towards equal distribution of money. Besides these direct taxes, there are indirect taxes as well; such as sales tax, purchase tax, manufacturing tax, excise, etc. All these taxes are levied on the rich and the poor equally. These are ways to create the income of the state or it can be said that some amount is taken from everyone for everyone. This is communism.

The Indian Government realises millions and billions of rupees through such taxes. You will wonder as to what happens to such a big amount. The government runs the army for our protection on the basis of these very taxes. The army, navy, air force, courts of law, etc., all the conveniences that you receive are paid for by these taxes. In this way through these laws the government only redistributes wealth. It takes a little more from the rich and a little less from the poor and nothing from the very poor and destitute and distributes it equally among all. No faults or virtues of the recipients are taken into account. Everyone receives properly. There is no legal restriction to this Act. The government has the right to distribute or use this accumulated wealth in the way it desires. There was a time when the income-tax in India was at the most one anna in every rupee. Even the person with the largest income was liable to pay only one-sixteenth of his income as income tax. At that time no one had even thought that a day would come when taxes would become a means to the equal distribution of wealth.

Besides, the nation has to pay a large amount of money yearly as interest. Ten, fifty or a hundred years ago, when the nation must have faced some calamity like warm, etc., then the government must have borrowed money from the people. You know that the people who lend money are the self-same capitalists who have money saved out of their income after meeting their expenses. When the nation took this loan it agreed to pay interest at the rate of 3 or 4 percent. Today a large part of the money which the government collects as taxes goes towards paying this interest. This interest goes into the pockets or coffers of the rich. When the amount was borrowed then all the things purchased from this money were supplied by the capitalists and each item included their profit as well. Later, these very people charge interest also.

In short, because of private property even today a large part of the taxes finally turns up with the capitalists.

31

TAXES LEVIED BY THE MUNICIPAL CORPORATION

General public convenience
Extra income
Responsibility of the lesser income group

A part of the house tax, etc., levied by the municipal corporation, likewise, finds its way into the coffers of the capitalists.

The municipality runs many schools, many hospitals, builds roads, maintains public gardens and does many acts of public service and convenience. Roads, etc., are also its responsibility. House tax is collected in ratio of the rent of every house concerned. The rent of a particular house will determine its house-tax. In this manner every tenant of that house has to bear a part of the house tax levied on it, whether he is rich or poor or in a position to pay or not. House tax is a communist tax which is levied equally on the people. Suppose an old couple live in a small room whose rent is Rs. 10 per month. They do not even get to eat two times a day. They say, 'what have we to do with the public parks or gardens? Why should we bear the burden of this tax?' But no, this is mandatory. Every person among the general populace has to bear this house tax and other municipal taxes.

The municipalities purchase goods worth hundreds of thousands every year. Those who sell or supply these goods are the same private property owners. The profit goes in to their chests. The municipality also has to borrow money. The lenders are those very people of the rich class. The interest on these loans is their income.

There are many services where the income of the nation is more than expenditure such as the Post and Telegraph Department, Electricity Dcpartment and the Railways. All this profit is accumulated in India's composite fund. As a result the need to increase income tax or super tax does not arise.

Similarly, if the municipal taxes and taxes levied for public services which every person has to pay are realised in excess of expenditure then they go towards lessening the responsibility of the government.

32

ALL THE LAND BELONGS TO THE STATE

Who owns the money and the land? Property declared by the Mughal Emperors and the British Zamindari Abolition: India's revolutionary step

The rent you pay for your house is a clear and direct way by which the capitalist landlord realises money from you. As has been explained earlier, income tax, house tax, etc., are indirect methods by which the rich and capitalist class of people exploit the general public, but the rent of the house or the revenue on land is a direct way of exploiting us.

There were no proper records or lease-deeds to determine as to which land belongs to whom in India.

Upto the 10th or 11th century the general rule was 'might is right'. Money and land belonged to the mighty. Whosoever had power in his hands could take possession of land and become its owner. When the power of a person waned, then other persons from outside or his neighbour attacked him and snatched away his land. It was in the 16th century that the Mughal emperors gave a legal status to land and lease-deeds, etc., were executed giving people ownership of land. Soon there were many landlords, jagirdars, zamindars, rajas, etc., all around.

When the British government established itself firmly in the 19th century then they usurped many lands of local rulers.

At the same time they gave shelter to many jagirdars, mirasdars, kings and barons and zamindars. As a result, the land which was in possession justly or unjustly with someone remained with him if he was favoured by the British. For a century or a century and a half, these zamindars remained owners of these lands and levied land taxes in accordance with their whims and accumulated untold wealth. Many zamindars spent fortunes on their hobbies and enjoyment. Whenever the owners of these lands wished they drove their tenants away or deprived them of their land. Every zamindar had his own manager and revenue collectors and a small force to realise the revenue.

The land was tilled by some peasant. He stayed on the land along with his wife and children in a hut which he himself built. If one day the zamindar would be struck by a fancy that a golf links he made on this land then the poor peasant would be driven out immediately. No matter how big a hue and cry he made, he would not be heard. If some other zamindar suddenly got an idea that rearing goats and sheep on his land would prove more lucrative, he would immediately order that all the peasants and labourers living on that land be thrown out. And they would be in fact thrown out and replaced by goats and sheep. Those people would not be able to find shelter anywhere. These landlords and zamindars would arbitrarily levy land taxes and revenue. These are the capitalists who have accumulated much wealth in this way.

How can land made by God be the private property of any person? Legally the ownership of land is never of any particular person. Land always belongs to the state. Land will always be held to be the property of whose ever rules the state. The people only have the right to live and settle on it and work on it. During British rule this right began to be bought and sold and this in turn was transformed into ownership of land. Even today legally no land is private property. Still the state has not thrown out anyone from it. So the landlords and zamindars who were so-called owners of land, sold the land to others or gave it to them on lease.

After Independence the biggest revolution took place in this direction. Those who had become the owners of lands and would whimsically extract land rent from the peasants, were forbidden to do so. At the same time it was declared that the person who has been tilling the land for a long time will become the owner of the land. The land would be deemed to be his. This was such a massive transformation that the whole outline was altered. The basis of capitalism was shaken.

In capitalism, as has been said, there are mainly two types of private property—land and money or capital. With the changed laws, land in most cases passed out of the hands of the capitalists. Although it was not nationalised but partly the aim of nationalisation was achieved. A ceiling was imposed on the land which a family could keep in its possession and the excess land was taken from it and handed over to the poor and other landless people. Now, out of these lands, especially lands in cities and towns are still in the possession and ownership of those very capitalists. They have built houses on them and their prices have risen considerably. In Bombay and Calcutta at places where the land is near the market place or commercial area it is valued at one thousand rupees a square yard. They build houses on this land and rent them out. Rent is arbitrarily levied by the landlord. Now, in the scheme of equal distribution of money, the government has put a ceiling on the increase of house rent.

Other private property should also not be allowed to exist like land and housing because by this, capital will accumulate at one place and these will be impediments later in bringing about socialism.

33

THE MAGICAL ILLUSION OF EXTRA MONEY

Rent for money besides land and housing
Money earned without working
Cost capital

Land and buildings are not such private property on the income of which a person can stay at home and live life without doing any work. More than the money accrued from the rent of buildings is the extra money with the capitalists accumulated through 'rent' accrued on this extra money and that is their second private property. The rent of this extra money is called interest.

Whosoever earns more than he can spend on his food and clothes and other necessities of life can save some money. It has been said earlier that real money, that is, food, clothes, etc., cannot be saved and kept for long. The second wealth is gold, silver and currency, etc. These are also things which cannot be saved. Their value exists only till you can buy the necessities of life with them. If for some reason you cannot buy four pounds of flour with the money you have, then of what use is all this saved and accumulated wealth? Gold and silver are metals whose price has fluctuated through the ages, but still they cannot be saved and kept. According to the principle of capitalism, if a person has more than his need and he can neither spend it nor save it then what should he do? He loans it out to someone who is in need of it at that time and takes some rent every year or every month for it. In this way the capitalists realise not only the rent of land and buildings, but also of the money in their possession and all this income is called unearned income, that is, income accrued without doing any work. For instance, you have some extra money saved. You cannot keep it intact because it may not remain the same after 10 years or 10 months or even 10 minutes. The reason is clear. During the past 20 years seeing the decrease in the value of money or let us say seeing the increase in the cost of things, one cannot possibly say that the value of money today will remain the same later. So, it is necessary to spend the money right now. And while spending this money if we can earn from it, then spending money in this way is called investment or cost capital.

34

Various Forms of Cost Capital

What is to be done with extra money?
Who believes in increasing money?
The rich need money the most

If you have food left over after eating and you give that food to some hungry person on whom you can rely on condition that he will return the same quantity of food after one year, then it can be said that you have spent your money and also saved it, that is, you have arranged for freshly cooked food at a date one year hence.

You will immediately say that hundreds of people would be available who would be ready to receive this food but it is difficult to get a trustworthy person who would return us food of the same quality and in the same quantity. How can we trust someone? It is not certain that those who are hungry today would be able to get something to eat themselves, what to say about returning our food. It is absolutely right, but there is a solution to this problem. You may not get trustworthy people among the hungry but your bank or your share brokers or your lawyers would certainly be able to find such people who can be trusted and they would be rich persons in most cases.

You will ask that if they are rich why they would need leftover food. Yes, their own stomachs are so full that they do not need food at all for themselves, but they do need food in large quantities, because that food is needed to feed those hungry people who cannot be trusted to return that food next year. They, too, do not trust these people but in lieu of that food they get some work done by the hungry people which will earn them money later. Any intelligent or talented man can do this 'business'. Suppose you have a large farm in the village and the road which goes past it, passes through very rough terrain. Now, you get the people, whom you give food, to build a good road through your farm. After building that road they may search someone else and get food from him.

Furthermore, you begin to charge some money as toll tax from the people who come and pass through that road. Suppose the passers-by are large in number then in a few days you will accumulate much money. You use the money thus accumulated to feed the hungry and have work done in return and have some houses built alongside that road which you rent out. And from the rent collected in this manner you again feed the hungry and have other work done. Gradually, you will build houses, factories and mills and employ thousands of people in those factories, and you give them saved food in return of their services. As you go along the quantity of saved food would also increase and with it you can do many things and there will be no end to this business.

Sometimes these plans fail, but still since you cannot keep your extra food saved up, you have to spend it and give it to those whom you can trust, and you will have to bear this risk. For the big industrialists or for their companies extra food or share capital is always available. In this strange way the very rich people become increasingly rich. And in no time shops, factories, railways, mines, steamers, huge buildings, palaces, telephone, aeroplane and cars, etc., come into existence.

So the saved food works like magic. Many lazy persons who have land and property, and capital become very rich and do not even know how they became billionaires. Their children are born millionaires whereas the workers, who are instrumental in achieving this, by whose labour all this is made possible, wander about in search of work to appease their hunger. They remain poor like before.

35

THE NATION AND CAPITAL

The aim of capitalist economy—more profit What is or is not in the nation's interest Capitalist mentality in nationalisation

It is to be noted that these capitalists only invest their money or the money taken from others in things where they see the most profit in the shortest period of time.

Suppose a light house is urgently needed to show the way to the ships at sea. There is a great need of ports for import and export. Similarly, hundreds of such basic things are needed for the progress of the nation. It is possible that the risk in making them is greater or perhaps they do not yield as much profit as other factories. So, the capitalists will never spend money in making them. All these essential things have to be done by the nation, by the government. Of course, if a cigarette-making unit is to be set up, the capitalists would be ever ready to set it up. They will increase the output of cigarettes to the maximum and through large advertisements and hoardings declare that smoking cigarettes is good for health and smoking their brand, the special type of cigarettes made in their factory, is almost like a religious work full of blessing. Although all this is a total lie, they do not hesitate in setting up many such factories and companies.

The second major thing is that the capitalist tries to earn the utmost profit in the shortest time. Suppose you send a telegram through the Post and Telegraph Department of the nation. You have to pay a charge of 10 paise per word. Suppose it was reduced to 2 paise per word then would the nation suffer an economic loss or not? If the rate of telegrams becomes so cheap, then it is natural for the number of people sending telegrams to increase greatly. Suppose the charge is increased to 1 rupee per word then the number would accordingly decrease. The Telegraph Department would earn the same amount of profit because with the increase or decrease in charges the number of persons sending telegrams would increase or decrease accordingly.

Increase the cost of some service or thing and it will rarely have an effect on the profit. Had the same Telegraph Department been the private property of the capitalists they would definitely charge 1 rupee per word and the profit from that would go into their pockets. In the industries run by the state, charges are naturally lower and cheaper because their motive is not profit. In case there is profit, it is spent for the benefit of every person of the nation.

In India in the industries nationalised till now, the mentality has been that of capitalists. Hence, the people at large have not been fully benefited by them. This happens everywhere where transformation takes place. The habits of people used to doing things in certain specified ways take time to change. Unfortunately, today's nationalisation is merely the change of the capitalist emperor. Where the capitalist ruled arbitrarily and whimsically, now there are new government officials who are perpetrating as much or even more injustice and exploitation. The only satisfaction is that where the capitalists were in power on their own accord these officials are in power because of the people's will. With the change in the will of the people, the biggest officer would be vanquished in no time.

For example there are the large projects of electricity-producing companies. In the companies which are under the capitalists the charge for instance is 10 paise per unit. In many places there are State Electricity Companies or State Electric Boards where the cost of producing electricity is at the most 2 or 3 paise per unit. Still, the government officials think it a matter of pride to sell electricity at the rate of 10 or 12 paise per unit and thus show that they also know how to run businesses like the capitalists.

It is a pity that they are not aware of the true spirit of nationalisation. Whereas the electricity plants of the old capitalists were set up at a cost of 100 million rupees, new plants of the same specifications today would cost more than tenfold to the government. Then why should you compete with them and how? It should not even be tried. The nation should provide electricity at cost or even less at places or all such industries and electricity companies should be nationalised. The present policy of the Government of India which 'wishes to do so and also does not wish to do so' leaves the people wonder struck and this certainly is uncalled for.

The capitalist mentality of profit-making, infused in all the industries which have been nationalised, should be removed root and branch. The nation's property belongs to the people for the benefit of the nation. Why should an officer consider it his personal legacy? The ideals of the government officials, too, are of the capitalist kind. They too have become millionaires by exploiting the people through their actions. Today they also think that now when the chance has come, why should they not become millionaires? It is certain that the bureaucrats can never become millionaires without resorting to illegal methods. As a result, intense corruption spreads. The same rule of the rich applies here.

36

THE SARVODAYA ECONOMY

Going back is impossible
Excess of resources
A national need

The form of economy propagated by Mahatma Gandhi has been given the name of '*Sarvodaya* economy'. Perhaps, this system may be considered appropriate and ideal for India seeing its pitiful state. But if India has to live honourably with other nations of the world while defending its independence, then how will it be possible to have *Sarvodaya* economy whose very basis is sacrifice.

Thousands of years ago the population of India was 100 million. At that time hand-spun and hand-woven clothes from the cottage industry would have been enough. But with the present population of 400 million people how it is possible to provide clothes for every one without machines.

Suppose they could be clothed, still would that condition be acceptable and proper when other nations of the world and even the neighbouring countries (which were part of you till yesterday) would adorn themselves with fine muslin and brocade while the people of this country would be wearing coarse hand-spun clothes. Undoubtedly, the people here would encounter a sense of inferiority and the desire to get all those good things would be kindled forthwith. The result will be same as it was 200 years ago in India at the time of the establishment of British rule.

Pats made out of cow dung are used as fuel in this country. If we say that till now the Indian economy has been a cow-dung economy it would not be an exaggeration. *Sarvodaya* offers no inspiration to bring the country out of that situation. Non-violence or Ahimsa is being inculcated into the people where the meaning of non-violence becomes only cowardice. If someone invades your house and you quietly accept it or apply a veneer of logic and keep running then it is not non-violence, it is cowardice. But where is the sense to understand even this? The slogan of non-violence is being raised from the hilltops. The tenets of panchsheel are being recited day and night. In such a situation when and in which circumstances can one say that now there is no need for non-violence? In this pseudo-mantle of non-violence the power of the nation is waning, its mental and intellectual level is declining. But how can one explain all this to the people?

In order to defend the country's freedom the biggest need is of richness of resources. This resource material will not be made in cottage industries or by handlooms. To generate these resources, modern and large machines would be required. Big industries would have to be set up. Large factories would have to be built.

The Planning Commission is working in this direction. There is enough stress on industrialisation. Industries have been developed properly on nationalist as well as capitalist lines but full success cannot be achieved till there is a definite path and policy.

Even learned secretaries and ministers are harping on *Sarvodaya* economy or cow-dung civilisation or cottage industry. Then how will it be possible to stand among the other nations of the world? This question is worth considering.

In the Mahabharata, Lord Vyas while giving instructions said that if you do not have resources you cannot he successful in war.

To gather all these resources and material, it is necessary that factories and workshops be opened in even larger numbers and everything be produced copiously. Then there should be no dearth of the necessities of life and alongside these, other things should be available in plenty.

This is only possible if the nation progresses industriously and forgets about the revered economy of sacrifice of 500 years ago.

The biggest objection to nationalisation is that foreign capital cannot be obtained by it in adequate quantity. If the industrialisation of India is desired, then the country will have to establish large industries on its own resources and capital. If not, then aid from foreign countries will have to be taken.

37

FOREIGN CAPITAL

Foreign aid inspired purely by self-interest
India's non-alignment
Export of Indian capital

Any capitalist industrialist, no matter which country he may reside in, will only invest his capital when he is sure of its security and of making a profit on it. If the policy of the Indian Government persists with the nationalisation of industries, no foreign capitalist will invest even a rupee in India. This apprehension is not entirely baseless, but stems from a wrong premise. No capitalist will invest money outside his country as long as he has an opportunity to do so within his country. No country is perfect and self-sufficient in the world and can never be so. In every country there are many areas which are open for investment and profit-making; still, it is a subject worth examining why these foreign capitalists invest in India and other backward countries.

Ever since the beginning of creation there has been struggle between man and man. This struggle is unavoidable. Whether the struggle is waged due to the greed for expanding the empire or for the propagation of one's religion and principles or whether it is economic, the struggle goes on and will continue in the future.

In the last World War, terrified by the tyranny and dictatorship of Hitler and Mussolini, the remaining countries waged war against them. The result is there for everyone to see. Fourteen years have passed since peace was declared but can it be called peace in true terms? Today the terrible friction between two major camps is about to explode into a new World War.

On the one hand there is communism adopted by Soviet Russia in which many countries are included, such as China, Czechoslovakia, Poland, Hungary, etc., on the other hand there are democratic republics under the leadership of America in which countries like Britain, France, Germany, etc., line up and are known as supporters of the capitalist system. Between these two so-called 'isms' there is a struggle, that is, between communism and capitalism.

Both the camps endeavour to increase their power. All the destructive molecules of the human mind are being utilised to the utmost. Terrible weapons like the atom and hydrogen bombs are the products of this mentality. Both the camps are betting on who can destroy this earth and how soon. In this competition there is also a race to invade other planets.

If Soviet Russia has sent its rocket or sputnik to the moon first, then America is trying its utmost to reach Mars before anyone else. This seems like a fairy tale when you hear or see it, but nevertheless it is true.

The population of China is much larger than that of India. The economic or social condition there was not better than India's. Still, when that country joined the communist camp of Soviet Russia it achieved development and industrialisation within a very short span of time. Did the Soviet Union extend all this help as a gesture of pity or generosity? No. A country provides economic and military support to another country only out of self-interest.

Ever since India has achieved independence, both the camps are trying to get India to join their bloc. Fortunately, the constitution of this country is that of a democratic republic and its Prime Minister is Pandit Jawaharlal Nehru. He has joined neither camp nor has he allowed his country to do so. This is the reason why India is not getting economic cooperation adequately from either of them.

Germany was defeated and destroyed in war and it is still divided into two parts. One part is under the Western power-bloc of America. It has progressed to an extent which it had not achieved when it was on the pinnacle of its power. In the other part of Germany which is under the Soviet Union, there too great progress has been achieved. Such are the claims although no proofs are fully available.

This is to say that if some foreign power gives economic aid to India then it does so not because of any benevolence towards India; they only do so to increase support in their favour and if India were to openly support one camp then it can receive enough economic aid. America and other Western countries have developed to quite an extent. Their principles are those of democratic republics. So they are closer to the principles laid down by India's constitution. But whether India joins any power bloc or not is a political question and discussing it may lead us astray.

We only wish to say that if India does need foreign capital for its development, then whether we get it or not, there should be no compromise or suppression of our principles in any case. If it is in the interest of the Indian people to nationalise industries then it must not be stopped because of the apprehension that we may not get foreign aid if we do so.

Another important question arises during this debate. And that is of transfer of capital to foreign countries by the capitalist industrialists of this country. When the rich and wealthy of this country were alarmed that the government may perhaps nationalise all land and property because it has given approval to a socialist economy, these rich people became fearful and apprehensive.

The support of the British was no longer available. Now what was to be done? They panicked. Many rich persons, especially the erstwhile maharajas and potentates, sent their wealth to foreign lands in the shape of gold and diamonds or by other means. This wealth must be lying there idle or earning profit or interest, but this much is certain that sending money abroad like this is adverse to India's interests.

It is the duty of the nation to not only declare such acts contrary to law, but also impress upon these people to bring back the wealth they have taken away to other countries. Strong steps should be taken to check such tendencies.

38

THE PROBLEM OF UNEMPLOYMENT

Immediate terrifying question
Legal and illegal methods
Initial impediments of every progressive reform

Leave aside equal wealth distribution, the biggest problem in India today is of unemployment. After the successful completion of the Third Five Year Plan, the Planning Commission is hopeful that 200 million new jobs will have been created. But by then many more adult workers would emerge.

The state of unemployment would remain as it was before.

India is an agricultural country. Those who have work have it for only 4 months a year. They remain idle for the remaining 8 months.

The nation will have to actively work towards eradication of unemployment. Once the goal is set on equal wealth distribution after adopting socialist economy, then it is a meaningless and weak hope if in the next 5 years only 200 million new jobs or employment opportunities would be created, and such a hope is rendered futile.

Just as it is the nation's duty to provide food and clothing to every person in the country, it is also the responsibility of the nation to see that every person does some job or another.

The problem of unemployment is not one that can be solved gradually.

The parliament of India and the state legislatures will have to enact laws to punish those persons who are idle and do nothing. Every adult should do some work or another on his or her own part. They must do some work whatever it may be whatever may suit their tastes or abilities, but labour should be a must.

The country which has not developed as yet, where there is a dearth of everything, there at least lack of jobs should not be a problem.

You will say that by this we are only propagating the communism of Soviet Russia.

This is not so. As far as work is concerned, perhaps the communists also want that every able and healthy adult should work. But instead of the terror of the Communist Party which has everything done by force, I am a supporter of having things done democratically by explaining to all and sundry their respective duties and responsibilities.

Otherwise how can unemployment be eradicated by its roots.

39

DIRECT TAXES AND THE TAXPAYER

Weakness of law
Wilfulness of related officers
State of the honest taxpayer

The main legal way of achieving the goal of equal distribution of wealth is through direct taxes.

Although there are many ways to do this, but the best way is through taxation. All the direct taxes, such as income tax, multi-income tax, wealth tax, gift tax, death duty, etc., are instrumental in eliminating inequality of money and wealth.

Ever since the Indian Government or the ruling Congress Party has declared socialist economy as its goal, there is some movement in this sphere. The results are worth considering. Laws have been enacted as per schedule but it is worth watching how they are being followed and will be followed in future. The law makers are so oblivious of the real difficulties that either they make a law on the principle of 'locking the door after the thief has committed the theft and run away' or a law which will not let a person live if he honestly adheres to it. Either authorities of the government, especially the officers are so engulfed in corruption that they have fallen low or they enjoy inflicting pain on others. In spite of the fact that a citizen has been paying his taxes honestly and correctly, he is harassed to such an extent that eventually he, too, realises that it is the greatest folly to be an honest taxpayer in India.

Suppose your income is Rs. 1 lakh yearly. You have paid your tax in accordance with law and in time. Within the stipulated time you submitted your return and placed all the books of account before the officer. For some unknown reason it occurred to the officer that your statement or return and the book of account are false and since he has unchecked power he declared it so and raised your income estimably to Rs. 3 lakh per year and placed a demand note of the corresponding amount of tax before you.

It is certain that if you are not into black marketing, you cannot fulfil that demand. The officer is empowered to levy penalty of an equal amount on you and use all the means by which government revenue is collected. It has been seen that in case you have some personal property that too can be attached and auctioned off. Needless to say that such confiscated property sells at highly reduced rates. With this you lose all your respect and credit in the market.

The irony is that an appeal against this order is not heard by a judge or a legal expert but another officer of the same income tax department who has probably done many such assessments till recently in his capacity as an income tax officer. Here again either corruption plays its part or the appellate officers are also sadists. Suppose you even win the appeal (which is going to take no less than 4 years in the entire process of hearing) and it was accepted that your income was actually Rs. 1 lakh as shown by you in the return and on which you have already paid the tax, what then?

Your shop has been auctioned off, your personal property has been sold at a fraction of its actual value, and you have forfeited all credit in the market. Equally you can get no compensation for all this loss and there can be no disciplinary action against the officer who had taken such an unjust action. He goes on perpetuating this injustice against other citizens and degrading them.

40

CORRUPTION VS. SADISM

The attitude of government officers
Excessive laws
Disregard of regulations

In India, the moral character of government officers, from top to bottom, has sunk very low and it will take a long time for it to be reformed.

But as long as this does not happen the scheme of equal distribution of wealth cannot stop. Therefore, those in charge of making laws should make such laws, giving them proper thought, which may be followed and observed without any difficulty and the government official are not able to either use corrupt methods or resort to sadism in having them promulgated. Many such laws can be enacted.

Almost 500 years ago the famous novelist Cervantes wrote a novel *Don Quixote*. It is the story of a ridiculous fool. Don Quixote, while preaching to his servant Sancho Panza, says that anarchy is far better in comparison with the state which enacts so many laws that one gets no time to ascertain whether they are being followed or not.

Many such groups of laws exist in India which are confined only to the covers of books. If the people observe them then there is no appreciation and if they are violated there is no punishment. What is the use of making such laws?

Laws may be few in number yet they should be observed by the people properly and at all times. To oversee this is as important for the administration as enacting these laws. But it seems that the state even now believes only in enacting laws and perhaps is not very bothered or concerned whether they are being observed or not.

The police department is only recourse which the government has to make the people observe laws, or where the events may take a serious turn it has the armed forces.

The police are so notorious that any good citizen is afraid to go and file a report with them although he may have been robbed of something invaluable or hooligans are laying hands on his honour and dignity.

The Indian citizen will bear all this quietly and will not have the gumption to go and call upon the police department. Why is this so?

The reasons are clear. If he goes to the police, they shall not hesitate to deprive him of whatsoever money and honour that remains with him, and in case he escapes this fate, then he will have to face the same capitalist poison which permeates everything. It is quite possible that he may not get any relief or justice without the support of money.

When this is the state of things then the question how will the equal distribution of wealth be successful becomes all the more serious.

If laws are enacted to make it imperative for workers and officers to observe their duties and the utmost punishment and that too easily and immediately is given to the erring workers, then the situation may improve.

It is necessary to improve and raise the moral standard of government workers. It is extremely important to make people observe the laws. Before enacting laws all the difficulties which may come in its way should be properly studied and after the enactment there should be no discrimination in its observance.

At present the state of things is such that there is one law but its implementation is different for a millionaire business man and entirely different for the common people. Other laws are expected to eliminate this disparity.

41

THE FIXED LIMIT OF CAPITAL

Ceiling on the ownership of land
Unlimited capital is more fatal
Competition—the biggest boon of capitalism

Private property is the basis of capitalism. It will have to be eliminated. In this category mainly two types of property are included: (1) Landed property and (2) Capital.

Ever since the Indian government has declared socialist economy as its goal, there have been many amendments in the first section.

No person can own and possess agricultural land beyond a certain limit. This law has been enacted in all states of almost the whole of India. Where a zamindar previously had thousands of acres of land, now legally one person or one family cannot have more than 30 or 50 acres. All extra land has been taken away from him in lieu of compensation or even without it. This is a great revolutionary change. Capitalists, who possessed thousands of acres of land since centuries, are now terrified. This is a positive and concrete step towards equal distribution of wealth.

Legally a ceiling has been fixed on the possession of land, but no such limit has been fixed on land other than agricultural land as well as houses and other landed property.

Any person can own hundreds of houses in a city and charge any amount of rent from his tenants. Fortunately, the government has passed a Rent Control Act. Under its provisions landlords cannot raise the rent to any extent they wish, still the period of raising the rent of new houses is unjust and inappropriate. The prices of houses are also exorbitant. As a result those zaminadars who had sold their agricultural land 15 or 20 years ago and built or bought houses in cities, consider themselves cleverer and more fortunate.

The reasons are clear. Still, the policy of the government in this matter has not been decided.

They have to establish a socialist system and also adhere to the principle of 'let everything go on as before.' A limit on agricultural land was fixed but no limit was imposed on capital. Why?

Why is it considered more dangerous for a capitalist to own more land than for him to own more capital? The reason is that rajas and zamindars were few in number. Hence it was easier to acquire their land directly or on the excuse of establishing a socialist government.

If the nationalisation of capital would take place or a limit imposed on capital then quite a large population would be affected. Mainly it would also not leave many of the members of parliament or state legislatures unaffected.

India is such a poor country that if a limit is fixed on capital in ratio of the ceiling fixed on land then the upper limit would be something like Rs. 50,000.

Then the law will dictate that no person can have more than Rs. 50,000 for himself or his family. He would have to surrender the remaining to the state.

This is step for which not even the staunchest socialist would be ready, because as far as acquiring land was concerned it was not a personal loss to these zamindars. Besides, the compensation given for acquiring the land has affected an increase in the ranks of the capitalists. Therefore, the zamindars can get more rent for money received in compensation. At least this is the hope that they have. So, besides the capitalists, these erstwhile zamindars, too, are an impediment in fixing a limit to capital.

Whenever the question of setting a limit to capital comes up, half the population, especially the educated class, will be up in arms saying this is the communism of Soviet Russia. It is very unjust. This should not happen.

Therefore obstacles arise in equal distribution of money.

Suppose the nation agrees to make such a law. Can it work properly? Would it be applicable?

Why not? Every person has to render details of his entire property these days and submit details of property tax yearly. So, an upper limit to total property can be easily fixed. Of course, all the obstacles which arise in the beginning when new progressive steps are put into practice will arise even now.

There may be a possibility that people will keep some money on the sly but when accumulated capital cannot be kept over a certain limit then that money will also have to be spent. The fixing of a limit to property does not mean that a limit will be set on the earning power of the people. The areas of business and profession will even then be open for people to earn money. The result of this ceiling would be that people will not be able to hoard that capital and transform it into 'dead money'. If the earned money is spent continuously it will pass through many hands and with this speed, money will gradually divide into many parts and reach many people. As for the fortune or misfortune of successors, one should not worry about it. There is a saying in Rajasthan and elsewhere too that 'if the son be good why hoard money and if the son be worthless then too what need to save money.' So, there will be a way for these successors.

When money cannot be put to use beyond a certain limit then this tendency of hoarding money will also disappear.

The biggest disuse of money is to hoard it.

If a person has money and it comes into use then even if there is unequal distribution of wealth, still, that capital will serve some purpose for the nation; but those people who save money and gold and precious stones and bury them in vaults or treasuries are the biggest enemy of the nation. Hoarding of gold and silver or precious stones, etc., should be declared a crime by law. It is indeed very unseemly considering the poverty in India that people should hoard gold and silver in this manner.

One part of private property, that is, ownership of land, is being abolished gradually or at one stroke. The other part can only be eradicated through direct taxation which has yet not been put into practice properly.

Capital, too, can be brought to an equal status at one stroke. By doing so possibly many people might not feel the pain.

The biggest headache of man is competition. If someone has nothing to eat and he knows that another person also has nothing to eat then he would not suffer as much as a millionaire who sees that his neighbour possesses a billion.

The basic instinct of man is competitive. Therefore, he feels a sense of relief with a co-sufferer. Wherever he comes across a prosperous and rich person he becomes unhappy at once, thinking himself to be inferior.

If a rule was made that no one would be allowed to keep more than Rs. 50,000 then those having a lac or a little more or less would not feel regret and discomfort as much as is now being felt because of the unequal distribution of wealth. That is to say that all can be made to leave their own extra wealth provided no one else is allowed to keep more than what they have.

The end result of all these extra taxes is equal distribution of wealth. If this purpose seemed successful to even a small degree then the people who have to pay taxes would not be so discontented. The basic reason for their dissatisfaction is the thought that huge amounts are being charged from them and are finding their way into the pockets of government officials and authorities directly or indirectly, and they will become the industrialists of tomorrow in our place. It is because of this jealousy and ill-feeling that the capitalist of India, even if he were a soldier of socialism, would oppose the measures taken.

The laws are such that room for their violation is left open even before they are finalised. In present-day India, only those lawyers and barristers are considered most capable who can show the easiest way to violate or disregard the law. The terror of income tax is such that lawyers, doctors and other professionals wish to take their fees in black without any compunction. Even esteemed people lay down this condition.

Where the upholders of law are in such a state, how can one think of the nation's progress and prosperity? At the base of all this erroneous thinking lies the self-same unequal distribution of wealth.

Once equal distribution of wealth becomes a possibility and socialist economy is established in its proper form, then there will be no charm in taking money or violating the law.

42

THE COOPERATIVE MOVEMENT

The ancient tradition of cooperative economy
The pitiful state of present cooperative institutions
The insistence of leaders and weakness of the people

Another call being loudly given these days is of cooperation. The cooperative movement is beneficial for the country, for every citizen of the nation provided that path is honestly and truly followed. But, if the same capitalist system is to be established on the name of cooperation then why lay the blame on cooperation. As it is, from the beginning of creation every man has supported cooperation in some form or another. The Indian family system, the village system and the city system, all are based on cooperation. Even today the shadow of that rudimentary and customary practice exists.

Suppose a man's daughter is getting married and Rs. 1,000 is the estimated expense. Almost all the people of the village will contribute according to their respective capabilities. Someone will give Rs. 1; another will give Rs. 2, a third person will contribute in kind, all according to their custom. The girl's father will have to spend only Rs. 100 or 200 from his pocket. All the remaining arrangement will be done through this village cooperative or cooperation. At every juncture of life, from birth to death this cooperative was present in every Indian house. Later, this sense of cooperation permeated into the caste system. People struggled to obtain essential goods to sustain their lives, so whole communities organised their own protection in an appropriate manner. Everywhere there were tribes or groups of people. All this was none other than a cooperative movement.

Even in illegal activity, it was impossible to succeed without cooperation. If four thieves would get together in order to rob a house, they too would cooperate with each other in taking the responsibility for various actions and the money got by thievery would also be distributed later in a cooperative manner.

This is to say that cooperation and the cooperative movement are nothing new as far as India is concerned. Now, it is being regarded as a legal way of obtaining the necessities of life or division of labour or distribution of money and by giving it the form of a movement, cooperative institutions have been set up at various places.

If the leaders and authorities of the cooperative movement perform their tasks honestly, the goal of a socialist economy can be achieved very soon equal distribution of money would become easier. But do these cooperative institutions run on the right path or can they do so? This question is worth considering. Just think over it.

In the various states of India various laws have been enacted for the regulation of cooperative institutions. For example, the principles of the Bombay Cooperative Institutions Act are alright, but in this century where there is a miserable lack of education, can a person understand his rights and vote properly?

Suppose you opened a store under a cooperative institution where all necessary items of daily life will be available to each member at a proper price. This is possible because a lot of foodstuffs, clothing and other necessary things are obtained from the government at controlled prices or bought from the producers direct without paying commission to the intermediaries. No wholesaler can sell things at that price.

Now, if the authorities and workers of that cooperative are all of them honest, then decidedly the members of that cooperative will be able to get all the things easily and at cheaper rates. But in accordance with his character and propensities man cannot resist thinking about personal gain and profit.

A pair of dhotis is selling for Rs. 15 in the market and is available to one's institution for only Rs. 10. Even after including all expenses, it cannot cost more than Rs. 12 to the members. Then it is possible that a person may think about taking some dhotis from the store at the reduced price and earning a profit by selling them in the market at a higher rate. Similarly, while purchasing goods for the store many such occasions arise when it becomes difficult to curb one's greed. Even when the inspectors deputed to oversee the regulations of the cooperative institution come on tours, only those people can serve them better whose pockets are full and who have the means to do so.

Besides consumer cooperatives, now even industrial cooperatives are running at full swing. Big workshops and factories are being set up under this department. The policy of the government is to help them as much as possible in this endeavour. As a result some clever persons, who by their manipulations control the institution, run it according to their wishes. These workers or authorities of the cooperative begin to think of the facilities provided by the cooperative as their own inheritance just as a capitalist considers his personal property to be entirely his own. How can this situation be rectified?

The people should be enlightened so that they become fully aware of their interests and rights. They should be made to realise that their vote is the power by which these authorities play their game. It takes a long time to educate people in this manner. No political party feels that it is in its interest to remove such ignorance.

So, the entire strength of political parties depends on the ignorance of the people. No matter how loud the political slogans are, no political party is interested in making people aware of their rights and privileges. This attitude of the leaders is to a large extent responsible for the nation's misfortune and the misery of the general public.

The supporters of cooperation are so enthusiastic about its success that they consider the cooperative path to be the best path for each and every thing. The left wing of the Congress party has in its Nagpur conference appreciated the inclusion of agriculture in the cooperative movement.

Many difficulties arose in implementing this scheme and that is why agriculture has not been able to be run on the lines of the cooperative movement. What will happen in future one cannot say.

43

MECHANISATION OF MAN

The tradition and devclopment of production
Mutually dependant production
Limited reservoirs of mankind's knowledge

In a way the capitalist economy has transformed human beings into machines. For example, 200 years ago, the needles and nails used in homes were made by the village smiths. From the material needed to make that stuff to the selling of needles and nails from door to door and the collection of money there from, all this work was done by him and his family. If we examine carefully, we shall find that almost 18 processes are involved in the business of getting the raw material to the sale of needles and nails. More than a dozen processes go into the making and polishing of needles and nails. In those days all these processes were executed by just one person.

Later, in order to increase production division of labour took place is such a way that each of the 18 processes were handed over to 18 persons who performed one task each. It is but natural that when a person does only one work regularly and continuously his workmanship will become better and better and production would be quicker and larger in quantity.

As a result the production of nails and needles increased by more than tenfold, but one man did not have the knowledge of the entire process; rather, he was confined to just one process.

The capitalist who organises all these people working at different levels in the interest of his own profit, makes sure that everybody does his work to the best of his skill and abilities. In time, machines run by electric power replaced human beings. So, the production of nails and needles was now confined to only one or two persons. The same applies to the weaving of cotton or woollen cloth. There was a time when the families of weavers would do all the work from shearing the sheep to making rugs and blankets from the wool thus obtained and also selling the same. In present day society no single person or family can do this job alone.

As a result, there was a decrease in the reservoirs of knowledge of mankind and property grew gradually with the increase in production. This is a peculiar shortcoming of the capitalist system.

Man has been made a machine and the reservoirs of his knowledge have decreased.

44

DEPENDENCE: SERVITUDE

Man caught in the claws of capital
Its fatal consequences

According to capitalistic logic, if a man does only one work, then after working for a few hours during the day he gets the remaining time as leisure. Then he can entertain himself. If all work from beginning to end will be his responsibility he will never get any free time, so he will not partake in any entertainment. There can be no objection if people get free time for entertainment. Perhaps leisure and entertainment are as necessary for man as work. The only objection is that the system of work in capitalism makes every person dependant on the other to such an extent that single person is able to do any work on his own.

It becomes an impossibility to get a man who knows one job fully from beginning to end.

Fortunately, the villages of India are still free of the obnoxious effects of capitalism and so, the carpenters, smiths or cobblers in the village are well-versed in all aspects of their profession or fulfil them with the help of their families.

In the cities this is not the case at all. The carpenters or cobblers in the city are totally dependent. Until they get sawn wood from the saw mill or tanned leather from the tannery they will not be able to do any work and even then they do not know as to what work will follow after they have completed their part of the whole process.

This dependence is not limited to the poor or the labour class. In household chores also it is the same. If the cook has not come then either one goes to restaurant or has to go without food.

The situation of the rich and wealthy is all the more pitiable. If their servants do not come for a day, they are rendered helpless. They cannot even get dressed by themselves or perform other personal functions. They are so inured to servants that for every little thing they have to rely on them.

This economy has made them so dependant and devoid of self-reliance that they are almost like invalids. If the cook or a servant doing house jobs gets no work he will soon face starvation as he does not know any other work. So, if he does not get the job he is versed in, his situation is bound to become precarious. But the situation of the masters would be worse without servants. They cannot perform any personal functions by themselves because they are used to having all the work done for them by servants.

In short, one of the fallouts of the capitalist system is that people become slaves of each other, they become helpless and dependant. They are afraid and apprehensive about breaking the chains of this dependence. Whether it the master or the servant, although they wish for equal distribution of wealth they are afraid to change the old order, the old economic system.

The reason is loss of freedom.

As a result the capitalists who have property and money cannot do anything by themselves. On the other hand the servants, labourers and peasants also are rendered helpless and lack freedom and so cannot do anything on their own accord.

45

THE MIDDLE CLASS

History
Channels of capital development
Rising numbers of the middle class

In order to manage and organise the system and also to set the conditions right between the other two classes, a third class of people naturally came into existence. This third class is known as the middle class and it, too, has a strange history.

The kings and zamindars were owners of land. Suppose a land owner had 10,000 acres in his possession. Because of the revenue obtained from the land he was considered very rich. With the passing of time he had five sons and each of the five sons produced four or five sons. So you can well imagine that by the third generation, with the division of land, only a few hundred acres would come to each of these landowners.

As a result while the grandfather was considered a rich man, his sons and grandsons or great grandsons were placed in the category of poor people.

In India, girls or women previously did not have a right to property. Therefore, only sons are mentioned in the example. Observing this situation, the society of that time made a rule that only the eldest son of that landlord would be the successor of his land and property and the remaining younger brothers be given a fixed amount yearly for their expenses. This system was devised by the sociologists of that time to prevent the property from disintegrating. As for the younger brothers, they were highly educated and at the same time their living standard and habits were all like the property owners and capitalists.

Now, when these landlords and capitalists needed people to manage their affairs, they became so helpless and confused that all of it seemed beyond their capabilities. But the younger brothers could do all such work. They would provide servants and workers to the capitalists and landlords on the one hand and on the other arrange work for the labouring class in lieu of food and clothing. This was the main function of these 'younger brothers' or the middle class of people who were neither landlords nor labourers. As their remuneration they would receive a part of everything. As the capitalists and landowners grew rich they became all the more lazy and laid-back. They would leave most of their affairs in the hands of their managers and stewards. All these people of the middle class were well versed in accounts and other managerial work. The entire management would be in their hands. The collection of revenue, the distribution of land to be tilled among the tenants, all this was under their supervision. Besides, they also kept accounts of rent from property or interest on capital invested, etc.

All decisions about where it would be more profitable to invest capital, where and how much capital is to be utilised, etc., were made by them. And so, in time, this middle class of people became more important than the other two. Without their help, both the rich and the poor considered themselves helpless. It is but natural that people would like to join the class of people which is more important is society. In time, many people of the working and labouring classes began to try and enter the middle classes.

It was not possible for the poor to obtain education. A century ago if a labourer or peasant in India learned how to read and write it was well-nigh considered a miracle. The rich would not tolerate nor allow the people of these lower classes to be educated. So, in the beginning many difficulties arose.

During British rule when the Christian missionaries began converting Hindus, Muslims and people of other religious beliefs into Christianity and educating them and teaching them in schools, the feeling that the poor have no right to education slowly waned. When these people gradually became educated, most of them naturally and automatically joined the middle class.

They were able to get clerical work, work related to reading and writing in the office instead of work involving manual and hard labour. On the other hand the rich class too, which had begun feeling helpless, began to do their own work themselves and in many ways they, too, began to consider themselves as members of the middle class.

The middle class workers have always competed with the trading class. The businessmen work independently and despite being very low down in the ladder as far as education and art is concerned, they can earn far more than middle class intellectuals. They can endeavour and by various devices earn millions. This has been a cause of envy.

The position of women in India 20 to 25 years ago was absolutely deplorable. They were ignored and exploited. India was a leader of male-dominated societies. Women were seen as hand maidens or dependants. Even today, women suffer quite a lot. Men have exploited them to no end.

46

THE BIRTH OF COMPANIES

The next chapter of middle class capital development
Company law Fall of the middle class

These people of the middle class who had emerged initially as younger siblings, stewards or secretaries, began to enter into trade and business for themselves. One the one hand there was the capital of the capitalist class and on the other the labour of the worker. By using both and exploiting them, this class established its rule over all trade and industry.

Small or big firms and houses were managed and run by them.

Even the Mughal Empire in India needed these office workers. Later when the British assumed power, they too needed these moneylenders.

These very people collected revenue on behalf of the government and also made available all necessary goods to the administration.

In this way firms and trading houses and local gaddis and kothis became extremely important. They would run small workshops and factories on their own resources. Capital was available to them from the capitalists and they would employ labourers on contract or wages.

With time their ambition grew. They began to form larger companies in place of the smaller trading houses. Many firms would get together to form a large company. In the beginning these larger companies only dealt in banking and money-lending. Later they began to trade in other things and areas too.

The companies started running industries, factories and workshops and this middle class community managed and operated the business of all those companies as managers or directors. And in this way this middle class became ever more powerful than those capitalists with whose capital they were running their businesses. The special facility with the companies was that any capitalist could join that particular business or industry with a small or big capital and become an owner in ratio to the money he had invested in that industry.

In the beginning there were shares of Rs. 1000 each; later, their value was fixed at Rs. 100 and now we have shares of Rs. 10 or 2. Any person by investing Rs. 10 or 1 can become a share holder in the company's profit. The basis of this business producing joint capitalist system is company management. In India management or industrial employment through companies was first established by the British.

Today, after many alterations and amendments the form of company law that is presently applicable is a certain indication that the government or administration holds the view that Indians who run companies are not at all honest. They are being forcibly made honest by the law.

However, I do not have to add anything about company law here. It is enough to say that all the nine rules and regulations in force today cannot help in the socialist economic set up. One can see the same bureaucracy established by the British in its distorted form.

Yes, every person began to try and become a part of that middle class which gradually overshadowed all the industries of India. Boys in lakhs began to go to school to pass the matriculation examination, some even went to colleges. Every person wished to join an office, sit on the chair under a fan and work on a table. He wished to work and rule as the middle class of that time worked and ruled. During the past 50 years the middle class has increased to such an extent that they have become quite cheap because of the law of demand and supply.

The situation now is that nobody asks for them. They have begun to get lesser wages than good and skilled artisans and labourers.

Similarly, it is easier to get female typist than housemaids.

The middle class needs good clean clothes to wear in accordance with their situation, good food and living. So, their expenses exceeded their incomes and in this manner the all powerful class declined. People of the middle classes are the most pitiable today. They form the largest part of the unemployed millions. They cannot cut-down on their expenses because they have to maintain their position in society. The other shortcoming of this class is that there is only one wage earner in the house while there are five, seven or ten mouths to feed. The middle-class women cannot work. The question of children working does not arise because their education is a must.

Initially these middle-class people did independent business. Even though they used the capital of capitalists yet they were masters in their own right. As the system of joint companies flourished, they gradually became workers. As a result, where formerly their children would become owners of the businesses automatically, now they had to serve according to their ability and qualifications. The middle class which was glorified 50 to 60 years ago and on whose shoulders the British had established their rule and who had become big agents and merchants, now declined. Their life became full of pain and misery. Some out of them were able to join the land-owning zamindar class or capitalists. They became rich and wealthy but the rest of the people in large numbers remained in want and misery.

47

THE SUFFERING WORKING CLASS

The great contribution of Karl Marx
An outline of the working class
Trade unions and communism

We have discussed the middle class. Now, let us see the poor classes, who are hungry, poor, labourers, workers and form the mob, whatever you may wish to call them. All those people are included in the working class of any country or castc or religion. They may be male or female, but they do not have any property of their own nor capital and they have to work for wages under someone and sell their labour at a price.

The movement of the working class was inspired by the great revolutionary Karl Marx.

Karl Marx was himself not a labourer. He was born in 1818 in Germany and died in England in 1883. He had as a matter of philosophy studied the working class to such an extent which was not equalled by any other person of his time. When he gave the call 'workers of the world, unite,' then he clearly meant that everyone get together to do away with private property and every person should have to do his share of work. This should be made compulsory. And the workers of world should capture the wealth and produce of the world without paying any revenue or rent to the lazy capitalists.

At that time in the whole world especially in the industrially developed nations middle class owners were prominent. Without them neither the capitalists could do anything nor the labourers. So the teachings of Karl Marx did not have much effect.

The book *Das Kapital* written by Marx could not become the Bible of the working classes. Marx was himself the son of a rich Jewish lawyer. As it happens, rarely do great men get the recognition they deserve in their lifetime. The greatness of Karl Marx, too, was revealed before the world 25 to 30 years after his death. Time had changed, industries had gradually come out of the hands of the middle class into the hands of big cartels and corporations and the middle classes had also begun to call themselves middle or working classes.

When the revolution took place in Russia the First World War was about to end. The entire working class was discontented and miserable. The rulers in Russia at that time—the Czar and Czarina and their spiritual leader Rasputin—were in disrepute. There were many other reasons why this revolt which started as treason, soon took the form of a movement of the working class or revolution. It was successful and a great new 'ism'—communism—was born in the world.

The teachings of Karl Marx impressed Britain and other nations, too. Socialist parties were formed in many countries. In Germany the Nationalist Socialist Party was established to bring about socialist system but it degenerated into a dictatorship later under Adolf Hitler.

In India of course the awakening of the working class was not possible. Under British rule who would talk about socialism of the labour classes where there was no individual liberty in the first place. When the leaders of the All India National Congress geared up to fight for complete self-rule only then their leaders formed a plan to organise the working class. All India Labour Union was established at that time. In the beginning some voices were raised after forming a labour union. Since the basic aim was India's independence there was no question of dealing separately with the problems of the working class. That organisation too was used in the Non-Cooperation or Civil Disobedience Movements of Mahatma Gandhi.

The situation of the labour class remained as it was. Even after India achieved independence there were many states and native rulers who forced the working class to work as serfs. They were made to work by force without receiving any wages.

The present form of the organisation of the working class is separate workers' unions. Every labour union is affiliated to some political party.

This method of labour organisation is not socialism or communism in any sense of the term. It is purely capitalism of the working class.

To explain this situation an independent chapter has been given later because the labour unions are very strong today one has to understand the labour market and their bargaining system to fully understand their working.

48

CLASS STRUGGLE

Struggle between worker and employer
The mentality of employers
Improvements in the conditions of workers in India

In the way a trader of the middle class considers it necessary to sell his goods for his livelihood, similarly a labourer also cannot live without selling his goods. His 'goods' are his 'labour'. The better value he can get for his labour, the better it is for him and the lesser he gets the worse it is for his welfare. And if he can get no price for his labour, he will be reduced to penury and become a pauper.

Therefore, a worker always tries that he should get the utmost price for his labour or that he should have to work the least for the money he is getting. It is his endeavour to work as little as possible and get as much money for his labour as he is able to get.

On the other land the employer always tries to get labour as 'cheaply' as possible. All employers have the tendency to get as much work out of the worker as possible by paying the least wages.

So, the interests of the worker and the employer run contrary to each other. As a result bitter and acrimonious struggles take place between the two which are called 'class wars'.

Our civic sense is outraged as the situation transforms into a vicious atmosphere. The administration has to intervene continuously so that the seller and buyer of labour can be kept within limits.

The employers only wish that the labour needed by them to manufacture their goods should be available to them at the cheapest rates. Whether the labourer is a man, woman or child, their interest is cheap labour. Their only aim is profit.

For example take any horse driven carriage. Most of these carriages belong to the contractors who are owners of the same middle class. They keep people on wages to drive the carriages. Now, suppose that if the horse is fed properly and pulls the carriage for 4 hours every day, then one horse can last for 12 years. But if he gets bad food and that too half of what he needs and has to work for 8 to 12 hours a day, he will die in 2 years' time.

When the carriage owner can get a horse very cheaply then why should be feed it properly and take only 4 hours' work from it instead making it work 10 to 12 hours every day.

The same is true of human labour. No employer who has an eye on his profit will hesitate to make a person work till he is thoroughly tired and his life span is reduced to 5 to 7 years provided he can get other workers in plenty.

In the capitalist system of the world this has always been the case.

In India, too, under the capitalist system this happened till recently. There was a time when small children were made to do hard labour under the whip. The work of breaking stones was assigned to women. The thinkers had invented sayings that during the life-span of one employer 5 or 7 generations of workers come and go. That is to say, workers died in such numbers and were born in such numbers that there was no dearth of labour. But today when we think of it we are horrified at this monstrous state of things.

That is the reason why Karl Marx raised a cry against these middle-class employers in his book *Das Kapital* and even advocated violence in no uncertain terms to do away with this system.

Much before the publication of this book by Karl Marx the foundation had been laid in Britain and other countries of this struggle between employer and worker and many laws had been made to stop this injustice which is known as Factory Acts.

49

THE FACTORY ACT

Initiated by the British
Opposition by workers as well as employers
Human reform is not based on personal generosity

In India, too, in the beginning of this (20th) century, Factory Acts were made on the initiative of the British and they were promulgated. Thanks to these laws no work could be taken from children under a certain age. Women would do only certain specified work. It was necessary to provide specified food for expectant mothers and mothers of small children. Working hours were fixed for all workers. The hours for running factories were specified. Adequate protection was made compulsory with regard to machines or implements by which there was a threat to the safety of workers. Contrary to earlier times when employers would give the workers rotten food or clothes in lieu of their labour, now payment in cash was made compulsory. Proper latrines and urinals were to be provided for workers within the factory. It was made compulsory to carry out white wash and repairs of the factory premises at regular intervals.

To see that these provisions are not violated many factory inspectors were employed. They had the right to enter any factory on behalf of the government to ascertain whether the factory act was being properly implemented or not. If the act was being violated somewhere, a case could be registered against the factory owner.

These laws were not made because of the socialist movement but were made voluntarily through the development of the natural tendencies of man.

These improvements took place during British rule although they were staunch supporters of capitalism. While examining these laws, it might seem that all factory owners must have protested and every worker must have welcomed it openly. But this is not so. Even among employers there are various kinds and so it is with the workers. Many factory owners wished to take care of the welfare of their workers even without the Factory Act. You will ask then why did they not do so? There is a reason behind this also. The goods they manufacture must sell at a profit in the market. If the cost of labour in making a product is more than that of the employer who is not generous and who is making more profit by manufacturing and selling goods at a cheaper rate, then the generous factory owner will not be able to compete against him. It is possible through cheap labour not only to earn more profit but also to make goods at a cheaper cost. If a good-hearted employer takes proper steps with regard to good wages and health of the workers he cannot produce at a lower cost as compared to the other manufactures that are hard and exploitative.

Therefore it is not possible to bring about these rules and regulations on the personal generosity of human beings. The reason is that he who is generous is wiped out. So, for such social reform laws need to be made by the government.

The opposition to the Factory Act initially by the workers is more remarkable than that of the factory owners. When the poor or destitute subjects came to know that because of the new law their children would no longer be able to earn the small amounts of money they could earn earlier they were definitely dissatisfied. For a person with means and money this may not matter but the man whose family earns Rs. 4 daily will decidedly suffer inconvenience if eight annas (half a rupee) are taken away from it. Although the parents are aware that it is not at all proper for their children to work at such a tender age, but hunger and poverty force them to do so. Therefore, at first the working class opposed the new law.

Poverty is such a terrible thing that it destroys even the parental feelings of love and they are forced to send innocent children to work in factories and workshops. The temptation of these small earnings by small children was such that they did not think it necessary to educate their children and considered it futile to do so.

50

WOMEN WORKERS

The evaluation of wages given to women
Labourers: the slaves of factory owners
And women the slaves of slaves
The road towards decline and fall

With the proliferation of factories and workshops the number of women workers also increased. But their wages in the labour market were always low than those of men. The argument of factory owners was that if they needed labour then why should they not employ only male workers; besides, women are ready to work for lesser wages. The logic behind their argument may be faulty but it forced women to work for lesser wages because of the need to earn their livelihood. The labour-purchasing factory owners profited from this sense of inferiority among women workers. Since women were ready to work for cheaper wages, their large numbers added another chapter to the problems of the working class. And along with it, the reduction in the family's income because of the restriction placed on child labour was extenuated to some extent.

The main work of women for the nation and society is the care and management of the home and family, upbringing of the children, etc. Although women are no less than men as far as work is concerned, still, because of their physical and bodily state and the responsibilities of the home and family the nature of their work is not related directly to industry or workshops. It is usually believed that women are subservient to men. The family runs on the earnings of the male member. It is his responsibility to provide for the family and the woman and children. Therefore, the labour of women is not valued at par with that of the men. The result of this was that male labourers became slaves of the factory owner while the women became slaves of these men. In the social condition where a woman became the slave of a slave, it is no wonder that her miserable condition only gives rise to pity. The benefit of earning is ascribed to men and so the women have to bear their oppression. In spite of physical and mental torture she has no recourse by which she can be free. The cause of this horrible condition is this painful economic condition. The moment a woman raises her head in protest, she is accused of tarnishing the name of the family and banished. To avoid this stigma woman does not find any opportunity to rebel against this condition. Sometimes she takes to the path of sin because of poverty and social oppression. All the laws made to eradicate prostitution and uplift social thinking and behaviour are inadequate because all these laws do not penetrate to the core of the problem and solve it. By these laws, only the ailment can be treated, not removed and then even there is doubt and uncertainty. Till women are not made free and self-reliant economically, they will have to remain slaves of men and a victim of social oppression.

The wages of sin are very attractive in comparison to earnings from labour. In every age or period religious preachers have preached against following this path. The future of the country and society depend on the future generation. The future of the children depends on women. On examining the problem carefully and in depth the poverty of the present condition as well as the inadequacy of the law related to it, is more than evident. The situation is that normally women observe that those who are free as far as sexual behaviour is concerned are in a comfortable and happy position, while their husbands are in a dire strait. So it is quite natural that women are drawn towards the path of sin.

It is said that youth lasts only a few years. So, the attraction of men towards it is momentary. As a result after her youth passes away the woman who sells her body for money will not be worth a penny. Therefore, it is better to earn one's livelihood honestly and live with one's husband and family. The answer to this argument is that after working hard the whole day, breaking stones and eating meagrely and bearing the burden of home and family when the woman gets old then, too, there's no one to care for her. In the manner a prostitute loses her charm and good looks, similarly the strength of women workers also cannot stay forever. In both cases they end in poverty, dying piecemeal.

Anyway, in the present capitalist economy, free sexual relationships help in obtaining more money with ease as compared to chaste living and working honestly. Again, since the number of women is more than that of men so they are available at all times at cheap rates.

In this way the capitalist system forces normal women to sell their labour cheaply and poor women to sell their bodies. Such is the pitiable state of women workers in India.

Even in other parts of the world the status of women is no better than this. Having no right to ancestral property, their condition in India is worse.

They are really hand maidens.

51

LABOUR ORGANISATION

The need for trade unions and their history
Strikes and their various forms
The other aspect of labour organisations

There is a history behind how the labour community could face and confront the capitalist industrialists.

It was evident that any man or woman who worked and was employed could not singly go and argue with the employer, nor could he confront him. The employers have a pet answer that if you are not ready to work in these conditions and on these rates there are hundreds of your brothers and sisters who are willing to do so. This answer would shut the mouths of the workers as long as they were alone and unorganised. So, in order to make the employer listen to their complaints and pleas it was necessary that there should be an organisation of some kind, because only a unified voice could have an effect on the employers.

It was not possible for every kind of labourer to organise in any way whatsoever. For example take the servants working in the houses. One servant does the cleaning, sweeping, fetching water and washing clothes etc. The cook does the cooking. Everybody more or less is confined to their own spheres. They work alone. Nowhere do they work as a body or a group. They rarely find opportunities to meet and confer among themselves. It is very difficult to have an organisation or union of such workers. Or like the farm workers. There, too, hundreds of workers are engaged in tilling, sowing, harvesting, etc., but their getting together and discussing their problems and demands is also not possible; so, a union among them could not be formed. Even if a union is established then it is extremely difficult to keep it well-organised, because they live at such distances from each other that they do not often meet. This is the condition of all workers of this type. A union can only be formed of workers of factories, workshops, mines, railways, trams, buses, etc. The organised form of such workers is already in vogue.

There are many trades where everybody works together, still, there cannot be a union among them; for instance, take the artists of a film company. A famous heroine gets lakhs of rupees while hundreds of other extras do not get remuneration enough to even eat properly. The difficulty of forming their union is evident. An organisation or union is possible only among people of equal status, behaviour, work and tendency.

Where there is a major difference in income or social status, no union is possible. There can be a union of the workers of a factory, workshop, railways etc.

There was an endeavour to form unions of masons, ordinary labourer, carpenters, smiths, etc., and the unions were partially successful.

Unions are formed as a revolt against the unbearable oppression of the masters or owners. By uniting, the workers can compel the mill owners to see things from their point of view.

When they succeed in their aim or if they fail, the union breaks up.

Then again, when there is some calamity or trouble, unions are formed afresh.

Thereafter, they collect some money to temporarily keep the unions working.

When difficulties arise on being dismissed from service, then out of that collected fund money is withdrawn in order to survive.

Organised unions which were formed temporarily, later became permanent labour unions. This is the history of the labour unions.

What is to be thought is what such labour unions can do to stop the unjust interference of the capitalists.

When a labour union is established fully then they tell the employer to his face that they will not work for such and such a wage.

The hitherto ever-present risk that 'if not you then others will work' is no longer there.

If all the weavers of a city form a union, then if the employer reduces the wages they can say that we will not work at this lowered wage because there will be that fund to bail them out. By refusing like this they will force the factory owners to close their factories for months. At the same time when the organised workers strike, the factory owners will not be able to get other workers in their place immediately. One or two workers can be threatened to be thrown out and being unorganised their fears are natural but when all the workers strike then it is well-nigh impossible for any factory owner to handle the situation.

Workers strike like this not only to protest against reduction of wages but also for increase in wages, reduction in working hours or any other demand which they wish the factory owner to accept.

Their success or failure depends often upon the situation of trade and commerce at that time. If trade is normal then the owners do not accept the demands and close down the factories. They keep the factories closed till the workers having depleted their funds start starving. In the end when the strikers are brought down to their knees, then the factory owners agree to run the factories and workshop on their own conditions.

But if trade is flourishing at that time and closure of factory or lack of production for even a day can cause large differences in the company's profits, then the owners concede to the worker's demands and keep the factories going at all costs. They make their own calculations that they are comparatively better off in accepting the workers' demands than closing the factories.

But later, the factory owners again look for opportunities to even out the score. Whenever trade declines or is running at an even pace they calculate that they will not be affected if the factory is closed down and they will not lose much profit or lose very little and then they once again resort to cutting down the wages.

If the workers do not accept these cuts, they close or lock out the factory. This is the strike by the owners.

Whenever trade flourishes workers' strikes take place and they are largely successful.

Whenever trade declines there are lockouts and they, too, are successful.

The situation of trade is changeable all the time.

Five to seven years after the First World War there was a big recession and as a result poverty and unemployment reached such levels that the whole country had to suffer because of their fallout.

During and after the Second World War trade flourished and reached new heights. Consequently, wages rose considerably. All movements by workers were successful.

These strikes and lockouts are a type of class struggle.

52

STOP WORK

Class struggle
Struggle to increase or decrease production
Opposition to renewal
Malicious policy

It is true that people have to go through difficulties and suffer because of strikes and lock-outs.

At present, we are analysing the system of workers' unions. Many aspects of it yet remain untouched. There are ways of civil disobedience for the success of the agitation of workers, such as tools-down, stay-in strike, etc.

It is the aim of every union that all workers or at least the workers doing that particular job become members of that union, so that the management cannot break the strike by taking on other workers. But sometimes differences between men or workers extend to such an extent that they are ready to even kill each other.

Those workers who agree to take the place of striking workers are called 'traitors', 'stool pigeons', etc., the worker community banishes them socially and morally.

When the labour unions call for a strike, picketing starts at the factory gate and workers squat in front of it. They try to stop any worker who is not their member and wishes to enter the factory and work. In India unions have not reached such low levels as are seen in other countries.

In Britain no one can become a worker without becoming a member of the union. And, it is unimaginable that a worker can work without joining the union. Wherever there are strikes and the management arranges to get labour from outside to work in the factory, there are physical confrontations.

When they go near the boiler of the mill they find an explosive planted there which explodes causing the death or injuries to workers. The machines, buildings and other things in the factory are always at risk at the hands of these striking workers. The strikers become so violent that life and property are all at their mercy. There are a hundred such actions which an everyday worker resorts to, which may cause bodily harm to the 'traitor' or 'stool pigeon' who has gone to work in their stead.

Besides this, the greatest problem which the striking workers create in Britain is that they will not let any worker from outside to work in the factory. This is not only at the time of the strike. Even ordinarily if you are in need of a carpenter or a fitter you cannot keep one of your wish; you will have to apply to the union and the union will send you who it wishes at its own terms. A carpenter, smith or fitter will not work even a little bit more than the work assigned to him.

For instance, if only two screws have to be tightened to complete a job which any carpenter or fitter can tighten easily, still, because of the rules laid down by the union, the work will be held up till the screw fitter comes and tightens the two screws.

If you employ an outside worker no worker of the union will work with him and immediately a strike will take place, resulting in disturbance and discord.

There the owners have no freedom that they may employ workers of their choice; neither is the worker free to work where he wishes. No matter how hard he tries, he cannot get work on his own in some factory or workshop. He has to first go and seek protection of some union which charges him an ample entrance fee.

Later, he has to work in the union for 2 to 3 or 5 years with many apprentices on minimum wages. There he has to serve various masters and then only he becomes a full member. Then the union informs him that his name has been selected for some factory or workshop.

The leftists have made this custom into some sort of a spiritual cult. In English it is called 'close shop'. The sense is the same. Like the spiritual cult no other worker can enter and work as a labourer.

In America, these unions have a more terrible form. Although the above said cult is not there, still, without the membership of the union it is unimaginable for a worker to obtain work.

The union bosses often employ persons of a criminal character for this work. They have orders to remove those workers who oppose them or vote against them in the meeting of the general body. Only a few hours later the bodies of such workers would be found floating in the sea or lying on some roadside.

There are also other ways by which these unwanted members of the union are removed from the way. This is done by putting an end to the unwanted worker in an organised manner by the endeavours of other workers. This is done so that the blame may not rest on anyone.

Besides difference in opinion about the wages given by the factory owner or the demand laid down by the union, a worker is put to much trouble in ascertaining the amount of work he has to do.

In each field it is necessary that the average work to be done by the worker is fixed.

The owners, for instance, choose a worker who is very diligent and works hard or is especially skilled and then taking him as an example they wish that every worker should be given the same amount of work. If a worker works less than this he should be paid less.

As opposed to this, the workers' unions try to have the quota of work fixed at the lowest level determined by the least work a worker does in a day. If an ordinary mason lays three rows of tricks in a day or a weaver weaves 2 inches of cloth they wish that only that much work should be fixed and whosoever fulfils the quota should he paid full wages.

When the owners sell houses made by those masons or cloth woven by those weavers, they charge the highest price for them. Similarly, the unions also wish to get the highest price for the lowest amount of labour.

Both these are incidences of pure capitalism. Although laying three bricks in a day or weaving 2 inches of cloth may be ridiculous examples, but this much is definite that is the fixing of workloads it is the attitude of the unions that for work which can be done by one man, three or four workers should be assigned.

You will ask, then how do things go on, this should result in a breakdown, and they do happen. Prime Minister Pandit Nehru and President Rajendra Prasad have to keep exhorting the people through slogans like 'increase production', 'do not let the work slacken', 'work is worship'.

Why did work not stop totally in India, why did the production did not stop completely?

The reason for this is that man's common sense has not yet been defeated by the workers' unions. Even now there are hundreds and thousands of workers who work as part of their duty. Even if they suffer they ascribe it to God's will and bear it. In India even today the total membership of labour unions is not one-third of the total number of workers. Workers have mostly joined the unions only in cities. The National Workers' Union, the unions controlled by the communists or unions affiliated to other political parties—all of them together, fortunately, are not able to leave their poisonous capitalist impression on the Indian workers.

If it was only this we could have laughed and ignored it but in the discussion about socialism, this is a serious situation.

The real danger is that capitalism is spreading.

Whenever a person begins to believe that poverty is not given by God but is man-made, resulting out of the economic system established by him then the working class immediately sets out to form their own union and build up their strength.

Just as the landlord tries to earn as much profit as he can from the land or house which he owns, similarly the capitalist tries to obtain the utmost profit from his capital. The middle class entrepreneur or professional wishes to gain as much as possible from his education and ability. And so also the workers' union tries to get as much advantage as possible from their united strength.

The economy of the country is being crushed between these two equally strong forces.

Who is the bigger of the two? The capitalist or owner or the worker? This has not yet been decided, nor will it be easily done in the near future.

When open struggle broke out, the capitalists who were rather strong at the time managed to have laws made according to their wishes. For many years such laws were in force which declared any kind of workers' agitation or organisation to be unlawful and punishable.

The main aim of the nationalist movement was to end foreign rule. The real struggle between owner and worker did not take place in India in real terms. This feeling has arisen after India achieved independence. Even now according to the labour laws in force, it is not necessary to recognise workers unions in all but a few industries.

The law-makers have understood the results which have been seen in other progressive countries. It is very like the self destructive boon given by Shiva to the demon Bhasmasur.

The conflicts between owners and workers have arisen because of differences between them regarding wages or conditions of work. Sometimes the conflicts arise out of workers being laid off on some charge or another. But mostly they are of wages and money.

Whenever for some reason or another, the pockets of the owners or the workers are affected, disagreements and conflicts take place. Wages are paid in three forms: daily, based on time and on the piece-rate system. You must be thinking that the employers must always favour the piece-rate remuneration while the workers must be demanding daily wages.

This is not the case. In India most of the workers oppose piece-rate work and are interested in daily wages. But this is not the case everywhere. In the Western countries things had changed long ago. Even in India, with the ever new inventions in machinery or their use, workers are now demanding piece-rate work. They feel that there is a possibility of earning more by piece-rate compared to daily wages.

Besides, with the new machines it is possible to have more work done with lesser workers. This leads to retrenchment or laying-off of workers. Those who are being thrown out will never opt for new machines to be installed by which those who are kept in service are paid more while a section is thrown out. The biggest opposition to mechanisation is this tendency. Retrenchment of workers is the biggest hurdle in modernisation. At times the disturbances reach such heights that new machinery cannot be installed in the first place and if it is installed in some places the workers cause damage to these machines. Why does this happen?

Because there is an apprehension that unemployment will rise and the membership of the union declines. Everybody is naturally worried about his livelihood. In their selfishness they forget that the prosperity of the country lies in increased production.

In whichever way the production increases it will benefit everyone. But none of them will tolerate it if there is any possibility of a fissure developing in their capitalist system.

The conveniences and rights which the workers received from workers' union's through strikes, were nullified after some time by lockouts as soon as the owners got a chance. In this manner the circumstances of the workers remained the same, rather deteriorated in the bargain. To rectify this situation it was necessary to enact laws. It was made a legal offence to violate any agreement between the labour unions and the mill owners. No matter how dissatisfied an owner may be with regard to the agreement arrived at with the labour union, he cannot disregard it. Such is the law.

Just as the workers adopted wholesale bargaining, the entrepreneurs also thought that unless they, too, formed organisations they would not be able to run their industries in the face of the workers' demands. Consequently, they also got together and established mill owners organisations or employers unions. There were chairmen and secretaries in the workers' organisations; so it was with the mill owners federations. Rules were laid down that whatever the agreement it should be equally applicable to all industries. Although our workers unions, unlike the Western countries have not become vicious and a condition akin to those countries has not developed here, in which no worker can seek employment anywhere without being a member of the trade union nor an employer employ any worker without the willingness or assistance of a labour union, still the capitalist tendency of labour unions is also operating here. In India the movements of civil disobedience and *satyagrah* which Mahatma Gandhi so successfully adopted to uproot foreign rule are an example for every small or big agitation. When Mahatma Gandhi gave this new and wonderful weapon to the people, there was a moral force behind it, a spiritual feeling of devotion. He considered *satyagrah* as the last recourse when all other means had been exhausted. It was never an instrument to force somebody to do something.

Unfortunately, some so-called Gandhians made it their business to use this infallible weapon rightly or wrongly and on proper or improper occasions to serve their selfish ends. They do not know that it was the aim of Mahatma Gandhi to bring about a change of heart and mentality through this medium. Today even if you force other people to accept your demand by means of hunger strike it has such an adverse effect on the heart and mind of the opposite party that centuries of love and good feeling are wiped out and seeds of discord are sown instead.

In India the leaders of the workers' unions have on many occasions perpetrated such horrifying acts using *satyagrah* as a camouflage that are completely illegal in the eyes of the law. Those actions of theirs are no less than the open hooliganism of the Western labour unions. This new form of capitalism of the labour unions is India's own. It is as dangerous to shake up the tender sensibilities of mankind as it is to inflect physical pain on it.

If somebody robs you of your money by showing you a dagger or a gun then he is a dacoit. You do not happily give him anything. Whatever you give him you give him under duress. There is a law against such a robber and the police protect you.

But if the same person tells you that you should give him your money otherwise he shall sit in front of your door and sacrifice his life by shunning food, then no matter how unjust his demand may be, thousands of sympathisers will gather and ask you to give the man half, if not the full, amount he is demanding. They will say, what if he loses his life?

Within a few days there will be a hue and cry that the hunger striker is growing weak and is on his death bed. The goons of society seeing the chance will not hesitate to create a disturbance. What was clearly a hold-up has become the symbol of justified demand only by virtue of obtaining the sympathy of the people. Anarchy and mob-rule operates in just this fashion.

If somebody is running on the street and a few people start running behind him shouting 'thief-thief' then passers-by will catch him immediately. Suppose you start beating him, all the others will also do so. At that moment no one will ask with discretion—'what has he robbed?'

No decision will be considered necessary at the time.

This sort of anarchical behaviour is called mobocracy.

The labour unions have developed out of such mob propaganda and are alive today only on the basis of such propaganda.

Whenever man has used his discretion and intellect, these unions were exposed. After all they are indulging to a larger extent in those very vices which they had set out to eradicate.

What else can they do? The epidemic of these unions has spread to such a degree that today unions are being set up for each and every thing. If some demand has to be presented to the administration or the government, send a memorandum from the union, send a delegation of the union. The government will promptly agree or pay more attention. An individual's plea, no matter how proper or worthy it may be, will not perhaps be heard.

The reason is that the government has also been formed by unions and organisations. Without the assistance of the unions, without their votes, the government of that party could not have come into power. Therefore, it is the duty of the government to pay a special attention to the applications of all those unions and also with special favour.

In India, the situation has come when the government is run by the Indian National Trade Union Congress. Whatever work its authorities wish will be done immediately, whether it is legal or illegal. A mill, workshop or factory may be running at a loss for years, workers have exceeded the required strength and even if all the capital of the factory is depleted gradually, the unions will not budge.

But if perforce the owners of the mill have to close down the mill because of sustained losses, then there will be a terrible hue and cry that the poor workers have been laid-off, they are suffering. The tale of their helplessness and miserable condition will be told in such exaggerated terms that ordinary people will not get a chance to think about any other aspect. It is also necessary to think about the condition of the mill owners. They have no source of economic assistance if the mill or factory suffers loss or is closed down. Even after mortgaging all their property or offering all kinds of security it is difficult for them to receive any loan from the government in order to run the factory or workshop in question. If there was some sort of assistance to overcome this lack of capital, the mill or factory could perhaps be again put into gear. But without any assistance mills close down, the owners are declared bankrupt. In the absence of capital there is no other way out.

As soon as the mill closes down that very government which showed such apathy in giving assistance will agree on the behest of the labour union to invest money without any security. In this way that factory, workshop or mill will be handed over by the government to the union to run under the 'unemployment assistance scheme'. Many such incidents have happened and are happening. Although under the Indian constitution the government cannot acquire any body's private property without paying proper compensation, still there are many persons whose private property, wholly or in part, has been usurped by the government. One reason could he that the owners of these industries do not have a strong enough voice which may reach the people in power. When the Supreme Court of India declared this action to be illegal, then the government immediately amended the constitution itself. The fourth amendment of the constitution was passed only in order to enable the government to take into its charge without ascribing any reason or paying any compensation any private property if it is an industry and run it for 5 years under its own supervision and control. Industrial Control Regulations were made for this very reason. Many other Acts were passed by which the fundamental rights of Indian citizens were taken away.

It is because of the unions that every wealthy person has to surrender to the government 70 to 75 percent of his income. It is possible that one day all these capitalists, too, form their union and go on strike saying that we now shun the responsibility which we are fulfilling by paying taxes to the government. Neither will we earn, nor will we pay taxes.

Seeing the power which the unions command at present, it would be no surprise if one day a situation such as this arises.

A time is soon coming when all workers will together feel this and becoming anxious implore that they should be saved from the strife of these unions which are under the spell of capitalist elements; that they are more afraid of these unions full of capitalist elements and tendencies than capitalists themselves.

In developed nations like America this voice of the oppressed workers is steadily gaining strength from day to day.

53

THE VALUE OF MONEY

Analysis of personal capital
Valuation of property for taxation

We have talked enough about corporate capital.

It is also necessary to assess the importance of personal capital and capitalism.

Suppose you have a little personal capital then it is necessary that the condition should be analysed from that angle, too. Property does not have a fixed form or shape. If you have some wealth the question is in which form you have it; whether it is in the form of a car, stock of food, gold ornaments or jewels or currency notes. Whatsoever wealth or money you have, its value exists only as long as the world accepts it as such. Money in the form of a stock of food is always valuable, but the problem is that it is perishable, it is easily destroyed and does not last; so, you cannot keep it in its original form for long, you must alter its form in some way or another whether you change it into silver or gold or a motor car. As a sewing machine or a typewriter its value lasts longer than as food. All other things can be categorised as money only if there is a buyer for them. Suppose there is a country where all the citizens have gold and silver as extra money and all the people wish to sell their gold and silver together. If there is no buyer, what value will all that money have? No one will value it at all. It is no laughing matter. It has happened in many countries of the world. Even after giving countless currency notes or pounds of metal a cup of tea was unavailable.

So what should be considered as money?

For humankind food and clothing and all things essential for life can be called money and because they are not permanent, no one can really become rich by possessing them. So, in the end, it is only a question of social recognition. Nowadays, all of us accept gold silver and currency notes as money and because all people or at least most people believe so, it is money. You have some extra money, you change it into gold and silver or use it in some way in which it is safe and secure till a time in the future. Such money is called investment capital. If you have such capital then all income that you earn from it is your unearned income.

In socialism unearned income is considered improper. The Indian government having accepted the socialist system has imposed income tax which levies different rates for earned and unearned income.

To attain the socialist goal, the Indian government has imposed another tax called the wealth tax. Many people say that such a tax has not been imposed in any other country of the world. This is not true. Although, Britain, America and other such countries do not have a wealth tax such as ours, still in Holland, Scandinavia, Germany, etc., such wealth tax has been quite successfully levied. As a result of this, socio-economic equality has been achieved there to some extent. Death tax or death duty exists in many countries of the world. Its form is certainly of equal distribution of wealth in the social order.

Now the question is what your property is, how should it be valued for these taxes? As I said your wealth is wealth only till people accept it as such or it is used as such.

If you have to sell your cycle or typewriter to pay your property tax and likewise all the taxpayers in the country have to sell off their cycles and typewriters to do the same then you can very well ascertain what would be its value.

A thing costing Rs. 500 will not be touched for even Rs. 50 and no buyer will be available for it. In order to pay taxes, you need money. As a result you will have to sell your cycle for Rs. 50 or even less.

Palatial houses and estates of eminent lords and barons, which had been built at a cost of millions, were sold for a pittance in this manner and when the age of socialism came in the social order, the situation deteriorated to such an extent that no matter how rich a person was he could not retain even a house. Its upkeep and maintenance began costing so much that it became difficult for them to bear it.

In India also the situation is similar or is soon going to be so. Huge palaces and manors of the rich were sold cheaply or have been put up for sale because in future it would be well-nigh impossible for any person or family to keep such big establishments while paying wealth tax or death duties.

54

THE CAPITAL MARKET

Real form of investment capital
Rupees bought for rupees
The witch craft of extra money

Still, suppose you have some extra money, then have a look at how you are going to invest it and what is the prevalent system for this. The place or institution where your money is put to use or is invested, is called the capital market. Yet, it is not like a bazaar where things are lined up for sale. But there are certain peculiarities of this bazaar which many people do not understand. A bazaar to buy and sell money! When it is addressed by this name, it seems nonsensical. You can buy vegetables or fruit for Rs. 5 but how will you but money for Rs. 5? It seems ridiculous. What is the sense in buying rupees by rupees?

Rupees are not bought by rupees but rupees are taken on hire; that is, they are borrowed. Borrowing and hiring are not synonymous.

You borrow a bowl or a sofa from your neighbour and after your purpose has been served you return it to him with thanks. You borrow or lend on the basis of mutual love and affection.

But there is no place for love or affection in the capital market. There what you take, you pay for and what you give, you charge a price for it. It is such a pure business.

This everyone knows that you will not return the same thing that you are borrowing; you will immediately consume it, use it. If you had borrowed a candle or a loaf of bread from your neighbour you would have used up the candle and eaten the bread. So you cannot return the very same things. You can return a new loaf of bread or a new candle in their stead.

Similarly, when you borrow money or capital, you actually borrow the things you can buy with it, like bread, candle, etc., which are necessary for consumption at that very moment.

When you borrow a rupee, you do so because you have to buy something with that rupee which will be used right there and then. You cannot return the same thing. You will either return some other thing or the rupee you will earn by selling your labour.

In any case, till you return that rupee that person will not be able to purchase anything out of that rupee. So you pay him some compensation for waiting till you have returned the money borrowed from him. This extra amount paid to that person can be called the hire or waiting charge of the money borrowed. If you have paid this waiting charge then there is no obligation on you. But when you use your neighbours' bowl or sofa you are obligated. When you borrowed money and returned it with the hiring charge then no one was obligated. He served your interest and you served his. Money is lent because it is extra. No one gives or can give money which he needs. Now, this extra money is an authority letter to buy bread, clothes and other perishable commodities. If there are five loaves of bread left over with your neighbour and you borrowed them on the condition that you would return five new loaves the next week then just think that had you not borrowed them they would have spoiled and gone waste. By borrowing them and then promising to return them next week you have in a way done him a favour.

Had you not takes those loaves, perhaps he would have offered his five loaves in return for four or three or even two loaves after a week.

This too is right. If it comes to suffering a loss then it is better to agree to save half. This situation is called 'cutting the losses' by economists or 'reverse interest'. That is, the fee to keep the amount intact.

As opposed to this, if you loan out your money when you do not need it and wait for it to be returned then economists call it 'direct interest' which is a charge to use the amount.

Both the situations are the same. No one pays to keep his money intact. Only interest is received for the amount lent.

The reason is clear. There are very few who lend money while the number of borrowers is large. If everyone had enough money for his own needs, there would be no person desirous of borrowing money and a situation would arise where one would have to pay to keep his money intact. In such a situation it is quite possible that you may deposit Rs. 1,000 in your bank and the bank would say, well, alright, but we shall return only Rs. 500 or 700 next year or even less.

Because if you do not eat your bread, you can save and keep it for someone else only if there is another customer for it.

If everyone had excess money, who would be concerned about it? In such as situation you will think why not spend it right now, what is the need to save it and pay for keeping it intact.

But in a capitalist economy such a situation cannot arise. This situation can only arise when most people have enough money and the poor are in a minority or not at all. This is not possible in a capitalist system.

So, you should rest assured that you can lend or invest your excess money. You will, as a result, get interest for waiting.

The Capital Market is the place where earnings of a whole year are bought or sold for a lump-sum. How much of the earnings you can buy depends upon the market value which changes from day to day. If some day lump-sums are less then you can buy more earnings. Besides, its ratio also depends on the certainty of earning and its fluctuation.

You take Rs. 1,000 to the share broker and say that you have to lend this money to someone or invest it or out of these Rs. 1,000 you wish to buy yearly income in future, then he may buy you income worth Rs. 40 which will accrue yearly without making any effort on your part or he may buy you yearly income worth Rs. 60 which may fluctuate. Or approximately Rs. 1,000 worth of yearly income where there is a risk of losing the whole amount.

The poor populace has nothing to do with this capital market because they have no property or inheritance or security to offer. They have to take a loan and that, too, on the promise that they will work and when they will be paid for their labour, they will return the loan.

Assurances are more uncertain than the risky shares and therefore their rate of interest is higher. For example, a poor labourer can get money on loan only by paying one anna per month interest on Rs. 1. This is the prevalent rate and getting money on loan at this rate is considered alright by the labourer.

This means that on Rs. 1,000 that labourer pays Rs. 750 as interest in a year. No rich man can dream of paying interest at this rate. The poorer you are, the higher the rate of interest; the reason being that chances of return will be so much more unlikely.

So, when you read that the Reserve Bank rate of 3.5 percent has become 4 or 3 percent, do not think that you or I will get loan at this rate. At these rates only they will get who, like the government, are absolutely certain that they will pay that interest. Whatever be the bank rate, the poor labourer will have to pay interest at that rate of 75 percent.

The cost of money changes in accordance with the purpose for which loans are taken. This means that you are investing your capital.

No one calls investment a loan, but it is a loan all the same. It is a special kind of loan given to companies on specified conditions. The big businessmen and entrepreneurs of the city always establish big limited companies. For instance a company operates a large cloth mill; another company establishes a locomotive workshop, while a third company is into shipping. These companies do not borrow money in the ordinary way by which in a certain period with a certain interest the amount borrowed is returned to you. Their condition is that the mill or factory or workshop being opened is the joint property of all the shareholders that is all the people who have lent money and when the factory or business will start yielding profit, then all the shareholders will be given a share of the same in ratio of the money they have loaned or invested. Unfortunately, if losses are incurred and the amount invested is lost, then you can rest assured that you will not be asked for more than what you have invested. This business is like a chance. If you are fearful then these companies issue you with preference shares of fixed interest at 6 to 7 percent to encourage you. On these the fixed interest is paid before the shareholders are given their share. These are called preference shares while the others are called ordinary shares. In these two kinds there can be several types of preference and ordinary shares but they are all different methods to take extra money on hire, the difference being in their conditions of paying the hire or interest.

In this manner if you have purchased shares out of your money by which you have an yearly income and in case you need a fixed amount of money, you can always sell off these shares to those people who are interested in and willing to purchase an yearly income in lieu of a fixed lump-sum amount.

Where all these transactions take place is the department of the capital market which is called the share market or stock exchange. To buy and sell shares you would need the services of an intermediary or agent who is called a stockbroker. When you have to sell your shares then your broker meets another broker who has been appointed by another party who wishes to sell their shares. Thereafter, in accordance with the conditions and rates prevailing at the time he will fix the price after due bargaining and sell off the shares. The constitutional function of the share market is the sale and purchase of the shares of old and running companies.

Besides this, another strange business goes on in these share markets with great gusto and it is called 'speculation' or 'forward trading'. Speculation means that one buys and sells shares worth hundreds of thousands without having one share in one's possession and without actually investing or receiving any money. Well, a detailed analysis of 'speculation' will be made in the next chapter. Meanwhile, I have only this to add that in the share market, shares of only those companies are bought and sold which are old and established companies.

Now, it is in the nation's interest that new companies are opened, new factories and workshops are set up, production should be boosted and all money lying idle should be used in it. Suppose you have Rs. 5 lakh in excess of your requirements.

You buy shares of some company in the share market. With the money you have, it would not be possible to set up even a loom in a jute factory, nor will an extra gunny bag be made. Only your money will go to them who had already invested in those mills. Your name will be entered in place of their names. Now with the money which you have given, they will do as they wish. They may gamble it all away, have a house built or drink it away; they can do whatsoever they want. How did the nation benefit from this money that you invested?

You will say that if this is so we will never purchase shares of old and established companies, we will buy shares of only new companies and deposit the 'share application money' in their bank. The idea is sound, but if you do not have an experience of this and do not have full knowledge about money, you will certainly lose your capital.

Establishing a new company is such an art in which tremendous trickery may be practised. Regretfully, I have to say that in India there are more dishonest and crooked people among those who set up companies than honest people.

The government has changed and amended the company laws and other laws. It was thought that with the last law made in 1956, fake companies would not be set up now. But whatever may the law be, these malpractices have not abated. New ways are being shown how the people can be deceived by violating the laws. When those who are forming new companies are able to collect the desired capital, they conveniently become directors, managers, etc., and begin to take large amounts as remuneration and benefits, etc., and start having a good time. No effort is made to achieve the purpose for which the new company was set up and fulfil the assurances given. If the workshop or factory has to be newly set up, then often large amounts are taken as commission straight or through relatives and friends, etc., they lose all the accumulated capital and then declaring that the company has failed, they dissolve it.

It should be kept in mind that all these activities are carried out by them purely on a legal basis. You cannot do anything except raise some hue and cry in the shareholders' meeting. And if you are not careful then in the meeting you might say some things against the directors which may prompt them to file a suit of defamation against you, and it is possible that you might have to pay some more money as a punishment for having said the truth. So, there is no advantage in protesting.

In India lakhs of rupees of the people are lost in this way. Had they been invested properly in genuine companies, the national capital would have increased in a large way.

Even if you escape the shares of these dishonest or counterfeit companies, there are many companies which are not able to be successful because of lack of proper organisation. Fifty lakh are needed to establish a factory. The people who are going to establish the factory have prepared an estimate of 25 lakh. This error in the estimate will result in the factory not being set up. All the money invested has been wasted and the company will have to be reorganised. Many companies are able to yield profit after the second, third or fourth restructuring. The value of the shares of the initial buyers is reduced to a third or less or its original value. The experienced person knows when and where there is risk and when and where it is not. He waits patiently for the right moment to come and then purchases the shares. Seeing this, the ordinary share purchaser gets flustered. Even if after going through several ups and downs and bearing the burden of worries for several years if he's finally successful, the only inspiration he gets is not to indulge in this activity in future.

In order to avoid this doubtful situation and the suspense of 'what to do and what not to do', it is necessary that one does not go near the new companies and purchases shares only of the old established companies. These shares will be obtained for you from the share market by your broker. Rather, investing in securities and loan to municipal corporations is still better as the rate of interest on these may be low but there is no risk involved in them.

If there is a detailed description in the newspapers about new companies, do not fall into temptation. Rather, be careful and think that your investment is being put to risk or not.

55

SPECULATION

Gambling of the share market
No increase in national wealth

Now, suppose you are also a gambler to some extent. You may dislike gambling; still, the knowledge of gambling is also essential to comprehend the structure of the present economic system. A special kind of game is played in the share market which is called speculation. In it, imaginary shares are bought and sold. I will give you a glimpse of how this game is played. Later, you may like or dislike it, it is up to you. This gambling initiated by capitalism is very popular all around the world.

To understand it you must know that in the Bombay Stock Exchange or the Calcutta Stock Exchange you can buy the shares of any company without paying any money or even if you do not have any shares you can sell as many shares as you like, you do not have to give a share certificate till the next settlement day does not arrive. Settlement days come almost once every 15 days. You might suddenly think what difference it makes. But a lot of difference can take place in 15 days. Just think how the excess or shortage of lump-sums raise or lower the value of shares and how hope or apprehension affects the market rates of shares; how the news that cloth is going to be taxed or the export of jute is being stopped or the prices of tea in America have gone down or more coal is being mined has an effect on the market. The effect being that those companies in question have less or more profit to distribute among shareholders. As a result, the prices of shares do not change once a year, but on the other hand, there is daily fluctuation in the prices of shares. The shares which you had purchased many years back for Rs. 1,000 may have started giving you an income of Rs. 5,000 or 5 or none per year. Consequently, the shares you had purchased for Rs. 1000 may get you Rs. 50,000 or it is possible that you may have to sell them off for Rs. 100 or they may be totally worthless. As soon as you open the newspaper in the morning and see the share prices for the previous evening you immediately come to know how much your wealth has increased or decreased. It is rarely possible for it to remain the same. Yes, if you have not purchased shares, but Government Securities or Municipal Bonds then it is different.

Now, you must notice that there is a fluctuation in the prices of shares every day or every moment. If you purchase shares at the Stock Exchange you do not have to pay money till 15 days and likewise if you sell shares you do not have to give the share certificates. As soon as you know this you can immediately deduce that if by some method you can ascertain what would be the position of the shares of a company in 15 days' time, you can buy and sell shares and settle accounts on the predetermined date. If you have calculated correctly and the market has risen or fallen according to your estimate then you can definitely realise the difference between the relative prices of shares as profit without spending any money or handing over actual shares. This is the game which is called 'speculation' or *satta*. When you play this game, no one will blame you as it is quite legal. If you buy shares you are called 'bullish' or *tejadiya* and if you sell them you are called 'bearish' or *mandadiya*. On the day of settlement you have to settle the profit or loss of the shares in accordance with their value on that day. Thereafter, you can change the system of settlement for the next 15 days and it is not necessary that you have to even out the account that very day. By exchanging like this either you have to pay some fee or you receive something. Here, too, it is the same thing.

If the shares of a company which are bullish are greater in number then they would have to pay some fees for exchange and if the number of shares that are bearish are greater, then that will have to be paid to them. In this manner this 15 day cycle goes on. There is another circumstance which may arise. Sometimes a capitalist who is a great speculator buys or sells shares in such a large quantity which is more than the total shares of that company. Such a situation is called *khela* or 'playing the market'. In such a situation no person can surmise anything. So, one has to be careful at all times to save oneself from some speculator, some big capitalist, who may 'play' the market.

Because all this depends on the correct estimate and calculation by persons who are expert in speculation, so some speculators who can form correct estimate of the market trends, earn millions. They are able to understand this game and trade only through the share brokers. Sometimes they are tempted by the loose trading going on. Besides these, trade in bullish and bearish market is also done which basically is illegal. This means that any person can by paying a fee or premium choose an alternative to buy or sell shares at a certain price. For example, suppose the price of the shares of a company today is Rs. 160 per share. Your estimate is that the price will rise to Rs. 200. You do not wish to buy the shares and take the risk of the value decreasing to Rs. 100 or 125 per share, thereby incurring a loss. So you pay a premium fee of Rs. 10 or 5 per share and choose an alternative by which you have the right to purchase these shares even after a month at these very prices. If the market price goes down you only lose the premium fee and if the price increases during that month then your profit will be equal to the amount by which the price has increased, but your premium would be forfeit. Similarly, there is an alternative to sell by paying a premium. And likewise, there is the choice of buying and selling. You can opt for both the alternatives, which is known as 'hedging'. These transactions are illegal. If there is a large fluctuation in the prices, it is possible that the trader who has traded in these shares refuses to stand by his commitment and you cannot get your money back by filing a suit in the court.

Not only the share market, but the cotton exchange, silver and gold market, sugar and seeds and groundnut market, etc., also are markets where trading is done on assurances. They are also markets where speculation takes place. In these markets you can sell goods worth millions without actually possessing them and buy goods worth millions without actually having any money with you. You will have to settle accounts on the settlement day by paying or receiving the difference between the prices of the opening and closing days. This forward trading worth millions is done all over the world every day. And, in India, too, in big cities like Bombay, Calcutta, Delhi, Kanpur, etc., this forward trading amounting to millions daily is carried on. The basic form of speculation is that if your assumption turns out to be correct, you will earn. It is also understood that some other person will suffer a loss.

No money is generated in these businesses. This is not earned income. It is only transferred from one person to another.

The nation's wealth can never increase through these speculations.

It is seen that very intelligent or talented persons or persons who are knowledgeable about inner truths, often make mistakes in speculation and suffer huge losses.

So, you must have by now formed an idea of what speculation is. Many books can be written to analyse it fully but for a layman this information is adequate.

If half of the intellect, talent, energy and patience which is wasted in speculation were to be utilised in building or any other welfare project, all the slums of India would have vanished decades ago and poverty would have reduced considerably.

All the epidemic-like illnesses which engulf us again and again would never have done so because capitalist ability has the effectiveness by which those jobs which would have taken years to finish are completed in a matter of days. It is India's misfortune that talented people are so addicted to this civilised gambling called speculation, that they are not aware of their proper duty.

56

BANKING

Banking system
Control of the hire of money
The Reserve Bank

The share market is only a section of the capital market. The real place where excess money is received and given out is the bank.

You can open your account in the bank and when you require funds you may go to the manager of the bank who will readily give you a loan, provided they are fully assured that you will return the borrowed money in time.

This is the real business of banks.

Banks can give you loans in many ways. They can, for instance, allow you to withdraw money in excess of your deposits, what is termed as an overdraft.

Suppose you have to get some money from some person. If he executes a *hundi* or a note in your favour for a certain period, then, if the bank is assured that the person executing the note will honour the *hundi* and pay the amount at the predetermined time, the bank will pay you the amount of the *hundi* immediately and charge you a fee. This is called discounting the *hundi*.

So, there are many ways by which you can take money from the bank on hire.

The bank charges also fluctuate with time and circumstances.

You must have heard many times that the bank rate has increased or it has been cut and has decreased. This means that the Reserve Bank of India, which controls the entire banking system on behalf of the Government of India, has declared an increase or decrease in the rate at which money is given on hire in ratio of the excess or shortage of extra money.

Whenever the bank rate increases, businessmen feel the pinch. All the money that they have borrowed is either called back or they are not able to raise fresh funds easily even if they need them. They are not able to expand their businesses.

This is the reason that when the interest rates go up there is much worry and panic in the trading and business community. Similarly, when the interest rates go down a wave of delight engulfs them.

Bank rate, in fact, is a measure of how much extra money is available at a given time.

Now the question is from where all this extra money comes from that the banks lend out.

Whenever you open an account in the bank you have to deposit some amount in your account. No less than Rs. 300 or 500 are required to open the account and you have also to keep some amount in your account at all times.

Whatever the nature of your business, you deposit one thousand, ten thousand, a lakh or even more in the bank. Whenever you need funds you issue a cheque and withdraw this money and at the same time you also keep depositing money in the bank.

In this manner every trader if he is into active business has to keep at times a deposit of a thousand, five thousand, one lakh in his account with his turnover running into lakhs and crores yearly.

In this manner there are thousands of accounts with more or less amounts of money permanently deposited in the banks.

The banks do not charge you anything to keep this amount safe and sound.

And so, the money which the people deposit in the bank is in fact extra money which they have.

Banks give you, us and everyone money as loan when needed. Bank charge a hire for this amount.

This hire or interest is the income of the bank.

You will ask that if the banks give our money on loan then how it will give us our money back when we ask for it.

The question is justified. When you ask for money then the bank gives you your money back out of the accounts of thousands of depositors. But if they all of them ask for all their money back at the same time, then no bank can never fulfil its responsibility.

The reason is clear. Every bank in a fixed ratio keeps ready money or liquid capital out of the money deposited with it in various accounts. If all the money remains idle then banks could never carry on the banking business.

The reason is if the banks earn interest on whatever extra money comes to the bank then they can meet all their expenses and also earn same money. If banks do not earn interest of that money then no bank can ever run its business.

How much money out of the amounts deposited with the bank can be loaned out is a question that has been settled by economic experts through a chart. This chart is tested through the experience of many years.

In spite of all this sometimes rumours or half truths circulate that the economic condition of such and such a bank is in a precarious position. Consequently, people whose money is deposited in that bank rush in large numbers to withdraw their money. This is known as bank run. With such a run on the bank sometimes established banks, too, are unable to honour their commitments and return the money deposited with them and an emergent situation arises in the capital market.

In order to save the banks the Government of India has made many regulations under the Reserve Bank of India, whereby whenever such an emergent situation arises the Reserve Bank immediately steps in to help that bank.

The Reserve Bank takes the certified loan certificates for the amounts given out by that bank on credit and in lieu of these, advances an equal amount to the bank so that it can fulfill its commitments.

If a bank loans out money without accepting any security to borrowers or has invested money in dubious enterprises or incurred loss in speculation, then it is difficult for it to get help from the Reserve Bank. Otherwise, ever since the Banking Act has been enacted in India, there are very few chances of banks failing in the country.

The calculations of economists are strange. Their computations are beyond imagination. If you carefully study their methods of computation, you may at times wonder why such people should not be sent to the mental asylum.

For example, if supposedly the amount deposited in various accounts in a bank is 5 crores, then the economists say 5 crores are deposits. So 5 crores is the credit value of the bank so it can give out 10 crores as loan or keep a reserve fund. Similarly, to ascertain the creditworthiness of people the economists say that the creditworthiness of the person who took 5 lakh as loan and deposited property worth 10 lakh in the bank as security, should be considered as another 5 lakh. This is a strange method of measuring the creditworthiness of an individual or an institution.

The interest rates on loans given out by the banks are also determined by the capital market.

When more people need cash money and less people have extra money then the interest rates will be definitely higher. As opposed to this, if less people need cash and more people have extra money then the interest rates would be lower.

This situation is dependent on the principle of demand and supply. Although demand does not mean that people need money at all times.

In economics the real meaning of demand is the ability of the borrower to return the loan at the predetermined time. This requirement is called demand.

You might be aware that a terrible storm came in the economic policies of India in 1947. The then Finance Minister Liaquat Ali Khan made many such laws and proposed taxes in the yearly budget. As a result, the need for cash suddenly arose among many people. People were forced to sell off their property which was in the form of shares or securities. When most people started selling, the buyers were reduced and there was a recession in the prices. The situation which went out of gear then, has not been set right even till today in 1960.

You may ask what terrible thing happened in one year's budget which altered the economic balance of the country to such a degree.

Just pay attention—suppose on the advice of a strange and ambitions businessman or some revolutionary minister, the government makes a law that the rate of property-tax imposed will be increased from 1 or 2 percent to 40 or 50 percent, at the same time it is also said that on every person besides the wealth tax a special tax will also be imposed by which 40 or 50 percent tax will be charged on the amount of money one can get as loan.

The result will be that every person will rush to arrange cash money to pay his tax.

Everyone will withdraw all his deposit in the bank and also try to borrow as much money as possible. But since all the depositors will withdraw their money, there will be no money left in the banks which they could loan out.

Then everyone will rush to the Stock Exchange to sell their shares and securities.

The demand for excess money will increase to such a level that a special meeting of the Board of Director of the Reserve Bank will increase the bank rate to 10 percent. Since all the depositors will rush to withdraw their deposits, almost all banks will put up a notice after paying out 25 percent of the deposited amounts that the rest of the money is invested in shares and securities and will be paid back after it has been realised in a short while. The banks will declare postponement of debts.

Even with 10 percent bank rate, money would not be available. To pay their taxes people will say that they have shares and so can pay only by giving the shares. The amount to be paid in taxes is hundreds of thousands whereas the shares will not even fetch a few thousand. When everyone is a seller then where will the buyers come from?

A so-called socialist-minded Finance Minister once said that if this happens, let it happen. When 99 percent of India's population is already in a miserable condition then what matters if one percent of the people have to suffer all this. No one could understand how the economic policy of the nation benefitted from this. A threat was issued in anger that he would come before the parliament proposing such taxes not once a year but several times.

If this mentality is taken to its logical conclusion the result would be that no one will have any cash left. People will not be able to arrange tea in the evening and food at night. There will be widespread unrest and when the whole populace rises up in clamour everyone will pray for a dictator like Hitler or Mussolini or Napoleon to come and set things right, improve the administration and deliver the people from misery.

In short, when the economic policy of the nation becomes a thing of play in the hands of such short-sighted people then the last resort is dictatorship so that people can be free of such disorder.

Democracy is suspended because class war takes place and because of it all mishaps occur like closure of bans, lock-outs of share markets, etc.

Such events have happened many times in many countries of the world.

History is witness that wherever unthinking people have done this, the economic system was ruined.

57

MONEY

Easiest medium
The power of money and its acceptance
Inflation

After discussing all this, a basic question arises. What is money? What shall we call money?

Some gentleman is very moneyed. So, what is this money? So and so has a lot of rupees. What is a rupee? That is to say, whatsoever we may call it—money, wealth, cash, finance, rupee, currency, etc.—what is its nature, its true form, its features?

As we have explained earlier all the things which are the daily necessities for mankind, can be called wealth, such as food, clothes, houses to live in, etc. All this is definitely money. The things from which these things are generated are also money. For instance cows, buffaloes and goats give milk and from milk many other items of food such as butter, curd, cheese, etc. are made. So these animals, too, are money. From the land we get food grains, cotton, etc., so, land also is money. Horses, Bullocks, asses, etc., are used as means of conveyance or cartage and they are also money. In short, all things of use for mankind which you possess are money.

Now, you have to go from one village to another. Before you board a train you cannot tell someone that I have one sheep or goat, take it and give me poultry or eggs or vegetables in its stead.

It is possible that such a situation existed in ancient times, but now another very easy medium has been invented since centuries and millennia and that is coins made of metal.

Gold and silver were accepted as costly metals. One reason being their shortage and rarity, the other being their universal utility and use. So, civilised society made these metals as an easy medium of buying and selling things, of trading. Three to four thousand years back, even before the reign of Emperor Ashok, gold coins were in use in India.

Ten thousand years ago in Lord Ram's age or even before that in the Vedic age, mention is made in many places of 'swarna-dhenus'. The swarna-dhenu was a kind of coin made of gold with the form of a cow embossed on it. Many such coins have been excavated at *Moen-jo-daro* and other archaeological sites. Indications of gold coins are also observed in the caves at Ajanta and Ellora.

That is to say that perhaps in India metal currency had come into usage before other countries of the world.

In the 16th century when the Mughal Empire was in its prime, coins made of silver were widely prevalent in India. There was no rule like that of Akbar the Great in India. So, various kings and regional rulers had their own currency. Coins were made of silver and their value was determined by the quantity of silver in them. If a king wished to be unjust towards his subjects and desired to spend the wealth of his kingdom as a means of personal pleasure and comfort, he would instruct officers of his mint to adulterate the coins. Thus, without any effort, the value of his currency would double.

When the people would come to know of this, the prices of things would go up and so would the remuneration paid to labourers.

There was no fixed economic level in those days, so these falsifications could not produce a huge unrest. It is said that a king had made a water-carrier king of Delhi for a day and he had ordered that coins of leather be brought into circulation. For that water carrier his worn-out leather pouch was more valuable to him than those metals which were used in coin making those days. Anyway, leather currency may have existed for only a day but this leads us to a new point of view that if the administration or the government wishes something to be accepted as money even if it is not money in the normal sense of the word it can do so, as long as people have confidence in that government or administration. In the whole world gold or silver was considered as money. Later, notes made of paper were printed making gold and silver as the basis. If you see from the point of view of the value of the currency note of Rs. 100 or 1,000 is a rectangular piece of paper 6 inches long and 4 inches across and so much has been printed on it that it cannot be used for writing. Someone may not even buy 100 or 50 such notes for one paisa, but its value is Rs. 100 or 1000 because it has the power to buy goods Rs. 100 or 1000 for you. You may buy food, clothes, cattle, anything and give those pieces of paper in exchange. They will be accepted and valued accordingly. In the way centuries ago kings would adulterate the metals used in the making of coins and decrease the amount of silver and gold used, so, too, the present government by printing notes in excess reduces the value of money which is intrinsic to the notes.

In the First World War Germany was defeated. The victorious nations imposed debts worth millions as compensation for destruction during the war. Now, Germany did not have the wherewithal to pay off this huge debt. So, it printed notes worth the amount of the debt and settled it. As far as printing notes to pay off foreign debts is concerned, it could be termed as an act of patriotism, but since the government had borrowed millions of Marks from within the country, they, too, had to be settled by printing currency. Once this unreal currency came into circulation in Germany how could one differentiate. The notes of the German government had a uniform value whether the notes were with an ordinary citizen or the British. The situation deteriorated to such a degree that the cost of dinner in a hotel became 5 million Marks. Similarly, currency notes for a cup of tea were given in such a quantity as would fill a cart.

This is one way of deceiving your creditors. In the international order it has been accepted that the machines printing currency notes be speeded up. One cannot understand what difference it will make if mentally Rs. 1 was regarded and accepted as Rs. 1000. The thing which cost one rupee will now be available for Rs. 1000. The difference will be there in returning loans. Similarly, the creditor nations decrease their currency to realise more from their debtors. That means, they reduce the number of notes in circulation. The first situation is called inflation and the second situation is called deflation. Both the actions are not honest. They are dishonest tactics. The nation whose economy has inflated or deflated currency will be counted among those nations who resort to objectionable practices. Therefore, if a nation wishes to establish its true worth in the world it should stabilise the value of its currency. There should be no inflation or deflation. The value of the currency should be stable as far as possible.

It is now 12 years since India achieved freedom. In these 12 years, five or six finance ministers have come and gone. Every minister had a different currency policy. But it is a matter of satisfaction that the economic policy of India has not been discredited in the world. Although, lately the government has had to adopt a policy of inflation to a certain degree but since it, too, has a definite principle that up to a certain point inflation is not only blameless but also needed for development, India has not crossed the limit and no danger has arisen out of it.

How much inflation and why is it useful—this is a separate subject on which there is much difference of opinion among the economists of the world. But the object of this book is not to go into problems of currency; so, it would be sufficient to say that the paper notes can also be more or less in number as the amount of silver was increased and decreased in ancient times.

To indicate how many notes are enough for India is a difficult matter.

First, let us see why notes are needed.

In the country where banks have not developed fully, much transaction is carried out through notes or paper currency, otherwise most of the accounting and trading is done through cheques. Suppose you have purchased goods worth Rs. 5,000 and sold goods worth 5,100. So, only one note of Rs. 100 should be enough as you have adjusted Rs. 5,000 through cheque. The same cheque you received back. This happens when you buy goods from the same person to whom you have sold goods. If you do not trade with only one person, but with different persons, even then the buyers and sellers will deposit the cheques in their respective bank. Nobody will have to give out money, nor receive it. Suppose all your accounts are in different banks. Still, through cheques the entire statement is tallied and with a couple of entries between banks, all the trading is done. Only the excess or shortfall is paid for. All this is done through the clearing houses of banks. Now the problem remains of paying or receiving the small excess amount which is to be paid.

To remove this every bank has opened its account with the Reserve Bank of India and every day entries of debit and credit are made in the books of the Reserve Bank. So, in city such as Bombay where millions are transferred from one company or person to another in trade, one does not need even one currency note of a rupee.

So, how many notes are needed? If only debit and credit entries have to be passed in the Reserve Bank of India, why print notes?

As we have said above, the banking business is not yet fully developed in any country and in India it is quite backward.

Before the Second World War currency notes worth Rs. 4 billion were considered enough for India and Pakistan together. Today 20 billion are not enough for India alone. Is this all inflation? Inflation is not just printing more notes but printing notes without reason and so, increase in paper currency without increase in wealth is inflation.

The economic and industrial system of India is progressing on the path of development. So, in accordance with the economic system it is necessary to increase notes.

Compared to other countries Indian currency is still at a minimum in the world.

Here, there is still not Rs. 100 worth of currency per person where as in America Rs. 2,000 worth of currency per person is in circulation. The same is with Britain.

As the standard of living of the Indian people will rise, so will the currency in ratio of this rise, and it will not be called inflation.

Even today many old people say that they do not trust paper notes. They only accept gold and silver as money. So they collect and hoard gold and silver.

This is the reason that the value of these metals increases day by day.

In my opinion this is also one of man's idiosyncrasies; otherwise why gold and silver are currency and why notes are not?

58

CURRENCY CONTROL

Nationalisation of mints
Valuation of cheque and hundi like notes
Inflation and banking
Nationalisation of banks

You must be well aware how important and useful banks, notes and coins are in today's world. In what ratio should metals be used in the coins and what national emblem should be embossed on the coins are all of them the work of government mints.

Coin making has been nationalised. No individual or private agency is allowed to mint coins at home or in its own workshop. It is illegal to do so. Before the Second World War, there was a proportion of silver in the rupee coin. Before that, in queen Victoria's time there was 91 percent silver in the Indian rupee. Now, there is no silver, but you cannot make silver rupees at your home or workshop and say that your coins are more valuable than the government coins; you cannot say that there is no underhand dealing in this, rather good coins have been made, so, why are you guilty? This argument will not be accepted, because according to law the right to make coins vests in the government mints, no one else's. As long as coins are considered money, they are money and so their account has to be kept by the coin issuing department of the government.

If a larger ratio of coins is minted, the situation will be the same as it was in Germany in the case of paper currency after the First World War. If every person or workshop was given the right to make coins then it would become extremely difficult to regulate them.

Before independence, many native rulers had the right to make coins bearing their emblems and marks.

Every native state that made its own coins would do so within its own area according to the prescribed system and regulation.

Similarly, no one can print government currency notes. Legally, it is a grave crime to mint or print currency with heavy penalties attached for transgressors such as 7 to 10 years of hard imprisonment.

The currency department has been nationalised like the postal department. You can see, the postal department has become so commonly acceptable that you do not even notice it. No one finds it objectionable.

Probably, you may know of the law that if you charge or accept money to deliver letters from one place to another, you may be taken to court for this and after conviction you can be sent to jail. Delivering letters is the function of the post office, you cannot do it, it is illegal.

Similarly, running a mint or printing notes is an offence.

Of course, you can put into circulation other types of paper currency. You can give your cheque or promissory note (*hundi*) in payment for value of any goods that you have purchased. This *hundi* or native cheque can be transferred by the payee to another person and so on. Of course, you shall have to pay the amount mentioned on the *hundi* to whosoever presents it.

The circulation and acceptance of *hundies* and cheques are dependent on the creditworthiness of the person issuing them.

Many millionaires issue *hundis* worth lakhs of rupees. People have full faith in them. The merchant who sells goods, gladly accepts *hundis* or cheques in lieu of cash, while there may be other people whose cheques worth Rs. 10 are not acceptable to other persons.

All this is dependent on personal prestige and the creditworthiness of the person concerned. The *hundi* or cheque is like a note, a letter of authority to receive the amount mentioned therein. The only difference is that currency notes are issued by the government and everyone knows that the note can fetch goods which are equal in value to the value mentioned on the notes, no one can refuse to accept a note, but a *hundi* or a cheque may not be accepted by everyone.

Only the government has the right to make legal currency.

If someone refuses legal currency he can be punished.

What are rupees after all? If one cannot buy necessary things with them, then they are worth nothing even though they may be made of silver or gold. If desired things can be bought by it, then whatsoever it may be made of—copper, tin, lead, leather or even paper—it is certainly money.

It can be controlled only by the government, no person can do it.

Banks have not been nationalised yet, but when people will come to know of the advantages of nationalisation then the banks will definitely be nationalised.

The Indian government has shown a very proper and easy method to the people. The State Bank of India is a bank that has been fully nationalised. Today, while other banks are under the control of capitalists, this bank is being run entirely on nationalised lines.

Those people who have the slightest experience of financial transactions, know that comparatively lesser interest is payable on loans taken from the State Bank of India.

Economic assistance is available from the State Bank of India in any city or town. Although, the management and running of the State Bank of India has not yet been organised fully on nationalised lines, still the officers and workers are part of the old empire, which was under capitalist control and therefore, they are not able to suddenly shed the habits formed during decades. But it would not be right to say that the purpose of nationalisation is to make the extra money available to people at cheaper interest so that it may fully cooperate in the development of production.

A peasant produces paddy in 4 acres of land. If he is not able to manage manure and irrigation in time or get hold of labour at sowing or harvesting time because of lack of money, then undoubtedly he will be able to produce only 15 maunds of rice instead of 25 maunds.

The primary duty of the State Bank would be to evaluate the standing crop of the peasant and give him proper assistance and aid so that he can arrange for manure and irrigation, etc., in time and the production may increase. In the present capitalist-oriented banking system those industrialists who require crores, get loans at proper rates of interest, but for the smaller units which need Rs. 50,000 to 100,000, the terms of interest are very strict. Therefore, if smaller businessmen and peasants need ten or five thousand as loan, they have no other course than to go to the money lenders, who charge exorbitant rates of interest.

These people have very strict rules and their rates are very high, so much so, that the Jew Shylock would seem to be a generous person compared to them.

The advantage of the nationalisation of banks will be that everyone would be able to borrow money at reasonable rates of interest for his industry or business. Banks have a lot of surplus capital with them which they must invest and give as loan for interest.

If some bank manager sanction a loan without adequate security by which the loan may become a bad debt, then that bank manager is in reality doing an injustice towards those who deposit their money in the bank. But a bigger injustice is meted out to the people by letting the deposits remain idle in the bank without giving out the money on loan or investing it.

The reason is clear. That extra money represents a thing which is perishable.

In other words if there is one lakh with him it means he has one lakh rupees' worth of wheat or cloth or shoes with him. Since all this stuff will decay and be destroyed as time passes, so will the money. If all that stuff is needed by people and they cannot buy without money then all those goods will go waste with time. If they were given to the people on credit they would use the stuff and the property could be saved from destruction.

The main benefit from the nationalisation of banks will be that money will be available at cheaper interest to all kinds of industries and trades. There can be no middle path in nationalisation. The critics say today that if the State Bank of India which has been nationalised proves to be more successful than other banks, then only can nationalisation be considered correct, but this is not so.

It is evident that the main feature of nationalisation is monopoly. If the State Bank of India has a monopoly over the entire Banking system, the aim of nationalisation cannot be achieved. While the aim and object of other banks is earning profits, the aim of the nationalised bank is to provide money at reasonable interest so that some people get the necessary capital..

For instance, suppose the Post and Telegraph Department allows, then many companies will be ready to deliver letters at cheaper rates that the Postal Department but would not even touch the letters to be delivered to a longer distance. The result will be that the postal department will not be able to function properly. It will not be successful. It will only get the post which is more difficult and expensive to deliver.

In the eyes of the critics, profit is the measure of success; so, they would immediately say that this department controlled by the government is incurring loss. They will present statistics that the private companies earned so much profit; so, the private enterprise should be considered more trustworthy and essential and be given more recognition. They will also say that nationalisation leads to loss; the nation's wealth is being wasted in idleness, mismanagement and corruption, etc.

During the past few years a new kind of banking system has been introduced in India. It is the cooperative banking system. Some peasants, workers or consumers open cooperative banks with their resources to create the fund and after recognition by the government, the cooperative department of the administration gives it adequate economic assistance. Government loans are available to these cooperative banks at very cheap interest rates. These banks loan out money to their members according to their regulations. The interest charged is slightly higher; still, it is not as high as these small borrowers have to pay elsewhere.

These cooperative banks can work in their own limited sphere. They do not have excess idle money in their funds. Therefore, it is necessary that all banks should be nationalised. This is the only remedy. The Indian government is slowly following this path. Many State Banks have been made subsidiary companies of the State Bank of India. Slowly, the way is being cleared for nationalisation. But the only proper way is to nationalise all banks at the same time.

Some critics while criticising the nationalisation of life insurance, give examples and say that because of the corruption in life insurance the hard-earned capital of the people is in the hands of corrupt officials and there is no guarantee for the security of their money.

This is true. There is, of course, the danger in nationalisation that if the managerial and operational authorities are not efficient and honest, such difficulties and impediments will have to be faced. In the nationalisation of these different companies those very managers, accountants and other officials who had been running the respective companies from which they had been inducted, were entrusted with the job of running this huge corporation. Originally, they were under the direct supervision of those capitalist directors and company owners whose money, in large amounts, was invested in the corporation. Naturally, their interest lay in the profit of their respective companies, so the managers and accountants and other officials could not resort to dishonest practices even if they wished. Now, suddenly, this control was removed. And so, it gave them an opportunity to do as they wished. Taking advantage of the security of their position, coupled with bureaucracy, they gave way to greed and corrupt practices. Hence, the criticism. But the whole blame cannot be ascribed to nationalisation nor can nationalisation be opposed because of this. It doesnot mean that nationalisation was a wrong step. The nationalisation of life insurance was very necessary.

Leave aside the few directors who were more concerned with the interests of their own mother companies. In majority cases the situation was that the directors of the insituations had embezzled the capital of life insurance which was in a way the money belonging to the 'trust fund' and this amount was in millions. The operation was carried out by them in such a cunning manner and the embezzlement was done so legally that this was not noticed by anyone. Only a court of law can pronounce judgment on this. But this is amply clear that if an amount has been entrusted to you and you use it for your own selfish interest, it is a breach of trust and a wrong act in principle. It may or may not come under the purview of law, but it is nevertheless a moral crime.

No political party can have any objection to the nationalisation of banks. The only objection stems from some capitalist bankers who say that the government will not be able to run the banking business. The reason offered is that it is not easy to do so. According to them it is a mysterious business. As a matter of fact it is not mysterious at all, but has been made so by these people, because they themselves do not know about the basic principle of banking—what it is and what it should be. So, in order to hide their ignorance, they have concocted this myth of mystery.

The excess funds of thousands of people lie in deposit with you. It cannot be left lying thus. It is also not possible to relegate this money to be wasted. Then how should it be utilised so that it goes towards producing and manufacturing the goods which can be made from this money.

When these large banks did not exist in India, the banking business was carried on by the native bankers—the *sahukars* and the banking *kothis*. They operated on a small scale but they fully understood the banking business.

If the bank manager, who works with responsibility and integrity for the interest of the bank, continues to do so even after nationalisation then it is certain that he will prove to be more successful and useful.

59

COMPENSATION

Compensation necessary for nationalisation Where will the amount of compensation come from? Nationalisation does not mean more profits More order and convenience

While discussing nationalisation one subject remained untouched. It is possible that you are a shareholder of banks. You may wish to know about the fate of the money that you have invested in the banks.

Naturally, if those shares are taken away from you without giving you their full worth, you would never agree.

You must get proper value for your shares according to the market rates or any other rate.

The government must pay compensation.

Actually, the method of paying compensation is very simple and excellent. If work is carried out with thought and in an orderly manner, then no one will have any objection to the nationalisation of a business or enterprise.

In fact, nationalisation is taking over the private property of individual property owners. Nationalisation is the removal of ownership and possession of an individual over his property or business.

Now the question is that if the nation pays compensation after proper evaluation according to the market value or otherwise, then where is the amount of compensation paid from?

The nation only has the capital received from the taxpayers. The compensation will be paid from that money. If the government does not have enough funds from taxes then by levying special taxes or increasing the number of taxes, more funds would have to be arranged from the people. Only then compensation would be paid. In reality the money of the people was returned to them.

Suppose, only 10 persons in the nation have shares in banks. Also suppose, that those 10 people are the only taxpayers. Then what is the difference? Whenever the government needs money it will realise money by extra taxes or increased taxation from those very taxpayers; they will get their own property as compensation and the banks would be nationalised and become the property of the nation. The government would be able to claim that it had paid proper compensation while nationalising banks.

It was said only for the sake of example that only 10 persons are taxpayers and only they have to be paid compensation. In actuality it happens that the persons getting the compensation may not have to pay any taxes and those who bear the burden of taxes do not receive any compensation. This is well nigh possible. It is nothing new.

This aspect of equal wealth distribution is attractive and excellent.

Suppose you have 10 acres of land. The municipality has chosen that land to make a public park or a road. If the land is taken away from you without paying you compensation, you would not be happy, even though public interest is involved; this much is true. You would say injustice was being done to you, you were being oppressed and you would not accept gladly. You will say this was highhandedness. But if the land was taken away from you after paying you due compensation then maybe you will not feel as offended and possibly you would agree to give it happily, keeping in mind the public good.

Now suppose the municipality has to give Rs. 25,000 as proper compensation for those 10 acres. If the municipal corporation gives you compensation by raising this amount through an increase in house tax from those citizens who are going to get the full benefit of that public park them the person getting the compensation is part of the people who have paid the extra tax.

Meaning, that a small amount was paid by all and the public park or road was built.

In case personal property is nationalised by paying proper compensation in this manner, then very few people would object to it, although it comes to the same thing—that the nation grabs hold of that property without giving anything.

Another argument which is presented against nationalisation says that when the State Bank of India has been nationalised, then why the result of this institution should not be compared to the results of the private banks. The institution which fairs well in this competition should remain. The answer to this is that in banking, capability is not measured in terms of profit. Unfortunately, in this age of capitalisation economic advantage and the largest profit has become the measure of success.

The industry or workshop which gets the full cooperation of the bank, it can fair well in competition. But any number of businesses suffers from this competition and it has led to the closure of numerous enterprises because of the acrimonious relations between traders and businessmen. To access the damage caused by improper assessment of capability and the constructive plans of the nations another chapter is needed.

60

AGAINST NATIONALISATION

Lack of efficient managers
Lethal tendency of capitalist profiteering
The disorder of nationalised industries is immediately evident

Only through the nationalisation of banks can the essential goods be made available to the people at proper prices.

This, too, has been made clear that the amount paid in compensation is nothing but a jugglery of figures.

So, in principle it has been established that the foundations of nationalisation are strong.

But while implementing it, it is all the more necessary to note that just declaring nationalisation of such and such industry by the government is not enough. This will not achieve anything; rather, that industry or business will close down. Therefore, before nationalising any industry, institution or service, the administration must have an array of able managers to bear the responsibilities of that business and carry on proper operations.

Without proper arrangements, the decision to nationalise is not devoid of pitfalls. For example no government can nationalise the postal service till it has arranged for a Department of Postal Services and appointed officers, post masters, the post master general and even postmen.

Without the Defence Ministry, the armed services department cannot be run.

Without appointing the mint master and other officials of the mint, the mint cannot be run.

Without the Home Department the police department cannot be run.

Without the Railway Board railways will not operate.

Similarly, before nationalisation, it is necessary to arrange for adequate managerial personnel.

Only a popular, stable and just government can make proper arrangements for these managers and officials. It cannot be done by revolt, criticism or dictatorial attitudes. In those Western nations wherever nationalisation took place, these managers were summarily changed, while in the next elections after other political party came to power, it subdued its opposite party through revolt generated nationalisation, but they could not run the nationalised industries and factories. Finally, having despaired they had to again bow down before the capitalist private property owners.

In nationalisation, whichever industry or service it may be concerned with, clarity and honesty in economic affairs is essential. At the same time this fixed resolve is also necessary that they will make nationalisation a success. If on the excuse of nationalisation rampant looting takes place or the general revenue is increased to earn profit or the whole plan is jeopardised by haphazard management so that the same capitalist profiteers obtain the nationalised industries to operate and run, then decidedly no good will be served by nationalisation.

The railways can be taken as an example. A large part of the profit which the railways earn from fares and cartage is transferred to general revenue. As a result, the loss which would have been realised through extra taxes is now compensated by the fares and freight charges of railways. When the railways were in private hands, the profit given to shareholders was given from the savings after expenses on account of the care and maintenance of all the railway property. The profit, after deducting expenses for maintenance, on whatever the work or service which an industry performs, is distributed among the shareholders; for instance, the profit on the work of carrying passengers and freight by the railways, after expenses have been met regarding care and maintenance.

And the object of nationalisation should be that all those facilities and services should be performed as before. Rather, better than before and still the profit after deducting expenses should be returned to the consumers.

This is the only way to run nationalisation successfully and honestly.

If you are entrusted with something and you start earning profit from it to invest in another thing then this action is worse than the wrongful acts of capitalist owners. At least this should not be the aim of nationalisation.

The result of such a faulty tradition will definitely be painful. The industries which have already been nationalised, such as railways, their wagon-making workshops, engines and stations, etc., will slowly deteriorate and decay. Their maintenance will become difficult. Since there is no other person in the competition, so whatever the fares or freights are charged, you are forced to accept them and travel or send your goods. You will have to take recourse to the railways. And if the people raise a cry to denationalise railways and hand them over to the private sector, it should not come as a surprise. The question is not that there is no mismanagement in the private sector. There is and on many occasions there are dire results, but their responsibility is limited to themselves. They are not answerable to anyone and so their misdeeds are not made public so widely.

The misdeeds of the government come out in the open quicker. Their faults begin to pinch the people soon enough, whereas the actions of the private capitalists almost lie hidden and so there is an illusion that as compared to private companies or industries there is much more mismanagement and corruption in the industries and services which are in the hands of the government.

At present, in the developed economic ambience of our country, this propaganda is rampant. To some extent there is also truth in it. The reason being that nationalisation is being done right and left, but without proper management. The people who are campaigning against nationalisation are mainly those whose vested interests are directly or indirectly adversely affected. For this reason, too, it is a very important question before the Indian voters that those, whom they vote for in the elections, should not be anxious to protect their own vested interests, but fulfil their responsibility having understood the needs and interests of the people.

61

IF COMPENSATION IS NOT PAID

What if land is acquired without compensation? Agriculture and industries will go on nevertheless Decline in the purchasing power of the people Difficulty in economic balance

Nationalisation after paying compensation has been discussed.

There is one section among the peoplc who say why any sort of compensation should be paid for nationalisation?

Why not take away their property just like that?

Their argument is that just as landed property was acquired from the zamindars and was given to whichever peasant who was willing and ready to cultivate it, if these industries or banks or shares are taken away, then what difference would it make. Whosoever produced crops previously is still producing crops at present; the only difference lies in this that the land revenue or rent which he then paid to the zamindar, he pays now to the government. Similarly, the industries and banks are also running. The owners of shares change in the stock exchange with the sale and purchase of shares. There is never a permanent owner; then why should the government not acquire all the shares without paying compensation? By this decision the work of the companies concerned will not cease. There is another argument which they put forward. That the rate of wealth tax should be raised to such an extent, that the private property owners are not able to pay the tax and perforce have to surrender their property to the government to pay the taxes.

When all the people will set out to sell shares, there will be no one to buy them, even at dirt cheap rates. At that time the government can say that it will purchase their shares in lieu of tax at such and such rates. As it is, the owners of those shares are not going to get anything in the market, the government is adjusting their taxes by giving them proper value for their shares.

When caught in such circumstances the capitalists say, alright, but what about the yearly income we received from these shares? How will we meet our own expenses? The only answer to this will be work brother, labour like the others and live your life like they live theirs. Rely on your earnings and do not expect income without working.

Although for the socialist initiative regarding equal wealth distribution, property tax or capital levy is probably right, but this proposal is foolish as far as increasing general revenue for the expenses of the administration is concerned, of course, the nationalisation of all mines and industries without paying compensation can be done in one go.

The plan is admirable no doubt.

But, there are objections in acquiring property without paying compensation. People say that taking away the property of the rich will lead to unemployment, because they give work to a number of people. It is true that they keep servants and so many people get employment. But this employment is like keeping people to manage mad people, or increase the members of the police to catch thieves or increasing jail warders when the number of jails is increased. Then the mad men, thieves and those being jailed can also claim that they give employment to many and keep them free from being unemployed or without work.

Such employment is worthless; it does not produce or earn anything. It is not work. It is like living on the mercy or alms of a person. Undoubtedly, when the situation of the person giving employment weakens, when he is no longer able to give work or pay his workers and servants, then the condition of those people will also worsen. The reason is that they only know how to do that work under slavery-like conditions. They were never self-reliant. They are afraid of being on their own.

Whatever may be the case, this much is certain that when those rich people are bereft of their property, all their dependants will also become redundant. They are large in number. It can be said that in India all together one-third of the population comes under these two categories. Besides, there are lakhs of middle class people who derive their livelihood from these people. The jewellers, goldsmiths, ornament makers, etc., constitute thousands of people engaged in professions and trades which are dependent on these rich people and they will all become unemployed once the rich lose their wealth.

So, they can never agree to the proposal of taking away all the property of the rich without paying them any compensation.

If the industries, factories, mills and banks, etc., become national property, then all those shares will be transferred to the government which gave the rich their yearly income and through which the purchasing power of the rich existed and was increasing.

If this purchasing power comes into the hands of the government then it is natural that it will not be spent. The national property will be destroyed and with it lakhs of people in the nation will face destruction. Tremendous unemployment and insolvency will spread all around.

Maintaining law and order would become a problem for the nation. Anarchy will spread, people will resort to violence, there will be food riots and the situation will come when some dictator will have to take hold of all rights and suppress the unrest ruthlessly. In such a tempestuous condition no government can run with stability.

Finally, the government will have to offer all the private property back to those owners from whom it was taken and with this apology that they committed a grave mistake by nationalisation.

Another suggestion is that all the money thus taken away from the rich be distributed among the people in small quantities.

But the problem is that the same flaw exists here, to remove which nationalisation is necessary. When the aim of nationalisation is to take away any and all the money or wealth which has been obtained without earning it, then the same malpractice will be undertaken by distributing the wealth of the rich among the people just like that. Those who will receive money by this distribution will get it without earning it.

If the nationalised banks give everyone more loans than are necessary and that too at cheaper interest then that also will result in people opening four or five factories instead of one and needless competition will arise. People in favour will say that their purchasing power will increase and so the fear of unemployment will vanish. This may come to pass. But it will take a long time for this situation to change. It will take so much time to balance the purchasing power of the labouring classes and poor people that the economy will crash.

Unemployment will increase in such an extent that inner struggle will begun and it would not be long before all the aims of nationalisation are lost; therefore, if there has to be nationalisation it should be done after paying compensation so that while the purchasing power of one class decreases slowly, the purchasing power of the people is increased so that unemployment does not spread.

62

THE NATIONALISATION OF THE MANAGEMENT OF INDUSTRY

Control of profit by the government without paying compensation
Policy of encouraging small industries
Loss in national production

The Government of India has adopted a new form of nationalisation. It is the nationalisation of the management of an industry without actual nationalisation. In the initial stages this proposal was considered as only an experiment, success was not certain. Let us see what difficulties may arise in this scheme.

The first rule of such nationalisation is that the industrialists, who own it, run the industry; only the protection of the consumers of the product of that industry is ensured by law? In the last World War when the production of everything was very little as compared to its need, production and prices had been controlled. When there is an emergency, all such laws are accepted by the people whether they like them or not. But when those laws continue even in ordinary conditions, then it is very natural for the people to be discontented.

The government servants and officers have become so used to corruption while continuing these controls that they are not willing to do away with these controls at any cost. Every time statistics are presented which convey that if the controls were removed, then profiteers and black-marketers will force the prices to touch the skies and the people would not be able to easily procure the things they need.

To remove this false shortage, father of the nation Mahatma Gandhi had in 1948 almost forced the government to remove the controls. But because the government officials had once tasted blood, they were not in favour of the removal of controls or the success of such a measure. Years later, the production capacity of the nation increased. And the then Food Minister Rafi Ahmed Kidwai adopted brave measures showing rare grit. It is because of those brave measures that the Indian population is getting food to eat and other necessary things.

The features of this new nationalisation are that you will get raw material at fixed rates and the goods produced will also have to be sold at fixed prices to the representatives appointed by the government. The industrialists will not have any freedom.

For example, take the electricity company. The government has fixed the price of electricity at a certain rate per unit. If you have set up a plant to produce electricity, the license which you will get will be on condition that you will not charge whatsoever you desire. Under the Electricity Act, keeping in mind what you spend on the production and your total invested capital, it is decided that you get appropriate interest and rent, supposing 5 percent per annum, and accordingly the price of electricity is fixed. If some industrialist is thrifty and spends judiciously and works honestly, he is going to get only this 5 per cent profit; and those who spend money injudiciously, they, too are going to get 5 percent.

The openly honest person has to suffer and the dishonest person is respected. On seeing this it can be said that in nationalisation dishonesty fetches you more profits.

Take the other example of the sugar industry. There are almost 150 to 200 sugar factories in our country. The government has fixed the price at which they have to buy sugarcane. To whom is the sugar to be sold and at what rate, this too has been fixed. In all the sugar-making processes the government chemical engineers have direct supervisory authority. For all intents and purposes the sugar industry is running in the private sector. But its nationalisation has taken place in reality. Similarly, there are many industries and professions and trades whose control is so strict and organised that it can be said with certainty that if the government takes over all these industries after paying proper compensation, the mill owners would he more than happy. They have to bear all the inconveniences of nationalisation without benefiting from the profits.

63

UNCERTAIN ECONOMIC POLICY

Encouragement to cottage industry vs. development of national production
Production tax

There is another adverse result of the twisted and half-baked economic policy of the Government of India. They are not able to correctly decide what exactly they want. Much is talked about increasing production, but at the same time no effort is made to rid oneself of the 'bullock cart age' or 'cow-dung culture'. The party holding office has kept Gandhiji's *Sarvodaya* economy as its aim. Clamour to give encouragement to cottage industry or village industry is also being heard. It remains to be seen to what extent national production is being adversely affected by this.

Take the sugar industry. On the production of all the sugar factories in our country production tax has to be paid at the rate of almost Rs. 11 per maund. If someone sets up small factories with a minimal expense on the name of cottage industry, they will not be required to pay any tax. The sugar produced thus is not clear and clean. There is no control on the rate at which that sugar is sold. As a result that sugar fetches as much or almost as much price as the sugar made in big sugar factories.

If sugar is made in the mills by pressing sugarcane in the factories, only 10 percent of sugar is obtained, whereas in those small raw sugar factories almost 6 percent. This means that 4 percent of the sugar is left behind in the cane refuse which has no other use except being used as fuel.

Even if we leave the economic analysis for a moment, still this much is certain that if 10 million maunds of sugarcane is pressed in these small factories instead of the normal sugar factories, almost 400,000 maunds of sugar are lost and this is a national loss.

Certainly, such a tendency should be criticised which causes indirect loss in the national production. There is another thing worth considering. The cottage industries which are being encouraged with such zeal by the rulers are also the private property of the capitalists. It would be a mistake to think that those small cottage workshops would belong to the poor peasants or labourers. No. The labourers and the peasants have to do the same labour and farming which they have done all along. They have to survive by the dictates of fate. All the profit from this tax exemption goes to the wealthy owner and immediately small raw sugar-making factories are established at many places and they derive huge profits.

The second necessity of man is cloth. In India cloth is produced in about 400 mills. There is production tax on every kind of cloth. On an average this tax is levied at the rate of 25 paise per yard on fine cloth. If these very looms which are in operation in the mills are removed from there and after being separated are run on the basis of village industry or cottage industry then they will not have to pay this tax. The number of such power looms in the various provinces of India is over 100,000 and they are run on a collective basis as well as individually. If the looms which run in the mills produce 50 yards of cloth every day after 8 hours of running, the power looms produce only 40 yards, suppose those 100,000 looms were in the mills, they would produce 1 million yards more of cloth per day. This loss of production would have to be borne after all by the nation. The same mill owners who had to pay almost 25 paise per yard as tax, run these looms or have them run in the name of cottage industry and earn huge profits. The nation is at a loss. The income goes into the pockets of the capitalists, undoubtedly.

In the progressive machine age of the present day, to exempt raw sugar plants or handlooms or power looms from tax as encouragement is a step in which every person sees a temptation and thinks why take the trouble? Why risk lakhs of rupees in investment? Why pay lakhs as tax to the government? Why not start small factories in the name of cottage or village industry and save oneself from tax? Why not earn lakhs of rupees?

Where the interest of the nation is in increased production, this policy of the government directly hampers increase in production.

This is the same condition of all industries.

Because of this half-baked scheme of the government, today there is profit and benefit in corruption, in not doing work, in destroying natural resources. In this situation no matter how many Five Year Plans are made, it is a matter of debate how the nation will be successful in attaining its goals.

64

THE PRESENT FORM OF SOCIALISM

The form presented by the congress
One wrong step fatal for the nсxt 100 years
Popular disagreement to nationalisation

It is very important to establish socialism.

The question is that is the work being done which is necessary in order to fulfil the goals set out by the idea of socialism which the Congress Party or the government may have formed (although it is very doubtful if they have formed any idea at all)?

The Planning Commission has presented before the nation two Five Year Plans. The outline of the Third Plan has been prepared and is going to come before us very soon. Can it be said honestly that any progress was made in wealth distribution in the country because of the last two Five Year Plans or was the aim of equality achieved?

Capitalism is bad. It will have to be eradicated, but have the steps which should be taken to do so, really been taken in earnest and in a proper manner.

This too is certain that if sociologists adopt a wrong measure while resurrecting the social economy, they push back their goal by many years.

The nationalisation of industry or trade should be done, but if this nationalisation is done without the arrangement of proper managers or without paying adequate compensation or by hurting the sentiments of the people or as a result of some sort of revolt, then it is certain that the work may be accomplished for once, but the people's mandate would not be won. The work which caters to the interests of the people to the utmost would in a short while earn the ire of the people. And the object of nationalisation would halt for many years to come.

The example of life insurance is before us. Life insurance was nationalised. It was necessary. But because the government lacked good personnel, the supporters of capitalism can very well publicise that corruption is rampant in life insurance. If some person wishes to insure his life for his own safety or for the care of his family, he is treated as if he has committed a robbery or an offence.

The people at large are coming to the conclusion that it is better not to have life insurance rather than undergoing such indignity.

Everyone, right from the Chairman to the General Managers and Branch Assistants of the Life Insurance Corporation are endeavouring to find out how to utilise the funds collected from life insurance and invest them in business and industry. They have no time to devote towards fulfilling the main object of the corporation. After nationalisation a big corporation called The Life Insurance Corporation has been made which includes all the companies, and the work of this corporation is being carried out like the work of the erstwhile companies. The difference is that formerly the personal interest of the directors and owners of the companies made them more careful and efficient, even if this efficiency and striving towards the basic aim of the company was undertaken because of their self-interest. Now, the officers and authorities of the Life Insurance Corporation have no personal interest. They do not care whether the corporation earns a profit or runs in loss. They are not ready to heed anyone. They know that whosoever wishes to insure his life will have to come to them. There is no other way. Everyone will have to come to them.

Advertisements appear in full force that Life Insurance Corporation is making great progress after nationalisation; people are now free of capitalist exploitation, etc. But people are aware how hollow these claims are. The fault lies not with any officer particularly; the fault is of this system where nationalisation has been done without making arrangement for good managerial staff. Wherever this has happened, the results have been the same.

There is a strange tale concerning the compensation paid to the shareholders of the life insurance companies. No benefit accrued to the companies which were managed properly, which were working honestly, which had built up a reputation during the past years and whose fund was safe. The companies which benefited most were those whose business was running dishonestly, who had spent or misappropriated their fund or who were on the brink of closure. Consequently, the public came to the conclusion that those who run their businesses honestly are always at a loss. Similarly, up to now in all the cases where the Government of India has nationalised industry, a setback has been given to public sentiment. And there is no relief for the common people.

Naturally, the government failed to get the support of the people. So much so, that there is a whole section which has begun saying that British rule was better. At that time the people were more contented than in the system of administration of the present democratic Congress Government. The oppression being felt at present indicates that much delay has occurred in putting into practice socialist economy and because of inexperience, many impediments have been created in the path of progress.

65

THE SPECIAL OPINION OF A SPECIALIST

Dr. Calder's opinion
Semi-acceptance of his suggestions
Failure

In order to successfully put into practice the idea of socialist economy, a few years ago, the then Finance Minister had invited a Western scholar and eminent economist Dr. Calder on behalf of the Government of India to study the situation here and give proper guidance and advice.

Dr. Calder stayed in India for many months. He studied the economic condition here and gave many suggestions for the achievement of the goal of socialist economy. The guidance he gave and the suggestions he put forth are worth study. His ideas have been published in a booklet issued by the Government of India. It is a pity that though the Government accepted a few suggestions of Dr. Calder, it kept silent about the others for some unknown reasons. If you consult a specialist about a specific problem and accept only part of his suggestions and ignore the other part, the result will be fatal.

Suppose you have invited a specialist for consultation on the treatment of a patient. He comes and tells you that the disease is not lethal, but it is malaria and quinine should be given for several days; also, that if quinine is administered through oral application and through injections, the bacteria will die; but since the patient is very weak, therefore, to offset the ill-effects of the bacteria and the heat generated by the medicine, it is necessary that the patient is given a cup of milk or glucose every 3 hours.

Now, what will happen if the person attending the patient thinks that the remedy in reality is quinine and there is no need to give glucose or milk?

If only quinine is administered, it is quite certain that the state of the patient shall deteriorate instead of improving.

Or suppose the diagnosis is tuberculosis. The bacteria of this disease are even more lethal. From the textbooks we learn that they are not affected even by acid and can be eradicated by a special acid called creosote which is obtained from wood. Specialists have informed that if such and such amount of creosote is taken, the bacteria shall surely die. But people with experience know that administering creosote in that quantity will not cure the patient and eventually there will remain neither the patient nor the disease.

What is the advantage of such medicine or treatment? Yes, a few years ago creosote was considered an appropriate medicine for a tuberculosis patient. But nobody could claim that the patient will be free of disease by taking that small quantity of creosote. Only some relief could be obtained. Now, if some inexperienced person looking after the patient without understanding this fact and considering creosote to be the best medicine keeps on giving the patient creosote in large quantities, the result again would be the same. The patient and the disease will together come to an end.

The Government of India also behaved in the same manner with the advice and suggestions of Dr. Calder. Dr. Calder gave several suggestions to achieve the aim of socialism in which the main remedy was property tax. At the same time, issuing a warning about the future economic danger he also said that the rate of income tax in the country, which is the highest, should not be more than 45 percent, there were other suggestions also which he put forth.

All the suggestions, by which the income from taxation would increase, were accepted by the government and put into practice. But wherever such suggestions were encountered by which the government would not lose the faith of the people, they were ignored and not put into practice.

Property tax is a sort of capital tax and capital taxes are usually a structure of illusions. Still, it is not necessary that they are always impractical or impracticable. If the Government so desires is can, through fixing the rate of property tax at 100 percent, appropriate the entire wealth of certain people. In lieu of the tax all the things they have—landed property, shares, securities, gold and silver or cash—can be deposited in the heads of taxes. Such a step would not be nationalisation but pauperisation. The only aim of such a measure would be to grab the property of a person with impunity. The government can issue a dictat that the economic condition of the country is facing danger; the government is unable to pay off its debts and so has been forced to adopt such dire measures. The main objection to such a measure is that this attack on personal property is nothing short of highway robbery. This is not the measure to convert personal property into national or people's property.

The main danger of such invasions is that the desire of people wanting some sort of security is shattered. People feel themselves surrounded by dangers. Whosoever is in a position to save some part of his earnings, will not do so at all. Keeping in view such uncertainty, they accept that life is momentary and believe that they have to live in the moment. Do whatsoever you can do today, leave nothing for the morrow; spend all your savings. People resort to the old sayings about the three fates of money—donation, consumption and destruction. The meaning of donation has been discussed and it is giving to others without their earning it. Consumption is spending in whatsoever way one desires and leads of all sorts of evils, since the main aim is to do away with money before the government takes it away.

If plague spreads in a city and uncountable people begin to die daily, then the people of that city will not consider their lives safe for more than a week or 10 days. All persons who come to this definite conclusion that their lives are going to end in 1 week or 10 days lose the capacity for character building and no one can predict with certainty what they may or not do. The invasion on private property is like economic plague.

There is one more consequence of such a measure. Once the economic plague has found its victim, once the governments are habituated to resort to legal robbery to pay off their debts, people are always in danger.

The eminent Dr. Calder was fully aware of these consequences and that is why he had suggested safety measures and precautions. Had the report of Dr. Calder been fully accepted, India would have become even more capable to attain its aim of socialist economy. It would receive more support from the people. The faith of the people would remain intact and the nationalisation of one industry after another would have been accomplished easily with the peoples' mandate.

66

BOTH SIDES

Good and bad aspects of everything
Administrative capitalism
Mixed illusions about socialism

The difficulties in the measures adopted by the government so far in order to implement successfully socialist economy for the purpose of equal distribution of money have been revealed and have been clearly understood. When the government has collected the entire nation's wealth for distribution through nationalisation of private property, industry and trade, then the question arises whether it can be equally distributed as well through socialist methods?

There are two sides to everything, with each good work there is an evil aspect. Lord Krishna has said in the Gita: 'I am both death and immortality.'

So, what is meant is that there are two names of the same thing. Every good which is life-giving becomes death and death becomes immortality. In the prehistoric times, when the gods and asuras whisked the sea in search of the elixir of life, poison came out before the elixir of life. This mythical story may have its source in some actual event or it may have a spiritual connotation. This much is certain that good and evil have the same source.

Workers and labourers formed their own organisations and unions because they were not ready to bear the injustice of capitalist industrialists. These very unions began to do the same unjustifiable things in a large degree which the capitalist industrialists were guilty of. This danger is always there. Capitalism definitely is a terrible thing, whether it is practiced by the industrialists or the workers unions or even by the state itself. Its evil and negativity remains the same.

What is to be seen is that whether in the endeavour to establish socialism, every man achieves full freedom and equal rights and the goal of equal wealth distribution or not. It is also to be seen that in this process we might not have created state capitalism.

There is very little difference between the two. So, there is always a chance that what has been misconstrued as socialism may be state capitalism in reality.

Such mistakes have been seen to have happened in the history of the world. Such a mistake has been committed by Lenin, the disciple of the great revolutionary Karl Marx and himself a great personality of Russia who had taken up the task of establishing pure socialism with the result that in the end the ideology which he took such pains to support was reduced in the end to nothing but administrative capitalism. Although, for all intents and purposes it is called communism, but what sort of a system it actually is, will be discussed to some extent in a separate chapter which follows.

The naked face of this Russian communism is none other than administrative capitalism.

Why only a little difference can be discerned in socialism or this administrative capitalism? Administrative capitalism and capitalist dictatorship, which is known as fascism, both go to serve the objects which are the basic elements of socialism, such as increase in workers' wages, decrease in the death-rate of the people, rise in average life-span, the proper utilisation of the ability of capable people, ruthlessly suppressing laziness or slackness in work, etc. But all these 'isms' suffer from this shortcoming: in the end that they become victims of the tendency towards inequality, so that no civilisation may be able to confront this inequality, nobody should be able to boycott it in any form, in any circumstance.

This is the reason that in whatever I have written up to now, I have, after giving just a hint about 'what is socialism,' left the question unanswered that what should be the form of socialism? What sort of parliament is required to safeguard socialism? What should be the economy like?

Care should be taken that we do not get entrapped by the capitalist types discussed earlier or some other forms of capitalist traps.

If today we too commit a mistake, then we shall embrace the poisonous communism of Russia or China in the name of socialism, and it is possible that our Indian philosophy, beliefs, civilisation and culture perish.

In order to properly think about all these circumstances, it is necessary to first thoroughly study this false socialism, so that we do not fall into the illusion of accepting some other ideology as socialism.

67

LEGAL LOOT

Grabbing money from a particular class by legal force Giving away the money thus obtained to another class What is the benefit to the nation from this?

During the last World War the government had declared emergency and the Defence of India Rules were promulgated, under which the government could take over the property of anyone at any time without paying compensation or after paying nominal compensation. The laws and regulations of those days are still very much in force at present. The need of that time has become a permanent demand now.

The property of anyone can be seized after passing an act in parliament. In a democratic state every work is done for the public welfare. So, the property of any class in particular can be seized and taken over. It can also be transferred to someone else. The law begins with the words 'in the public good'.

How easy is it for the government to seize the income of any class through the passing of laws and then distributing them to the people of another class? But it does not follow that this will lead to the achievement of the goal of equal wealth distribution. Nor is it necessary to nationalise any industry or trade because of this. If people of any class or any business class or a group somehow attain majority in the parliament then with that power they can loot any other class or business community or any group for their selfish interests.

Such looting is done in reality for one's own selfish ends, but it is done in the garb of national interest and the best way for public welfare and with the claim that it is being done in the public interest.

If, for example, we suppose that the landlords of the city have captured the parliament, then they will see that their interest lies in making such laws by which the slums of each city are invariably removed, roads are widened, parks and gardens planted, etc. Ostensibly, such measures may appear to be purely utilitarian or for the public good, whereas the real aim of such measures is that the rent of their buildings in the cities is doubled or trebled and their income goes up. The people of the other class who live in those slums will have to rent houses at increased rates and suffer added hardships. What would remain with them after providing for food would go towards house rent. In this way the government of house owners will serve its own interest while pretending to bring about social reform and public good and deprive the poor people of their income.

Take another example. There are many cloth mills in India. When the mill owners after sustaining much loss are not able to bear any further loss then the leaders of the workers' unions (it is said that the present government is theirs) have laws passed and take those mills and factories in their own hands and then run the mills with the cooperation of some other mill owner or manager of their choice or preference the profits are taken away by the government or given to the workers as extra wages. It is the same story. Legally one class (that is the working class) seized the income of the other class (that is the mill owners) and provided for its own income.

Not only one or two, but many such instances can be cited. In each instance the idea of protecting the interest of the country is present, but in reality the class in power, purely for the interests of its own class, usurps through law the income and property of the people of another class.

Where the question is of the mismanagement of industries because of which loss is being incurred, this situation can be tolerated or it can be excused but there are instances where subsidy is given only to stop unemployment. If, on an average, the workers of a huge factory of one capitalist get Rs. 50 per month as wages and the workers of another factory are getting Rs. 100, then the governments which support the working class give a subsidy to the first factory as compensation, so that its workers may also be given Rs. 100 a month as wages. Such huge subsidies are given in many countries. In this country also many such incidents have taken place where distribution of money has been done through subsidy.

As a result, when the working class becomes powerful in parliament it will encourage such subsidies and when the mill owners are in power the workers will not be able to oppose them the consequence will be that the ordinary taxpayer will have to bear the burden of the idleness, mismanagement and short-sightedness of these selfish elements. This means that the money of the tax payers will be siphoned off into the pockets of either the workers or the mill owners. In any way, these subsidies are a form of alms which the taxpayer gives either to the workers or the mill owners.

Those who say that subsidies are part of socialism commit a grave mistake. If this contention of theirs is right, then this too must be right that all the thousands of grandchildren, legitimate or illegitimate, of the Nawab of Avadh, Wajid Ali Shah, who are said to be present even today in Lucknow should be given state grants and this too should be said to be a part of socialism.

This is an open robbery perpetrated on the people by those bankrupt businessmen, capitalists or their dependent working class. Those who really know and understand socialism will never tolerate these subsidies. At the same time they will say that by giving these subsidies the idle capitalists are being made richer who neither work themselves nor wish to let others work.

All these things are not imagined. Such a system has existed in other countries. The American capitalism is based on this foundation. Big corporate bodies, huge workers unions, institutions, large workshops and factories are run on such modern scientific principles that one is astounded. Even in electioneering they take possession of the government and the path they follow appears to be the glorious path of socialism.

They are ever ready to improve the living standards of the people and apply plans of improvement and reform of all kinds and it seems inhuman to stop or oppose them. The organised workers live in good houses instead of slums. The big industrialists like Ford, Lord Leverhulme, Rockefeller or Cadbury, etc., become common instead of one among thousands.

The feeling of considering the working class as subservient to themselves is transformed into a sense of self-dependence. New slogans are taking the place of the old hullabaloo. Instead of the demand of a healthy city by raising the municipal taxes, the call for more wages and more profit by people such as Ford is gaining more importance today.

The ratio of profit is the root cause of all evil. These profits again lead towards unequal distribution of money. In spite of progress all over, the demand of the true socialist still remains that distribution of wealth and division of income should be equal and its management should be in the hands of the people. Without this, the big industrialists will collaborate with the workers' unions to obtain power in the government and endeavour to serve their own interests.

The indication of the real socialist is to transform private property into public property; whereas, in the illusory socialism which has been mentioned above, the money or the income of one class of people is snatched away to give it to another class; that, too, not because one party has less and the distribution is envisaged to bring it to an equal level. At the basis of this so-called socialism is the tendency to profit from one's own political power and give more to those are already wealthy.

The world is ever-changing and ever-moving and nothing in particular can stay in the same condition for much time. This is a natural law of the world. It is necessary to have a look at the history of past changes before trying to understand what changes are taking place and will take place in future.

What has taken place in India within the last few centuries? No reign or rule, no government or state remains static and unchanged. Those things or measures which were thought to be cruel, detestable or socially excommunicable are necessary today and not to do them would incur punishment. The whole times have changed. I have personally seen that the squire or zamindar of a village would summon by force any low caste person to labour and any refusal to comply with the order would result in punishment. There would be whippings and physical tortures of many kinds. Not only the landowner or zamindar but the moneylender who gave loan on exorbitant interest would also make these oppressed people do free labour. It was a normal practice to pick up vegetables from the fields of these people without paying any money. The landlord could banish his tenant farmer at any time without giving any reason or compensation. Sometimes the reasons for such a punishment would be strangely incredible. For instance, the field from which the tenant was being thrust out was needed for the landlord's pastime of cockfights. Or it was required as a walk for the wife of the landowner and she being in *purdah* needed full seclusion. Such instances of oppression were aplenty during the past rule of landlords and money lending businessmen. Such behaviour appears inhuman today, totally unacceptable, but there was a time when it was considered quite normal and all sorts of oppression were rife.

Even today we can find a few people who are inclined towards the old system. The age has undergone a change, but they have not changed. Even today they repeat their homilies such as *'dhol, ganwar, sudra, pashu, nari, ye sab taran ke adhikari'* (the drum, villager, person of low caste, animal and women deserve to be beaten), that is, they can be made to work only through castigation.

Intelligence and talent are not birth rights of any rich or poor man. A person with intelligence and brilliance can be born anywhere, in a rich man's house or a poor man's hovel. Now, it so happens that whosoever with his brains and talent and courage does something for the country and the people, he is given the credit of changing the course of history, no matter from which walk of life he may have come. Such people are rightful heroes of society.

These changes occurred in social thinking or the social order with the passage of time. In our country the credit of revolutionary change is attributable to a very large extent to Mahatma Gandhi whose fearless leadership brought about a huge transformation in society. Now nobody considers low caste people as untouchables. Human rights are clearly enshrined in the constitution and every adult Indian has the right to vote. Today, whether it is the Maharaja of Bikaner or an ordinary cobbler, both are equal when elections come. Undoubtedly, the Congress Party has done much to inculcate this feeling of social equality, but the aim of economic equality was never there in this political institution. If Mahatma Gandhi were alive today, he might have suggested some remedy for this social evil. But, at present, an average Indian is in darkness as to the aims and objects of the socialist economy which the government has proclaimed it is going to set up. It is still not clear what kind of system or order it is going to be.

68

THE DYNAMICS OF PROGRESS

The speed of capitalism
The controllers of national capital
People's mandate

There is action and speed in capitalism. Only an active man can progress. A static state or inaction is an indication of death. It is the biggest quality of the capitalist age that it gives speed and movement to all activities. There is a joy in progress. Man always desires change. He easily tires of an unchanging life. No matter what direction change may assume, still, the desire and expectation of change is the natural inclination of mankind. It is because in this change and activity man finds happiness. Remember the joy you felt when you and your family rode for the first time in a bullock cart or a horse-driven cart. With the passage of time, things changed, communication became quicker and now there were railways, motorcars and aeroplanes. The pleasure derived from this speed has made man endeavour for ever more speedy transport. If one country has made an air craft which travels at 600 miles per hour, another country is trying to get its aircraft to travel at 800 miles an hour.

If the power to control speed, to accelerate or decelerate it at will is in one's hands then speed is undoubtedly enjoyable, but when speed, in its acceleration, reaches such levels that it become difficult to control then accidents are inevitable. The same can be applied to the capitalist economy. If its movement becomes uncontrollable, there is bound to be a collision.

Whosoever runs the economic automobile, whether it is the kings and princes, despotic monarchs, chosen leaders of the people and their representatives, persons controlled by workers' unions or the communists controlled by Soviet Russia, no one can regulate its speed or totally control it. Its speed goes on accelerating. Similarly, the industrial cycle of capital also speeds up continuously. No one sitting in this economic car can feel that he is safe. Whenever the driver changes, the new driver takes hold of the steering with the confidence that he will be able to control this speed, but in the end he too fares the same as the other previous drivers. Those who never thought of this before, are worthy of praise. This ignorance is probably a heavenly gift. He who knows of the dangers can never be a guard of a railway crossing. Only a semi-educated man can fulfil that duty properly.

There are very few people in the world who are really concerned with how this economic cart of capitalism is being run or which people are running it in different countries. If there are 5 or 50 educated and enlightened individuals, no one reads their books or pays attention to what they say. They are often ignored by people who think them to be idiosyncratic.

Among the Indian electorate it is not possible that even one in a hundred understands anything about this. They are content with whichever ruler who rules them. The rulers are under the illusion that they think of Rs. 5 as Rs. 100. The voters have no idea of this. Most of them have no knowledge about capital, just as sheep have no idea of a woollen mill, although it is for that mill that the sheep is shorn.

So, there is a basic difference in the ruling class and the exploited subjects. Our rulers may not know how to rule, but they do know that there must be a rule. The voters think that the administration or government unnecessarily puts obstacles in their path, in their freedom. About taxes they know only this that it is oppressive loot by the authorities of the ruling class. If this was the situation of the ruling class in the past, it mattered little, for they did not require votes from anyone. In the reign of kings and nawabs and even during the British rule, people at large were asked not to interfere in the affairs of the State nor express any sort of opinion. If any person dared to say anything to oppose this he was arrested and put behind bars. The people rebelled against this repression; a political awareness came to the country and the people; and, in the end, foreign rule had to go from here.

The peoples' mandate brought democracy, although there is no dearth of people who profess that adult franchise and elections hold no meaning; they say, what do the people know about votes when they are in the clutches of poverty and are devoid of political sense and education and literacy; they should not have the right to vote till they become fully knowledgeable of the political system or do not attain enough education or ability to differentiate between good and bad. Iskander Mirza, the ex-President of our neighbouring country Pakistan, had openly declared that giving the vote to the people was like giving it to cattle and other animals. Whatever be the case, such a situation has arisen in many countries of the world, some elected representatives of the parliament have formed an opinion that this government by parliamentary system is not worthwhile. Instead, there should be a government of strong and able persons. The majority was not a part of this or any view because the majority has no political view. Those strong and able persons should have enough money at their command and corruption should be given free rein. When the supporters of democracy came to know of this mismanagement they raised an outcry for discipline. Consequently, governments like Mussolini's in Italy, Hitler's Germany, Franco's in Spain and Ayub Khan's in Pakistan were established after dissolving the parliament.

In Russia some minority groups established a government on the basis of a new ideology making Marxism as its focal point and deluding the workers and peasants. In this manner, one's own government is dissolved when the rule comes to a standstill and some autocrat or dictator like Mussolini or Hitler is proclaimed as the protector of the people and given total power in order to avoid anarchy.

In the search for good governance sometimes even one's dearest thing—liberty—is sacrificed.

To what extent human liberty suffers oppression at the hands of such governments is a matter of debate and thought.

69

FREEDOM WITHIN LIMITS

Total freedom is impossible
Economic freedom after political freedom

When we think about freedom it seems that man can never be completely free. Total freedom transforms into anarchy in the end and that is destructive and death-like. Today, you we or anyone else may not have to become slaves of some others, but there are certain rules of nature which govern us and because of which we can never call ourselves totally free.

There is a detailed description in ancient literature of man's freedom. When the word *'chhand'* (metre in English) was used, mankind was divided into four castes or levels. Every level had to live in a particular *chhand* which means there was a rule by which mankind had to live in limits. This was the rule. This can also be termed as subjugation. For the sake of reference, there was the *Gayatri chhand* for Brahmins, *Anushtup* for Kshatriyas, *Jagti* for Vaishyas. It is an intricate and deep and detailed matter as to how they were in practice in daily life and what their place was in the physical day to day behaviour. But the essence is that humans were limit bound or metre bound. Only the animals were free of limits or metres. It is another matter that later man's intellect made slaves of animals and subjugated them, but God may have in his creation made animals free, but not mankind.

Just have a look at the programme of a day's events right from the morning, you are bound by certain rules from day to night. For example you have to leave your bed and get up by 6 o'clock or 8 o'clock or 10 o'clock. You are not free. Nature has saddled you with the need to perform your daily toilet. Food is essential for the appeasement of hunger and so things have to be arranged for it. Food has to be cooked after lighting up the fire. Many functions like these have to be fulfilled. So how are you free? This is to say that man is subservient by birth. The only difference is that while the rule of nature is not unbearable to anyone, subjugation at the hands of another man or even human society is not acceptable to man's heart or mind.

Man worships freedom by nature. He desires to work according to his inclination and liking as far as possible. That is why socialism while supporting equal distribution of money also favours equal distribution of leisure.

There is much debate about this. Leisure or free time is not only dependant on physical labour. There are differences between the work done by different individuals and these differences will always remain. For instance, if a specialist in the field of chemistry, who is carrying out research in the field of new medicines, sits in his laboratory and thinks, even for hours, he would be working; but many people would say that he is sitting idle, he has leisure and no work, whereas the worker in a factory toils for 8 hours daily. If leisure and work are measured thus, man will be reduced to the level of a machine. Just as a machine is run for a number of hours, after which it becomes necessary to give it some rest, so man will also become a machine. In such a system there would be no scope for the development of intellect, talent or expertise. Socialism never envisages that the talent of the country is deprived of development or that the intellect be crushed. There should be a proportionate arrangement for the area of work of individuals along with their leisure and rest so that the social balance is not disturbed. This in fact is the basic aim of socialism.

After the independence of India, the question of political freedom is no longer there. When the British ruled or any other foreign power had control over the Indian people, there was a sense of slavery or subservience. At that time the question of the individual freedom of man did not arise. When the whole nation is in fetters, how can individual freedom be thought of.

Today, when we are free of foreign domination and every Indian subject is fully free and even the constitution has assured complete freedom, we must give due consideration to the ways by which we may be free of the biggest subjugation, that is, economic dependence. The main aim of socialist economy is to free every Indian from economic subjection.

At present, in the capitalist economy many people have an abundance of leisure. They are free the whole day. Without doing anything they have an income of millions, whereas innumerable people undergo hard toil the whole day long and yet find it difficult to earn enough for themselves and their families. Who should get how much, in what manner economic equality may be established has been discussed at length. Now, the question is who should get how much freedom and who should get how much leisure and free time.

The present progressive age can be called the era of the machine. New machines are being invented all the time. Consequently, the duration of man's leisure or free time, too, is increasing. But the problem is that this leisure is not being distributed equitably. Here, too, it is the same situation. The advantage of this leisure, born out of inventions, only goes to people of one class who probably do not need it.

The daily hours of work in factories and workshops are fixed at 8 hours. The workers unions demand that the working hours should be fixed at 7 hours per day, and every worker given 1 month's leave every year, etc. Leave alone the factory workers. Their working hours may be reduced to 4 hours a day, but what of the others, all the other people who work, what should be done about them?

There are many professions in which the nature of work is not of fixed hours or shifts. It continues night and day. Suppose there is a doctor or a chemist who is engaged in discovering new medicines. This is mental work and even during the night while going to sleep he thinks about this. A doctor or nurse are engaged in tending to a patient and endeavour to discover all the pertinent information about that particular ailment or its cure. If they, too, go by fixed hours and after completing that predetermined time, leave the remaining work to others, then could someone else take it up?

An astronomer works out calculations related to the planets and the cosmos. How can he have fixed hours of work.

There is a minister who has to keep an eye on various events and happenings throughout the country, all accidents and mishaps, famines, upheavals, disturbances, etc., and this too every minute, every hour; even he cannot be limited to set hours of work.

Similarly, all philosophers are also so industrious that they devote all their time to their work, night and day, all through their lives. Thousands of years ago the sages of that time underwent a strict regime of thought and penance in order to fathom out the secrets of creation and the creator. They worked night and day. Even in the modern age, thinkers work so hard that they often lose sense of time itself.

One may fix the hours of work at 4 or 5 or 8 with regard to those who work in factories or those who work with their hands, but it is not possible to fix the working hours of a doctor or lawyer or a writer, a poet and such people who do mental work. However, it is possible that they work for 8 to 9 months and then rest for 3 to 4 months. And, often, it happens like this. All the people who work thus have to take leave in this manner.

Besides, there is a great difference in day-to-day work and work involving creativity and inventiveness. When you come to know that so and so has achieved this and this from his hard-earned earnings after working 16 hours a day for 30 years, then do not think that he has done a great thing. In fact, you may think that he is not capable of doing something great. He may, in reality, not be able to do any work which has not been done before and he is not in a position to think original thoughts. If you are interested in such people, then there is no dearth of such people in India. Go to any village. You will find old persons who sit in the *chaupal,* the community centre, recounting glorious tales of their past life with nostalgic élan. No matter how great a history they may lay claim to, you will soon come to know that they have only worked in those areas in which their ancestors or their contemporaries spent their lives and energies and they, too, are not capable of doing something new.

The crux of the matter is that there is a difference between the two works. An accountant works for 8 hours maintaining his books of account and yet does not tire, whereas an auditor checking those accounts cannot work for more than 3 hours at a stretch. A person used to writing by hand can easily write for 6 hours in a day, but if a typist types the same material, he will tire in 3 hours' time. Similarly, taking into account various jobs it can be said that the work of all the persons cannot be measured by time alone. Doing this would be absurd. So, what is to be done? Can you give the same free time or leisure or rest to all equally? This, too, is noteworthy that the time needed to rest or recover from one's labours, is included in the hours of work and often these hours of rest are even more troublesome than the hours of work.

In short, total freedom cannot be obtained from nature, but practically and in reality the questions before us are: (1) How much leisure or free time can we give ourselves; (2) How far are we free in doing what we wish to do in our free time?

For instance, can we city-dwellers go out into the forest on a fox-hunt? Many of us will say no. Because firing bullets from guns so near to the city may injure someone, at least there is such a possibility; or that killing a fox is cruelty towards animals. So, in this way according to the opinions of people a law will be passed that in such and such an area near the city, or for that matter at any place it is not legal to hunt, it is an offence.

Can we play football on the morning of *ekadashi* (first day after full moon)? Thousands of years ago in the reign of king Ambareesh it was a crime for a person not to go to the temple and take part in the worship there on the morning of *ekadashi*. After the British came to power, Sunday was declared a weekly holiday so that people could go to church. In the beginning there were strict rules that you could do no work outside your house. There are many instances of restrictions on what a person can do in his leisure at his own free will. There are many things which we were free to do but now we cannot. A husband in prior times thought his wife to be his personal property and no matter what he did with her she had no right to object. The Pandavas lost Draupadi while gambling and she had to perforce become a servant. Today no husband, however evil he may be, can do this and if he does, he is punished by law. Law defends his wife. Even the children were considered personal property. Some evil parents in order to satisfy their self-interest would maim their children and put them to beg. But today no parent can force their children into crime and if they do, they are punished by law.

There are certain limits to what you may or may not do at your will during your leisure. You may not roam around naked on the streets or drive or park your car or vehicle on the street where you wish or how you wish; you may not hunt animals or have a band playing on the street without proper permission. You may not wash your clothes with detergents in any river or lake or well. Many more examples can be cited but these are enough. So, there is no freedom without leisure and no leisure without rules and regulations.

If every citizen is allowed to do whatsoever he wishes in his free time, most people would find themselves in jail, because they cannot do many things at will which would not appear unpleasant or distasteful to some other person. You cannot at your whim do anything which may prove directly or indirectly harmful to another person or which he may object to, that means you can do only what people have been doing all along, and which is acceptable to society and which does not impede the freedom of others.

You have an umbrella or a walking stick in your hand. You are also fully free to swing it while you are walking on the street but others, too, have the freedom to walk on the street without any obstruction or danger. If by the swinging of your umbrella some other passer-by is apprehensive of being injured then it would mean that your freedom obstructs the freedom of some other person. So, you cannot do any such thing at will.

It is a trait among people to disregard or oppose the prohibition of the government. You must have seen that on the streets wherever it is written 'please do not spit here,' people deliberately spit right at that place. Where dumping of refuse is prohibited, people will pile up waste material. But, in order that these prohibitions do not become oppressive, the spirit of opposing them also has to be kept alive.

Anyway, where people start opposing without a proper sense of freedom and its limitations and begin to declare socialism or any other social reform as slavery then the matter becomes ridiculous. The question is should people be handcuffed in order to assure that no one slaps another person?

70

INDUSTRIAL REGULATION AND POLITICAL POWER

Evaluation of extraordinary quality
The role of power and leadership

After deliberating on freedom and its natural limits, the question which remains is that socialism, equal distribution of wealth and equal distribution of leisure may not become a laughing matter because of the difference in the ability and intellectual capacity of man.

Suppose a woman's voice is very sweet, as sweet as a nightingale's. Crowds gather in theatres to hear her sing and that singer earns thousands daily. Or, an artist is in possession of his painting which is so artistic that he can put it up in a gallery and earn hundreds just by the tickets to see it. A doctor asks for whatever fee he wishes (it could be in thousands) to operate upon a patient or cure a person of his serious ailment because of his experience and skill. Beautiful and extremely attractive women ask for and even get an amount, only after a short acquaintance, which some other woman cannot earn in a day or even a lifetime. Famous heroines show their tantrums and because of their temperament film companies have to incur losses, yet because of their popularity among the masses they bear all these tantrums.

Many more examples can be given but there is no need to be afraid of them. The reason being, they are all in a minority, they are exceptions and can be counted on the fingers. If their number increases then they will lose their power to earn money. They may earn money but they will never achieve or obtain industrial rights or the facilities of political power. In the history of the world no nightingale voiced beauty has ever had an empire nor has an artist achieved special authority and power because of his talent. Such people or other professional intellectual persons may well earn money but even for that they have to work. As compared to a barrister, who earns Rs. 3,000 a day, the manager of a mill earning only Rs. 500 or the secretary of the worker's union who probably earns Rs. 100 may be more important. The reason is clear. The beauty that collects money like this has adorned herself with necklaces of diamonds and pearls and in no way can harm the woman who earns money through labour and who has decked herself happily with cheap red beads. But the mill manager or the secretary of the union can create impediments in the daily livelihood of the ordinary woman.

So, industrial control or political power is more important than money. Besides this, we know that the money spent by the richest person on his own personal needs is not very much more than that spent by ordinary persons. What can a person do if he becomes very rich? What can Tata, Birla, Dalmia, Mafatlal or Bangad in India or Ford, Rockefeller, Carnegie or Newfield or Rothschild in other countries do with their money? In any case they cannot spend a large part of their wealth on their personal needs. They are forced to spend a huge amount of their money in the service of the people, in building colleges, universities, hospitals, etc. As compared to an ordinary person what percentage of their income do they spend on themselves? No one has probably thought about this. People, dazzled by the great wealth of these persons, are never able to really compare themselves with these millionaires. Evidences of this will be seen in every nation and province in the form of various donations where these wealthy persons are compelled to spend a large part of their money on works of public welfare.

If the income of the people becomes equal then a little more income of a few people will not make any difference in their standard of living. They will also have to live like the rest of the people. Suppose a famous singer organises a programme in a theatre and is able to earn hundreds of rupees daily. But if she requires a maid or a servant she would not be able to get one because after the equal distribution of income, why would another woman wish to become a maid or a servant? The work of the maid or servant will be done by the socialist institution. That singer will not be able to leave some money in her will for her children nor would she be able to buy some shares or invest her money in a source of unearned income because the inheritance will also go to the socialist institution. The capital investment custom will also belong to the socialist institution.

Although it would be difficult for any government by levying income tax or multi-income tax to prevent those people with individual talent to become very rich, still what would be the end result? There is no harm if such persons endowed with individual talent are able to obtain some special facilities or things in their day-to-day lives? But we also see that many people profit from the intellect or talent of other persons. If a beauty or Gauhar Bano earns money through the attraction of her beauty then no one will object; but if a wealthy man collects a hundred or fifty 'Gauhar Banos' and hires them for Rs. 5 and then charge Rs. 25 per person from the people hiring them out then how will society view this?

If a thief or pickpocket picks your pocket by the sleight of his hand then his skill may be praised or he may even be excused, but what of the man who has bought stolen goods from that thief dirt cheap and who has made it a custom to buy stolen goods and sell them and earn money out of it? No one can sympathise with that person.

There are many more instances like this where individual skills and talents are utilised by people who buy and sell and broker or give loans to earn money. They cannot be forgiven.

Now, what is to be considered is who among these is the most skilful and talented? All can claim that they are skilled in their work and so why should it be said that one person is profiting on the talents of another? Work is not just making something, but all services related to that production are also valuable. Skill is also needed for them.

For example the cotton which was previously raised in the fields was carded and spun by the women of the house and the men would weave cloth out of it and the need for covering the bodies of the whole family were met. For that work now the following people are required: (1) The peasant who grows cotton, (2) The capitalist who lends money, (3) The carder who makes cotton for weaving, (4) The cotton merchant or agent, (5) The people who run the mill to make the yarn, (6) The people who weave the yarn, (7) The wholesale merchants or agents of cloth, (8) The retailers of cloth, and (9) The shopkeepers who sell cloth. God knows how many more people can claim that it is because of them that your family is wearing clothes. And it is true. Without any one of the above mentioned people, the situation would be of an army without guns or bullets.

But on searching you will not find one among these people who is accustomed to working on original lines. All of them do what they are instructed to do. Always, there are a large number of people who are accustomed to doing what they are told to do as compared to the people who can give instructions. It is for this reason that if the workers get Rs. 100 per month, those who give instructions get Rs. 400 to Rs. 500 per month as salary.

Where there is no need for original decision-making but a responsibility to make people fulfil jobs entrusted to them and general maintenance of discipline then such work has a special value. Everywhere the person, who can manage this, invariably becomes the head or the leader.

The people who have these qualities, such as the 'jobber' or 'master' in a factory, the 'warder' in a jail, headmasters in schools, etc., that is, those who lead and maintain discipline, are usually unpopular, but their work is significant. Consequently, the owners place people of their choice on such posts at a higher salary.

This is the situation everywhere. Every Indian member of a joint Hindu family knows that if there is a man or woman of a cruel or autocratic temperament in the house, he or she, is able to control and direct the whole family and if in this autocratic control there is a slight degree of selfishness or sadism then in time revolt is inevitable. But if such control is not there, then the situation may worsen. It becomes that of riding a horse without reins. Therefore, there are only a few who object to working under instruction and guidance. Also, many people are rather happy that they are freed of the trouble of thinking or planning. Very few rebel against discipline. Of course, if the people in authority misuse their power or insult and belittle the people working under them then every person will oppose this tendency.

So, our mental idleness always directs us to follow. And so, in every age only a few are great leaders or super human beings.

Parents ask their children about where they went without taking permission? Why did they do this or do that?

And it is these very parents who, after some time, say why don't you act according to your brain? Why do you pester us for every little thing?

In short, it is the tendency of mankind that some should work and others should bear the burden of making decisions of thinking, etc. No one is ready to shoulder the blame for a plan which has failed or has borne adverse results, because there is a danger involved in it. Everyone shies away from danger. It is not necessary that everything should succeed. Therefore, if the leadership and guidance is reliable then people prefer to work under it.

Only circumstances can properly analyse the mentality of mankind. Supposing that you enquired from a coolie or a railway official about the platform on which the train to Lucknow is standing? And that railway worker knowingly and just to vex you put you on a wrong train or even roughly told you that he does not know, you will always feel an antipathy towards all railway workers or coolies. As opposed to this if he informed you of the right train or behaved with you politely, you will tend to favour the coolies or railways workers. This example is relative. All of us, big or small, feel the need of some guide or another. This is quite natural, for everyone does not and cannot know about everything.

Now if you are benefited by the guide and your work is facilitated you will praise him, but if through his guidance you are put to trouble, you will certainly feel adversely inclined towards him.

All the ills in the world arise not because 'there is no good governance'; rather, 'why is there any governance?' This is our objection. It is true that no person likes interference of any sort in the work that he does. Everyone wishes to do as he feels like. But if the skill to do a certain job vests in five (or even less) people out of a hundred, then the remaining people will definitely be ready to work under the guidance of those five (or lesser) people. Rather, they will wish that those five lead us somehow and guide us. If someone opposes this they will be angry on him and will not stop short of violence. This is the reason that many ambitious false persons are able to get themselves accepted as leaders although their knowledge is very poor.

Even able leaders become unpopular very soon because of their cruel natures. No matter how able they may be if they unnecessarily scold their inferiors or insult them or make an exhibition of their exalted position, they lose the respect of their followers. As a result, dissatisfaction brews among the people who follow them and the leadership of those persons is endangered. But this much you can be sure of that the tendency among the followers to obey the leader and his directions is never permanently destroyed. After a momentary resentment it is again re-established among the people. There is nothing to fear in this. It is quite natural. But the thing to be afraid of is this that such new leaders who emerge and uproot the old leaders may not be accepted as ideal leaders. If this happens then dictatorship is bound to result. Nelson was worshipped by his men. In Russia Lenin and thereafter Stalin were worshipped as gods. Mussolini in Italy was regarded as a saviour and Hitler in Germany was regarded as the father of the nation. But any leader who advocated anarchy or the desire not to accept any governance and to revolt against it was never popular permanently, nor will such persons become popular in future.

Unfortunately, this is the biggest shortcoming of capitalism that whosoever assumes the reins of power rejects social equality. Another hierarchy comes into existence—leader and follower. The word follower is now attributed to persons who are subservient and bear all sorts of insults. In this system ability becomes a commodity, to be sold and bought in the market and everything becomes cheap or dear according to demand and supply. Since ability is in short supply it is priced higher in the market. Whenever the income of the leader exceeded that of his followers there was a class division. When a hierarchy was established in society then revolt was the natural outcome. The argument which prevailed was that 'why do you give us orders?' we are no less. And so bitterness brewed in society.

Such thoughts do not arise in the army. When the havildar gives an order to the soldiers, no one says 'why do you gives us an order, we are no less than you?' But in the day-to-day life of the soldiers or in the behaviour of their neighbours or their women such competitiveness or hierarchy can be easily noticed. Rather, it is seen that people are not so pleased by their own progress, than by the adversity suffered by their neighbours. If people say that they are very happy then 99 out of 100 people are lying.

Under the capitalist system there is hardly a person who would oblige someone or do some work on being asked politely or on request. But if on the other hand he is shouted at or threatened, then he wags his tail and hurries to comply. At the bottom of all this is nothing but economic weakness. The tendency to become slavish arises out of poverty and want. In time as the poverty of our country is done away with, this mental poverty will also vanish.

The tendency to shy away from leadership creates more difficulty. If you send five soldiers on duty, all of them will hesitate to take up the responsibility of leading the group and it may be that the person who finally agrees does not really comprehend the task. Similarly, in India we are not able to find judges for the High Courts or the Supreme Court. The reason being, no one is ready to shoulder the grave responsibility of the exalted position.

The remedy is that members of the general public are made members of the jury. But you must have seen how people avoid becoming members of the jury. And how, when their name comes up for jury-duty they have to comply willingly or unwillingly. In the same way the task of leadership has to be thrust upon the people whether they wish or not and this is the only true remedy.

Besides, there is a natural inclination in persons to perform or do certain jobs. Some people are born leaders. They can never work like servants or followers. On the other hand, some people are born followers. They can never lead a group or collaboration. Such people, who are talented by birth, will do the job entrusted to them successfully, no matter what remuneration they may be given. So, it is not necessary that the persons who make plans or give orders get lesser pay than the workers or followers. Unfortunately, the capitalist system has created such a benchmark by which those who are extraordinarily talented and capable, do not become leaders while incapable, ordinary persons are made leaders, just because of selfishness or for some interest.

So, one must not be afraid that the persons with special talents need to be paid more money in order to work. Leave alone incentive, experience tells us that even if they have to face obstacles and suffer hardship, they do not leave any stone unturned in following their natural bent to lead. The only fear is that after once becoming a leader they do not misuse their powers and become autocrats; and, taking refuge behind these powers and authority, entirely ignore the goal of equal income. This happens; this can happen.

In the socialist system, services become public duties. A civil servant, a judge, a captain in the army or a general or the principal of a college are not tempted by the prospect of getting more remuneration. These gentlemen only require this from the nation that they may live a life devoid of worries and for this they need only so much by which they can serve the nation properly. But in the capitalist system it is doubtful if people can remain gentlemen. When they come to know that work cannot go on without them, they begin to demand exorbitant wages and remunerations and get them and leaving aside national interest endeavour to become individually rich and wealthy.

Capable people are needed for leadership or management of factories, workshops and industries they, too, demand a high price for their expertise.

Whether they are talking to persons of the working class or leaders of trade unions or shareholders, they say without any hesitation that they will work on such and such condition and if this is not acceptable, they can do the job themselves. Some people of the trade unions may agree to this, but they certainly fail. There are many reasons. Because of their inexperience, the managers come into direct confrontation with the shareholders who invest the capital and a useless struggle ensues between two equal groups.

The managers again say that if they are allowed to manage the industry on their own then in spite of giving them exorbitant or improper remuneration, the workers will get more wages and the shareholders twice or four times the dividend. This is true. Those capitalist managers extort a large amount for their efficiency and capability.

So, in this fashion, in the capitalist economy people derive income from renting property, renting money and renting ability.

So when land and property and capital are nationalised it is necessary to nationalise human ability also. To some extent ability has come under nationalisation. All income taxes, multi-income taxes and extra-income taxes are levied on the profits from renting out capability. If not all, a large part of this becomes the property of the nation.

The word 'rent' is being used rather generally and freely. The word rent is usually used to denote the income derived from giving one's property to another for use by a landlord or zamindar. But in a broader sense all money given to people for the use of anything or for services is a rent. More crops are produced in one field than another or one man has more professional expertise than others; this will result in the difference of rent paid to each of them. We will have to accept the difference which nature has bestowed on the production capacity of one field as compared to another. Similarly differences between one man and another will have to be accepted. But they can all be nationalised. All these gifts of nature can be made the property of the nation through taxes or by other means. Until this does not happen, those persons will keep deriving benefits and with time their children will not only become idle land owners and wealthy capitalists but also suspend indefinitely the goal of equal distribution of wealth and social equality.

The condition will remain the same. Great philosophers, scientists, astronomers, chemists, mathematicians, inventors, teachers, preachers, social reformers and saints will remain so poor that their wives and children do not get enough to eat while the industrial managers and corporate executives become so rich that they earn millions and their wives and children roll in wealth.

Seeing these people, the poor of society will suffer from inferiority or revolt. They will not be able to even converse with each other leave alone establishing relations. The economic disparity will increase to such a level as to cause discrimination and hatred between the two communities.

No God-gifted talent can be wasted or destroyed, but differences in their respective incomes can be definitely brought to par. As soon as the incomes equalise, all the children and wives of these people will open out. They will inter-marry and form other relationships between each other and this will lead to all round improvement in the nation.

There is another problem in India. It is the difference in caste and creed or the hierarchy of race. Although in this modern age it has changed a lot, because of the laws passed by the government, still, this custom is so much a part of Indian culture that even today these differences matter a lot in marriages or social intercourse. In this respect only this can be said that there is no place of caste or creed in a socialist economy.

This discrimination will have to be removed.